CHRONIC HEPATITIS B AND C

Basic Science to Clinical Applications

Chronic Hepatitis B and C

Basic Science to Clinical Applications

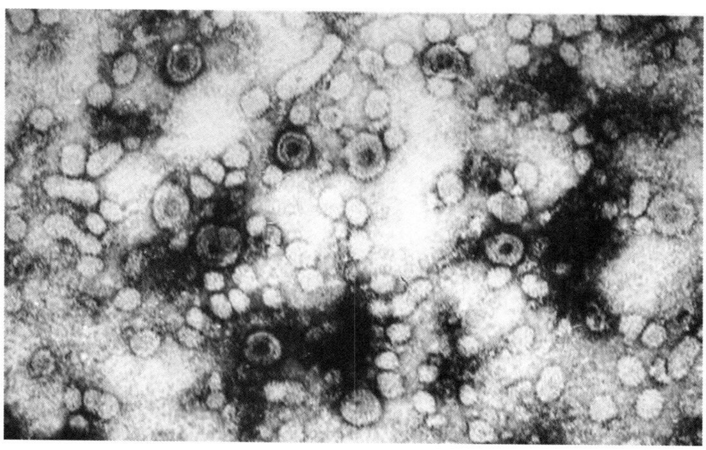

Editor

Chiaho Shih

Academia Sinica, Taiwan

World Scientific

NEW JERSEY • LONDON • SINGAPORE • BEIJING • SHANGHAI • HONG KONG • TAIPEI • CHENNAI

Published by

World Scientific Publishing Co. Pte. Ltd.
5 Toh Tuck Link, Singapore 596224
USA office: 27 Warren Street, Suite 401-402, Hackensack, NJ 07601
UK office: 57 Shelton Street, Covent Garden, London WC2H 9HE

Library of Congress Cataloging-in-Publication Data
Chronic hepatitis B and C : basic science to clinical applications / editor, Chiaho Shih.
p. ; cm.
Includes bibliographical references and index.
ISBN 978-9814299787 (hardcover : alk. paper)
I. Shih, Chiaho.
[DNLM: 1. Hepatitis B--physiopathology. 2. Hepatitis C--physiopathology.
3. Chronic Disease. 4. Hepatitis B--therapy. 5. Hepatitis C--therapy. WC 536]

616.3'623--dc23

2012012247

British Library Cataloguing-in-Publication Data
A catalogue record for this book is available from the British Library.

Typeset by Stallion Press
Email: enquiries@stallionpress.com

Printed in Singapore by Mainland Press Pte Ltd.

Foreword

This is an excellent book for use in teaching about chronic hepatitis to graduate students, medical and other health sciences students and as a reference for the professional research community.

Nobel laureate Baruch S. Blumberg, M.D., D.Phil., who discovered HBV, was originally going to write the foreword to this book. Sadly, Dr Blumberg passed away in 2011 before the book was ready for release. (I think it is terrific that the Editor has chosen to dedicate the book to him)

Since I worked with Dr Blumberg, during his time at The University of Oxford, and he then worked with us here, at the Hepatitis B Foundation (HBF) labs, until his death, the book's editor, Dr Shih, asked me for comments. Although I am no substitute for Dr. Blumberg, I can say with confidence that this is a book he would really have enjoyed. For one thing, it combines the practical with the basic science, putting discovery science into a clinically relevant context. That was of course, his life's work, having discovered the cause of chronic hepatitis B and then helped produce an effective assay for the virus and preventive vaccine. On the other hand, the book covers a wide range of topics regarding viral hepatitis B, C and D, from detection, prevention, pathobiology to therapeutic management. Dr. Blumberg's interests were broad, and he always felt clinical hepatitis and liver disease should be understood in the context of fundamental science. Finally, and of course, many of his close friends and colleagues are contributors, making the book all the more meaningful.

The discovery of hepatitis and development and use of an effective vaccine is one of the great accomplishments in medical and public health of the last century. The complicated natural history of hepatitis B disease

Dr. Baruch Blumberg (right) reunites with Dr. Harvey Alter (left), friend and colleague from the National Institutes of Health, at the Hepatitis B Foundation's Princeton Workshop in 2002 (photo provided by Joan and Tim Block).

in people made the discovery of its etiology, all the more remarkable. The disease pathobiology is thoughtfully covered by two leading clinician scientists, Drs. Chu and Liaw. The immunology of hepatitis B is central to prevention, pathogenesis and resolution, and the chapter on immune responses is by one of the field's thought leaders, Prof. David Milich. He systematically addresses the major known facts about the host B and T cell response to chronic HBV, as well as drills down to explaining the complex interplay between cytokines and chronic infection.

HBV is ubiquitous, and its epidemiology is stunning, with more than 350 million people chronically infected worldwide, and as many as 2 billion with exposure and resolved infections. One of the most serious outcomes of chronic HBV is primary cancer of the liver, and the chapter by Drs. McGlynn, London and Evans, shows the relationship between HBV and HCC, and provide excellent illustrative graphs of disease incidences that will be useful to anyone researching or teaching this subject.

The molecular biology of HBV and of HBV- induced disease has been among the most surprising in virology. HBV replicates via an obligate RNA genomic intermediate, despite being a DNA virus. HBV clearly causes cancer in people, despite having no clear oncogene. These fascinating paradoxes are summarized nicely in the chapter by Drs. Hodgson and Slagle.

There is a growing body of evidence that mutations in the virus play important, perhaps even in some cases, decisive roles in the course of infection in people. Mutants of HBV emerge that may frustrate treatment, vaccination, or contribute to pathobiology. The chapter by Dr. Shih takes a mechanistic and functional approach to this fundamental issue, particularly from the perspective of naturally occurring core and surface antigen variants. A virion secretion defect, or a compensatory effect on such a defect, appears to be a common feature in these naturally occurring variants.

Chronic HBV is treatable, but not curable, and long term therapy is needed. The molecular basis for the two categories of medications, the direct acting antiviral (DAA) polymerase inhibitors (small molecules) and the immuno-modulatory interferons (biologicals), and the emergence of clinically relevant and observed resistance, is well described in the chapter by a group that has been among the leading authorities of the subject (Drs. Nguyen, Desmond & Locarnini).

Hepatitis D, the extraordinary sub-virus that requires co-infection with HBV to complete its replication cycle and is associated with worsened outcomes, continues to be a major, although often overlooked health threat. It is not overlooked in this book! The chapter by Dr. Sureau shows how the molecular biology and natural history of liver disease associated with HDV are as fascinating as it is complicated and goes a long way to making it all understandable.

But, let's face it, these are times that are no less than revolutionary in chronic viral hepatitis C. Hepatitis C is curable. That is an extraordinary thing to write, and even seeing it in print, with my own hand, causes me to step back in awe.

The management of HCV clearly involves a combination of immunological and direct acting antiviral strategies, and this is covered in the chapters by Drs. Liu and Chen and Drs. Yu and Chuang. Development of tissue culture infectivity systems has revolutionized the study of HCV and development of antiviral. These discoveries and the interplay between the immune system, the therapeutic interferon's and HCV are also nicely addressed by the groups that help make breakthroughs in this area (Drs. Kato and Wakita and Drs. Horner and Gale).

The molecular biology of HCV is discussed in a chapter that focuses on the recently discovered HCV F protein (which appears to be the results of an internal start site), written by one of the groups making this discovery (Drs. Yuksek and Ou).

There are clear extra hepatitis affects of HCV, from influences upon the immune systems, to neurological and vascular diseases, and perhaps even non-hepatic malignancies. The chapter by Drs. Machida and Lai take a candid look on these possibilities, evaluating the relevant epidemiological and molecular evidence.

The book looks at an individual's host and genetic factors that affect outcome of natural HCV infection and treatment. HCV genotype plays a strong role in determining outcome of treatment with interferon and ribavirin. However, the infected individuals' genetics and other clinical factors also play important roles in natural infection and treatment related outcomes. Drs. Tanaka and Mizokami review both viral and host genetics of HCV. Again, how these factors influence, and can be used, in the use of the current and upcoming antiviral treatments is reviewed by Drs. Yu and Chuang, where the use of the kinetics of virological response to intervention is also explained.

So, I do think that Dr. Blumberg would have been one of this book's fans. It is a shame he is unable to write these words himself, and I feel frankly presumptuous, in the role of a substitute hitter. However, for teaching and for research, I think he would have enjoyed the contents.

Timothy M. Block, Ph.D.
President, Hepatitis B Foundation and its
Institute for Hepatitis and Virus Research
and
Professor, Microbiology and Immunology,
Drexel University College of Medicine

Preface

Viral hepatitis is a very ancient disease. Hepatitis B virus (HBV) and hepatitis C virus (HCV) can cause acute and chronic liver inflammation. Chronic infection with HBV or HCV can lead to cirrhosis, liver failure, and highly malignant hepatocellular carcinoma. Furthermore, coinfection with HBV and hepatitis delta virus (HDV) often causes more severe liver disease.

Diagnosis of HBV by ELISA or PCR is available, and vaccination of HBV surface antigen is successful in reducing the incidence of HBV infection and liver cancer in Taiwan and other parts of the world. However, the treatment of HBV chronic infection remains a challenge. Long term treatment with nucleoside or nucleotide analogs often leads to drug resistance. Interferon alpha (IFN-alpha) is more expensive and associated with side effects. The efficacy of pegylated IFN-alpha is low for most perinatally infected chronic HBV carriers. Similarly, at present, there is no effective treatment for hepatocellular carcinoma. Worldwide, there are approximately 350 million chronic HBV carriers. Therefore, there remains an urgent need to research more on both basic and clinical science of HBV.

Diagnosis of HCV by ELISA or PCR is also available. Treatment of HCV infection with pegylated interferon and ribavarin is successful for most HCV genotypes. Several new oral drugs targeting HCV NS3, NS5A and NS5B are being actively developed. However, to date, there is no HCV vaccine available. Despite the establishment of an *in vitro* cell culture system for HCV, the study of HCV innate immunity and liver pathogenesis remains a problem due to the lack of a small animal model for *in vivo* HCV infection. The recent finding of the correlation between certain types of interferon lambda (IL28B) polymorphism and the clearance of HCV, provides another opportunity for individualized therapy of chronic HCV carriers.

This book consists of 14 chapters, including 7 special topics on HBV and HCV, respectively. The scope of the book is very broad, and the selected topics range from epidemiology, immunology, molecular virology and oncology, to clinical therapy. I hope this book will serve as a friendly introduction for students, and as a resource of information for investigators in both basic and clinical science.

HBV was first discovered by Dr. Baruch Blumberg, who kindly agreed to write the Foreword for this book. Unfortunately, Dr. Blumberg passed away suddenly last year. Dr. Timothy Block, who is a close friend and a long term collaborator of Dr. Blumberg, graciously agreed to prepare the Foreword for this book. We also thank Joan and Tim Block for providing a photo of Dr. Blumberg. It was taken at a workshop held by the Hepatitis B Foundation, in Princeton. It appropriately includes Dr. Harvey Alter, who was Dr. Blumberg's long time friend, and helped in the discovery of HBV and is credited as being a co-discoverer of hepatitis C. This was a photo that Dr. Blumberg treasured, since it reflected the friendship and collaboration. For me, it shows two giants, representing a golden age of hepatitis research.

Last, but not the least, I would like to thank Szu-Yao Wu, Chih-Yin Lee, and Shu-Fan Chou in my lab for careful proofreading of book chapters, and Ms. Sook Cheng Lim at World Scientific for her outstanding dedication in editing this book. Finally, I thank all the contributing authors for their generosity in sharing with us their insight and knowledge in chronic hepatitis B and C.

Chiaho Shih, Ph.D.
Distinguished Research Fellow
Institutes of Biomedical Sciences
Academia Sinica
Taipei, Taiwan

Contents

Contributors

Dinn-Shing Chen, MD
College of Medicine
National Taiwan University
7 Chung-Shan S. Rd,
Taipei, 100 Taiwan
chends@ntu.edu.tw

Wan-Long Chuang, MD
Hepatobiliary Division, Department of Internal Medicine
Kaohsiung Medical University Hospital
Kaohsiung, Taiwan
waloch@kmu.edu.tw
fish6069@gmail.com

Chu, Chia-Ming MD
Liver Research Unit
Chang Gung University
Taipei, Taiwan

Paul Desmond, MB, BS
St. Vincent's Hospital
41 Victoria Parade
Fitzroy 3065
Australia
paul.desmond@suhm.org.au

Alison A. Evans, SCD
Fox Chase Cancer Center
333 Cottman Ave.
Philadelphia, PA 19111

Michael Gale Jr., PhD
Dept of Immunology
University of Washington
Seattle, WA 89195-7650
USA
mgale@u.washington.edu

Amanda J. Hodgson, PhD
Baylor College of Medicine
Mailstop BCM-385
Houston TX 77030
USA

Stacy M. Horner, PhD
Department of Immunology
University of Washington
Seattle, WA 89195-7650
USA

Takanobu Kato, MD, PhD
Department of Virology II
National Institute of Infectious Diseases
Toyama 1-23-1, Shinjuku-ku
Tokyo 162-8640, Japan

Michael Lai, MD, PhD
Institute of Molecular Biology
Academia Sinica, Taipei
TAIWAN
michlai@usc.edu

Chun-Jen Liu, MD, PhD
College of Medicine
National Taiwan University
7 Chung-Shan S. Rd,
Taipei, 100 Taiwan

W. Thomas London, MD
Fox Chase Cancer Center
333 Cottman Ave.
Philadelphia, PA 19111
E-mail: wt_london@fccc.edu

Stephen Locarnini, MD, PhD
Victoria Infectious Diseases Reference Laboratory
10 Wreckyn Street
North Melbourne-3051
Australia
stephen.locarnini@mh.org.au

Dr. Keigo Machida, PhD
Dept of Molecular Microbiology and Immunology
University of Southern California
Keck School of Medicine
2011 Zonal Ave.
Los Angeles, CA 90033
USA
kmachida@usc.edu

Katherine A. McGlynn, PhD
Fox Chase Cancer Center
333 Cottman Ave.
Philadelphia, PA 19111
Tel: 215-728-2204
Fax: 215-214-4053

David Milich, PhD
Vaccine Research Institute of San Diego
3030 Bunker Hill St.
Suite 205
San Diego, CA 92109
Tel: 858 581-3960
Fax: 858 581 3970
dmilich@vrisd.org

Masashi Mizokami, MD
Department of Virology & Liver Unit
Nagoya City University
Graduate School of Medical Sciences
1 Kawasumi, Mizuho, Mizuho
Nagoya 467-0001, Japan

Tin Nguyen, MD
Victoria Infectious Diseases Reference Laboratory
10 Wreckyn Street
North Melbourne-3051
Australia

James Ou, PhD
Dept of Molecular Microbiology and Immunology
University of Southern California
Keck School of Medicine
2011 Zonal Ave.
Los Angeles, CA 90033
USA

Chiaho Shih, PhD
Institute of Biomedical Sciences (IBMS)
Academia Sinica
Taipei, TAIWAN
cshih@ibms.sinica.edu.tw

Betty Slagle, PhD
Baylor College of Medicine
Mailstop BCM-385
Houston TX 77030
USA
bslagle@bcm.tmc.edu

Camille Sureau, PhD
INTS
6 rue Alexandre Cabanel
Paris-75739
France
csureau@ints.fr

Yasuhito Tanaka, MD, PhD
Department of Virology & Liver Unit
Nagoya City University
Graduate School of Medical Sciences
1 Kawasumi, Mizuho, Mizuho
Nagoya 467-0001, Japan
ytanaka@med.nagoya-cu.ac.jp

Takaji Wakita, MD, PhD
Director
Department of Virology II
National Institute of Infectious Diseases
Toyama 1-23-1, Shinjuku-ku
Tokyo 162-8640, Japan
wakita@nih.go.jp

Liaw Yun-Fan, MD
Liver Research Unit
Chang Gung University
Taipei, Taiwan
liveryfl@gmail.com

Ming-Lung Yu, MD
Hepatobiliary Division, Department of Internal Medicine
Kaohsiung Medical University Hospital
Kaohsiung, Taiwan
waloch@kmu.edu.tw
fish6069@gmail.com

Kamile Yuksek, MD
Department of Molecular Microbiology and Immunology
University of Southern California
Keck School of Medicine
2011 Zonal Ave.
Los Angeles, CA 90033
USA

HEPATITIS B VIRUS

Chapter 1

Natural Course of Chronic Hepatitis B Virus Infection

Chia-Ming Chu and Yun-Fan Liaw*

Introduction

Chronic infection with hepatitis B virus (HBV) is a global public health problem. Despite the development of highly effective vaccines against the disease since the early 1980s and the implementation of universal newborn vaccination programs in more than 168 countries, there is still a huge burden of liver disease due to chronic hepatitis B. An estimated 350 million people in the world are chronically infected with HBV and 75% of them reside in Asia-Pacific region.[1] Although most hepatitis B surface antigen (HBsAg) carriers will not develop hepatic complication from chronic hepatitis B, between one-quarter and one-third are expected to develop progressive liver disease, including cirrhosis and hepatocellular carcinoma (HCC) and 15–25% will die from hepatitis B-related liver disease. It is estimated that worldwide over 200,000 and 300,000 HBsAg carriers die each year from cirrhosis and HCC, respectively.[2] In Taiwan, HBsAg carriers are at 5.4- and 25.4-fold, respectively, increased risk of mortality from cirrhosis and HCC.[3]

* Correspondence and request for reprint: Prof. Yun-Fan Liaw, Liver Research Unit, Chang Gung University and Memorial Hospital, 199, Tung Hwa North Road, Taipei, Taiwan, Tel: 886-3-3281200 ext. 8120, Fax: 886-3-3282824, E-mail: liveryfl@gmail.com

The natural course of chronic HBV infection is complex and variable and has still not been completely defined. Substantial improvement in the understanding of HBV virology and host immune response to HBV, combined with the recent availability of highly sensitive HBV DNA assays and quantitative HBsAg assays, during the past decades, has led to new insights into the natural history of HBV infection. A better understanding of the clinical outcomes and factors affecting disease progression is important in the management of patients with chronic HBV infection.

Epidemiology of Chronic HBV Infection

Worldwide, an estimated two billion people have been infected with HBV, and some patients with acute HBV infection develop chronic HBV infection, defined as persistence of serum HBsAg for more than six months. The risk of chronic infection after primary HBV infection varies and depends on the age and immune status at the time of infection. Among infants born to hepatitis B e antigen (HBeAg)-positive mothers, and hence infected in the perinatal period, the probability of chronic infection approaches 90%. When infected at one to five years of age, 20–30% of children become chronically infected, while among older children the probability falls to 5–10%. The risk of chronicity among normal, healthy, immunocompetent adults may be as low as 1%.[1] The extremely high chronicity after perinatally acquired infection is presumably related to the immature immune system of the neonates. Another possible mechanism is that the fetus is tolerated *in utero* to HBV following transplacental passage of viral proteins.[4]

The global prevalence of chronic HBV infection varies greatly among different geographical areas, and can be classified into high-prevalence (Southeast Asia, China, Sub-Saharan Africa and Alaska); intermediate-prevalence (Mediterranean countries, Eastern Europe, Central Asia, Japan, Latin, and South America); and low-prevalence areas (USA, Western Europe, Australia, and New Zealand) based on the prevalence of HBsAg carriers of > 8%, 2–8%, and < 2%, respectively.[1] Associated with a wide range in prevalence of chronic HBV infection are differences in the predominant mode of transmission and age at infection. In high-prevalence areas, perinatal transmission is common and accounts for 40–50% of chronic infection in Southeast Asia. In contrast, inapparent parenteral spread from child to child is the major mode of transmission in Sub-Saharan areas. This difference is related to the higher prevalence of HBeAg in Asian carrier mothers (40%) than in African carrier mothers (15%),[5,6] as 80–90% of HBeAg-positive mothers will transmit the disease to their offspring, as compared with 15–20% of those seronegative for HBeAg.[7] In low-prevalence areas, hepatitis B is a disease of young adults, typically those who have high-risk behavior such as sexual promiscuity or drug abuse or are in high-risk occupations.

The worldwide incidence of HBV infection is decreasing as a result of vaccination and public health education on preventive measures against risk factors. After the implementation of universal vaccination programs in newborns, the HBsAg carrier rate among children in Taiwan decreased from 10% in 1984 to <1% in 2004.[8] The estimated total population of HBsAg carriers in Taiwan decreased from 2.9 to 2.6 millions.[9] In the United States, the incidence of reported acute hepatitis B declined by 81% from 8.5 to 1.6 cases/100,000 during the period 1990 to 2006. However, immigrants from high-prevalence areas are now responsible for an increasing burden of chronic HBV infection in many developed countries.

Clinical Presentation

Only a small percentage of patients with chronic HBV infection reported a history of acute or symptomatic hepatitis. In low- or intermediate-prevalence areas where HBV infection is predominantly acquired during adult or childhood years, approximately 30–50% of patients with chronic HBV infection have a history of acute hepatitis. In patients from high-prevalence areas, where HBV infection is predominantly acquired perinatally, clinically evident hepatitis is almost lacking. Most patients are incidentally identified to be HBsAg carriers during health check-up or routine screening, and a few due to non-specific constitutional symptoms. In high-prevalence areas such as Taiwan, as high as 40% of HBsAg positive patients with overt acute hepatitis are actually chronic HBsAg carriers who remained unrecognized until they present the episode of acute hepatitis. They were HBsAg seropositive but seronegative for immunoglobulin class M antibody against hepatitis B core antigen (IgM anti-HBc), so-called "previously unrecognized HBsAg carriers with acute hepatitis flares or superimposed other forms of acute hepatitis."[10] In these areas, a history of acute hepatitis in HBsAg carriers is more likely an acute hepatitis flare of chronic HBV infection rather than acute hepatitis B.

The patients with chronic HBV infection may present one of the following four biochemical and serological profiles: (1) HBeAg positive with normal serum alanine aminotransferase (ALT) levels; (2) HBeAg positive with abnormal ALT levels; (3) HBeAg negative, antibody to HBeAg (anti-HBe) positive with normal ALT levels; and (4) HBeAg negative, anti-HBe positive with abnormal ALT levels. These four patterns of presentation actually represent different phases of chronic HBV infection.

Phases of Chronic HBV Infection

As a result of the dynamic interplay of complex interactions involving HBV, the hepatocyte and the host immune response, the natural course of chronic HBV infection consists of distinct phases, characterized and diagnosed on the basis of HBeAg/anti-HBe serology, serum HBV DNA levels, ALT levels and liver histology. Typically, chronic infection acquired perinatally or during infancy has three phases: the immune tolerant, immune clearance, and inactive residual phases.[11–13] In a subset of inactive carriers, HBV may reactivate and trigger immune mediated liver injuries. This reactive phase can be viewed as a variant of immune clearance phase.[1,14] In adult-acquired chronic infection, there is usually no or very short initial immune tolerant phase. Otherwise, the clinical course is essentially the same as seen in patients with perinatally acquired infection. The clinical, serological, histopathological and virological characteristics of these dynamic phases of chronic HBV infection are summarized in Table 1.

Table 1. Phases of Chronic Hepatitis B Virus Infection: Clinical, Immunohistopathological and Virological Characteristics

Characteristics	Immune Tolerance	Immune Clearance (HBeAg-positive CHB)	Residual (Inactive Carrier)	Reactivation/Immune Clearance (HBeAg-negative CHB)
Age (years)	<20–25	20–40	>35–40	>35–40
Serology				
HBsAg ($\log_{10}$ IU/mL)	4.5–5.0	4.0–4.5	2.0–2.5	2.5–3.0
HBeAg	Positive	Positive	Negative	Negative
HBV DNA	Very high ($>10^{7,8}$ copies/mL)	High ($>10^{5,6}$ copies/mL)	Low or undetectable ($<10^{4,5}$ copies/mL)	Moderate ($>10^{4,5}$ copies/mL)
ALT	Normal	Elevated	Normal	Elevated
Precore/core promoter	Wild type	Mixed (wild type > mutant)	Mixed (mutant > wild type)	Mutant
Histopatholgy				
Inflammation	Absence	Presence	Absence	Presence
Hepatocyte HBsAg	Positive	Positive	Positive	Positive
Hepatocyte HBcAg	Positive (nucleus)	Positive (mixed nucleus and cytoplasm)	Negative	Positive (cytoplasm)
Disease progression	No/minimal	Yes	No	Yes
HBsAg seroclearance	No	No	1–2%/year	No

ALT — alanine aminotransferase; CHB — chronic hepatitis B; DNA — deoxyribonucleic acid; HBcAg — hepatitis B core antigen; HBeAg — hepatitis B e antigen; HBsAg — hepatitis B surface antigen.

Immune Tolerant Phase

The initial phase of chronic HBV infection is characterized by the presence of HBeAg and very high serum levels of HBV DNA (usually $> 2 \times 10^7$ IU/mL or $> 10^8$ copies/mL); normal ALT levels; and normal or minimal histological changes[11–13] with intrahepatic hepatitis B core antigen (HBcAg) expression diffusely and predominantly in the nucleus.[15] There is usually little or no disease progression as long as serum ALT levels remain normal and the immune tolerance is maintained.[16]

The absence of liver disease, despite high level of HBV replication during this phase, is believed to be a consequence of immune tolerance to HBV. Even though HBV does not cross the placenta, the HBeAg secreted by the virus does. Experiments in mice suggest that a transplacental transfer of maternal HBeAg may induce a specific unresponsiveness of helper T cells to HBeAg in neonates. Because HBeAg and HBcAg are highly cross-reactive at the T-cell level, deletion of the helper T-cell response to HBeAg results in an ineffective cytotoxic T-lymphocyte (CTL) response to HBcAg, the major target of the immune response.[4] Once a chronic infection has been established, persistence of high viral load and continued secretion of HBeAg (the tolerogen) are necessary to maintain the tolerant state. The viral population identified during the immune tolerant phase usually consists of exclusively wild type HBeAg-positive (e^+) HBV with little or no mutant type HBeAg-negative (e^-) HBV.[4,17]

Immune Clearance Phase

The transition from the immune tolerance to immune clearance phase usually occurs between age 20 and 40, but may sometimes start earlier and even occur in pediatric patients. During this phase, serum HBeAg is still positive but ALT levels become abnormal, accompanied by declining levels of serum HBV DNA. Liver biopsy demonstrates increased histological activity[11–13] and intrahepatic nuclear HBcAg expression decreases with a concomitant increase in cytoplasmic HBcAg expression.[15]

Little is known about the mechanisms that regulate the loss of immune tolerance in chronic HBV infection. The finding that immune clearance phase is accompanied by a change in the intrahepatic distribution of HBcAg from nuclear to cytoplasmic localization suggests that it may be triggered by a change in the presentation of viral antigens.[15] However, a more recent study suggests that the shift of hepatocyte HBcAg from nucleus to cytoplasm during the immune clearance phase may be secondary to liver damage and regeneration.[18] On the other hand, the immune clearance phase is associated with a decline in serum HBV DNA level and accumulation of mutant type HBeAg-negative (e^-) HBV,[4,17] and hence decreased secretion of HBeAg; both may trigger the loss of an immune tolerant state. But their causal relationship also remains unclear.

Hepatitis Activity and Acute Hepatitis Flares (HBeAg-positive Chronic Hepatitis) during Immune Clearance Phase

Most patients in the immune clearance phase are asymptomatic and have mild to moderate elevation in serum ALT levels. However, the clinical course may be punctuated by spontaneous acute hepatitis flare, defined as elevation of serum ALT over five times the upper limit of normal (ULN). These ALT elevations and acute hepatitis flares are considered to be the results of HLA-class I antigen-restricted, CTL-mediated immune response against HBV antigen(s) and its downstream apoptotic mechanisms.[19] The reasons for spontaneous acute hepatitis flares are not clear but are likely explained by subtle changes in immunological controls of viral replication. Several studies have found that acute hepatitis flares are often preceded by a sudden increase in the serum levels of HBV DNA[20] and enhanced T-cell response to HBcAg and HBeAg.[21] This chronological sequence of HBV DNA and ALT in spontaneous acute flares is similar to that observed in acute hepatitis flares secondary to therapeutic intervention with corticosteroids or cancer chemotherapy. Histological evidence of acute lobular hepatitis superimposed upon the changes of chronic viral hepatitis is frequently observed during these flares.[22] IgM anti-HBc may also appear at this time, but is generally in lower titers than in acute infection.[23]

The reported incidence of acute hepatitis flare during the immune clearance phase varies among different patient cohorts. In one hospital-based study, the annual rate of acute ALT flare was as high as 25% per year during the first three to five years of follow-up.[24] However, in another cohort involving HBeAg-positive asymptomatic young carriers with normal ALT levels at entry, the overall incidence of acute ALT flare is only 29% during a mean follow-up of 3.7 (range 1–8.6) years.[25] Most acute ALT flares are asymptomatic, but around 20% of patients present with symptoms of acute hepatitis,[23] and approximately 2–3% of acute flares may be complicated with hepatic decompensation.[26] In high-prevalence areas such as Taiwan, acute ALT flares of chronic HBV infection is the most important etiology of acute hepatitis[27] and fulminant hepatitis in adults.[28]

HBeAg to anti-HBe Seroconversion

The immune clearance phase has a variable duration and often lasts many years until HBeAg seroconversion occurs. HBeAg seroconversion is usually preceded by ALT elevation, followed by a marked reduction of serum HBV DNA to levels that are only detectable by polymerase chain reaction (PCR) assay and $< 10^{4-5}$ copies/mL, ALT normalization and resolution of liver necroinflammation.[22] However, abnormal ALT levels and high-level viremia persisted at the time of HBeAg seroconversion in about 5% of patients.[29] These patients progressed directly from HBeAg-positive chronic hepatitis to HBeAg-negative chronic hepatitis.

The average annual incidence of spontaneous HBeAg seroconversion is 10% (range, 2–15%), depending on factors such as ethnicity, mode of transmission, age, ALT levels, histological activities and HBV genotype,[19,30–34] as shown in Table 2. A higher rate of HBeAg seroconversion has been reported in non-Asian children with horizontal transmission than Asian children with vertical transmission.[35] The likelihood of HBeAg seroconversion correlates positively with

Table 2. Determinants of Spontaneous HBeAg Seroconversion

Horizontal transmission
Age
Genotype B > C, A = D
Immune competence
Higher ALT levels
Chronic lobular or active hepatitis (> chronic persistent or non-specific reactive hepatitis)
α-fetoprotein >100 ng/mL during ALT flare
Bridging hepatic necrosis

HBeAg — hepatitis B e antigen; ALT — alanine aminotransferase.

baseline ALT levels: HBeAg seroconversion rates at 18-months of follow-up are 0%, 3–8%, 17%, and 59–70%, respectively, if baseline ALT levels are <1, 1–2.5, 2.5–5, and ≥5 times of ULN.[19] In patients with acute flare, 72% undergo HBeAg seroclearance within three months if α-fetoprotein (AFP) levels increase to >100 ng/mL during the flare, as do only 18% of those without AFP elevation of this level.[30] The likelihood of HBeAg seroconversion also correlates with baseline histological activities: the five-year cumulative probabilities of HBeAg seroconversion are <25% and >65%, respectively, in patients with mild and more active hepatitis.[30] HBeAg seroclearance may occur within three months in two-thirds of the patients with more extensive hepatocytolysis, such as bridging hepatic necrosis.[30] In Eastern countries, patients infected with genotype B HBV — as compared with those with genotype C HBV — seroconvert earlier and more frequently.[31–33] Interestingly, HBeAg seroconversion is more frequently preceded by acute ALT flares in genotype C than in genotype B HBV infection, suggesting that a more vigorous immune-mediated hepatocytolysis may be needed to achieve HBeAg seroconversion in genotype C HBV infection.[33] In Western countries, spontaneous HBeAg seroconversion was similar between patients with genotype A, B, D, F HBV infection but much slower in those with genotype C HBV infection.[34,36] In Alaska natives with chronic HBV infection, the median age of HBeAg seroclearance is <20 years in patients with genotypes A, B, D, and F, but over 40 years in patients with genotype C.[34]

In Asia countries such as Taiwan, HBeAg seroconversion occurs at a median (interquartile range) age of 32 (26–36) years, with the majority of cases (90%) occurring before the age of 40 years.[37] In accordance with these data, the prevalence of serum HBeAg declines remarkably from 85% in carrier children (age <15 years)[38] to 5–10% in carrier adults over 40 years of age.[39] These findings suggest that HBeAg seroconversion usually occurs between 15 and 40 years of age in perinatally acquired chronic HBV infection. Persistence of HBeAg over 40 years of age is rare and can be considered as "delayed" HBeAg seroconversion.[37]

HBeAg Persistence and Its Outcome

In some patients, the immune clearance phase may last for many years without HBeAg seroconversion during a long-term follow-up. A recent cohort study from Taiwan demonstrated that the risk of progression to cirrhosis increased with increasing age of HBeAg seroconversion, with a hazard ratio of 3.8 per decade increase in age of HBeAg seroconversion.[25] Notably, patients with HBeAg seroconversion after 40 years of ages were associated with a remarkably high risk of progression to cirrhosis than those with earlier HBeAg seroconversion.[37,40] Clinical studies also show that persistence of serum HBeAg is associated with an increased risk for progression to cirrhosis, HCC development and liver related mortality. An earlier study from Germany involving 53 untreated patients with HBeAg-positive chronic hepatitis showed that hepatic decompensation or liver related death occurred significantly more often and survival without clinical complications was significantly shorter in patients with persistent serum HBeAg than those with spontaneous HBeAg seroconversion during a follow-up period of 38.5 months.[41] A more recent study from Taiwan also showed a significantly higher annual rates of cirrhosis and HCC (3.7% and 1.6%, respectively) in 147 patients with persistent HBeAg, as compare with 1.8% and 0.4%, respectively, in 86 patients who underwent spontaneous HBeAg seroconversion during a follow-up period of 7 years.[42] A study from Italy also showed high incidence of cirrhosis development and short survival in patients with persistent HBeAg seropositivity.[43] These findings indicate that prolonged HBeAg positive phase is associated with an increased risk of disease progression.

Inactive/Residual Phase

During the third phase of chronic HBV infection, serum HBeAg has been seroconverted to anti-HBe but still seropositive for HBsAg. The hallmark event of HBeAg seroconversion usually signals a transition from chronic hepatitis B to an "inactive carrier state."[11–13] Although HBV DNA is undetectable by hybridization techniques, it is often detectable by PCR assays in these "inactive carriers." These low HBV DNA concentrations probably reflect a low level of HBV replication, but do not associate with chronic hepatitis in most patients. During this phase, the patients are usually asymptomatic and have normal ALT. Liver biopsy shows no or mild necroinflammatory activity with variable degrees of fibrosis, including inactive cirrhosis, and intrahepatic HBcAg is absent.

The majority of inactive carriers had levels of HBV DNA less than 10^4 copies/mL, a level that has been used to discriminate inactive carrier state from HBeAg-negative chronic hepatitis.[44] However, in a recent study from Taiwan, 10% of inactive carriers with persistently normal levels of ALT for more than 10 years had levels of HBV DNA $> 10^5$ copies/mL.[45] These inactive carriers were predominantly female, more likely to have genotype B HBV and wild type basal core promoter, as compared with patients with HBeAg-negative chronic hepatitis.[45]

Most inactive carriers remain in this phase with sustained remission of hepatitis and a lifelong inactive state, particularly if this phase is reached early in the disease course. In one study from Taiwan involving 283 inactive HBsAg carriers who seroconverted and remained HBeAg negative, 189 (67%) had persistently normal ALT levels. Of these 189 "inactive carriers," only one progressed to cirrhosis and two developed HCC over a nine-year follow-up period. The estimated annual rates of cirrhosis and HCC in the inactive carriers were 0.1% and 0.2% per year, respectively.[29] The prognosis of inactive carriers from intermediate- or low-prevalence areas is even better, possibly due to the shorter duration of the infection. In 317 asymptomatic HBV carriers from Montreal detected upon blood donation, the vast majority remained "healthy" and asymptomatic after 16 years of follow-up and only three died from HBV-related cirrhosis and none from HCC.[46] In another study from northern Italy, no difference in survival was observed between 296 healthy blood donors with positive HBsAg and 157 uninfected controls who were followed up over 30 years.[47]

Spontaneous HBsAg Seroclearance

During the inactive/residual phase, HBsAg may disappear from the serum (HBsAg seroclearance) spontaneously. Short-term studies showed that the annual rate of spontaneous HBsAg seroclearance was 1–2% per year in Caucasian carriers in low endemic areas, and even lower (~0.8%) in high endemic areas where infection is usually acquired perinatally or in early childhood.[48,49] However, a recent 11-year (mean) follow-up study in 1965 asymptomatic anti-HBe positive subjects [age = 16–76 years (median = 34)] showed an annual HBsAg seroclearance rate of 1.2%, being higher (1.8%) in those over age 50. The cumulative HBsAg seroclearance rate was 8% at 10 years, increasing disproportionately to 25% at 20 years, and 45% at 25 years of follow-up.[50]

Factors significantly associated with spontaneous HBsAg seroclearance include older age at entry; normal ALT levels at baseline [48,51]; HBeAg negativity[48,51,52] or low viral load at baseline, especially if levels of HBV DNA < 300 copies/mL,[53] genotype A HBV (> genotype D)[36] or genotype B HBV (> genotype C) infection[49]; sustained remission of hepatitis during follow-up[50]; and presence of cirrhosis[48] or HCV superinfection.[54] Among these, advanced age was the most constant and important predictor for HBsAg seroclearance in most series.[55] The annual incidence of HBsAg seroclearance varied among each series, but correlated significantly with the mean or median age of patients at enrollment of each cohort.[55] Interestingly, in one case-control study from Taiwan, HBsAg carriers with spontaneous HBsAg seroclearnce

Table 3. Determinants of Spontaneous HBsAg Seroclearance

Advanced age
Normal ALT levels at baseline
HBeAg negativity at baseline
Low viral load especially if HBV DNA level <300 copies/mL as baseline
Genotype B (>C) and A (>D)
Fatty liver
Persistent inactive carrier state
Cirrhosis
Hepatitis C virus superinfection

HBsAg — hepatitis B surface antigen; ALT — alanine aminotransferase; HBeAg — hepatitis B e antigen; HBV — hepatitis B virus.

had significantly higher body mass index and higher degrees of fatty liver than those without.[56] The underlying mechanism by which fatty liver enhanced HBsAg seroclearance remained unclear. Notably, in two independent large series cohort studies recently reported from Taiwan, fatty liver[57] and obesity (body mass index >30 kg/m^2) [53] were independent factors significantly associated with HBsAg seroclearance. The deteminants of HBsAg seroclearance are summarized in Table 3.

HBsAg seroclearance is generally accompanied by loss of all serum markers of HBV replication, including HBV DNA by PCR assays, and improved liver histology.[55] However, intrahepatic HBV-DNA, mainly in the form of covalently closed circular DNA, may persist longer and up to four years after HBsAg seroclearance.[49] After HBsAg seroclearance, anti-HBs seropositivity increases over time to 59% at 51 months of follow-up.[49] Patients with spontaneous HBsAg seroclearance had an excellent outcome because virtually none of the non-cirrhotic patients without concurrent HCV or HDV superinfection developed any untoward sequelae during a median follow-up period of 62 (12–179) months after HBsAg seroclearance.[58] However, in patients who have pre-existing cirrhosis or HCV and/or HDV superinfection, clinical outcomes of disease progression such as hepatic decompensation, HCC or liver related death may still occur.[58]

Reactivation of Hepatitis B

Following HBeAg seroconvesion, a subset of patients ultimately undergo spontaneous reactivation of HBV replication, with the reappearance of high levels of HBV DNA (>20,000 IU/mL) and a rise in ALT levels. Only a small proportion of carriers with HBV reactivation is associated with the reappearance of serum HBeAg (HBeAg reversion) and the remainder are persistently anti-HBe positive, suggesting that reactivation of hepatitis B usually results from HBV variants with precore or core promoter mutations that abolish or downregulate HBeAg production.[25,29] In addition, HBV replication can reactivate in inactive HBV carriers as a result of immunosuppression or cancer chemotherapy.[59]

HBeAg Reversion

HBeAg reversion occurs in a minority of patients following HBeAg seroconversion. In a study from Alaska, 432 (80%) of 541 seroconverters remained HBeAg negative and anti-HBe positive throughout the study, whereas the other 109 (20%) developed HBeAg reversion. HBeAg reversion episodes are frequently accompanied by a flare of hepatitis, and HBeAg tended to fluctuate between seroconversion and reversion, commonly having two to three reversions over the six to seven years of study.[52] However, HBeAg reversion was relatively uncommon in other studies. In two studies from Taiwan, one involving 283 patients with HBeAg-positive hepatitis and another involving 240 HBeAg-positive carriers with normal ALT at entry, HBeAg reversion following initial HBeAg seroconversion occurred in 12 (4.2%) and seven (2.9%) patients, respectively, during a mean follow-up of 8.6 years and 6.8 years.[25,42] Most HBeAg reversion occurs during the early years (median 1.1 years; range 0.3–3.2 years) after the onset of HBeAg seroconversion.[29] Another recent study from Italy showed that only one (1.6%) of 61 seroconverters had HBeAg reversion during a mean follow-up of 22.8 years.[43] Despite the low frequency, HBeAg reversion is significantly associated with increased risk of progression to cirrhosis,[29] as well as development of HCC.[52]

HBeAg-Negative Chronic Hepatitis

The majority of patients with reactivation of hepatitis B are negative for HBeAg,[25,29] and these patients are said to have "HBeAg-negative chronic hepatitis B." The general features of HBeAg-negative chronic hepatitis B are summarized in Table 1. Patients with HBeAg-negative chronic hepatitis B tend to be older, have lower serum HBV-DNA levels and lower inflammatory activity, but more advanced chronic liver disease, as compared with those with HBeAg-positive chronic hepatitis B. Moreover, acute exacerbation occurs at an estimated annual rate of 10.6%,[24] and spontaneous remission is infrequent (<15%).

The prevalence of HBeAg-negative chronic hepatitis B in chronic HBV infection varies widely among different geographical areas. The relative proportions of HBeAg-negative cases among cases with chronic hepatitis B are 80–90% in the Mediterranean areas; 30–50% in Taiwan and Hong Kong; but less than 10% in the US and Northern Europe.[60] However, the proportion of HBeAg-negative chronic hepatitis among anti-HBe positive carriers in different geographical areas remains largely unknown. This issue has been addressed in a few prospective studies that followed up on the natural course following spontaneous HBeAg seroconversion. In the two studies from Taiwan mentioned before, 24% (68 of 283) and 12% (29 of 240), respectively, developed HBeAg-negative chronic hepatitis during a mean follow-up of 8.6 and 6.8 years, with the annual rates of HBeAg-negative chronic hepatitis being estimated to be 3 and 2%, respectively.[25,29] The cumulative incidence of HBeAg-negative chronic hepatitis B following HBeAg seroconversion was 25% after 16 years of follow-up, but reactivation of hepatitis B typically occurred during the first five to 10 years following HBeAg seroconversion.[29] In a study from Italy, 9 (14.8%) of 61 seroconverters developed HBeAg-negative chronic hepatitis during a mean follow-up of 22.8 years.[42] In pediatric patients, the rate is even lower. Only 4 (6.3%) of 64 pediatric patients developed HBeAg-negative hepatitis during a mean period of 15 years following spontaneous HBeAg seroconversion.[61]

The incidence of reactivation of hepatitis B among incidentally identified anti-HBe-positive carriers with normal ALT levels varies between Western and Eastern countries. In one study of 1241 anti-HBe-positive carriers from Taiwan, 211 (17.0%) developed HBeAg-negative chronic hepatitis during a mean follow-up of 12.3 years, the cumulative rate being 20.2% after 20 years of follow-up. Reactivation of hepatitis B occurred much more commonly during the first five to 10 years of follow-up and became extremely rare after 20 years.[62] A study from Italy showed that only 3 (4%) of 68 healthy blood donors seropositive for HBsAg developed HBV-related biochemical changes during a mean follow-up of 11 years.[63] A study from northern Greece showed that only 4 (2.1%) of 195 "inactive" HBsAg carriers encountered HBV reactivation during a median follow-up of five years.[64] A more recent study from Greece showed a cumulative incidence of HBeAg-negative hepatitis of 24% at the end of fourth year follow-up of 85 "inactive carriers."[65] These apparent differences among these studies may be related to the different ethnic origin, HBV genotype, mode of transmission and age of the inactive carriers, that account for the different outcome of inactive HBsAg carriers among these countries.

Factors predictive for reactivation of hepatitis B following HBeAg seroconversion have been identified, as summarized in Table 4. Male gender, genotype C HBV (>genotype B),[66] genotype D HBV (>genotype A),[36] and baseline HBV-DNA levels ≥2,000 IU/mL[64,67] are associated with an increased risk of reactivation of hepatitis B. Age of HBeAg seroconversion before 30 was associated with a particularly

Table 4. Determinants of Spontaneous Reactivation of Hepatitis B

Male gender
Genotype C > B, D > A (C and F > A, B, and D for HBeAg reversion)
HBV DNA ≥ 10^4 copies/mL after HBeAg seroconversion
ALT flare ≥5 × upper limit of normal during immune clearance phase
Ages of HBeAg seroconversion >40

HBV — hepatitis B virus; ALT — alanine aminotransferase; HBeAg — hepatitis B e antigen.

low incidence of HBV reactivation.[50,62] In addition, ALT levels ≥5 times of ULN during the immune clearance phase and age of HBeAg seroconversion ≥40 years are also associated with an increased risk of reactivation of hepatitis B.[40,66] These findings suggest that hepatitis B is more likely to be reactivated if more vigorous immune-mediated hepatocytolysis or a more prolonged immune clearance phase is needed to clear the virus.

Studies in Taiwan have shown that patients with reactivation of hepatitis B have a 20-fold increased risk of progression to cirrhosis as compared with those without.[25] The annual rates of progression to cirrhosis and HCC were 2.3% and 0.5%, respectively, significantly higher than 0.1% and 0.2%, respectively, in those with sustained remission of hepatitis.[29] Notably, among patients with reactivation of hepatitis B, the incidence of cirrhosis correlated significantly with male gender and the age of reactivation over 40 years.[68]

Concurrent Viral Superinfection as Part of Natural Course

In high-prevalence areas such as Taiwan, 50–60% of adult patients admitted for acute clinical hepatitis are previously unrecognized HBsAg carriers with reactivation of hepatitis B or viral superinfection as they are positive for HBsAg but negative for IgM anti-HBc.[10] Viral superinfection was demonstrated in as high as 30–40% of these patients, with hepatitis C virus (HCV) being the most common, followed by hepatitis D virus (HDV), hepatitis A virus (HAV), and hepatitis E virus (HEV).[27] Non-B viral superinfection in HBsAg carriers tends to increase the severity and case fatality rate during the acute phase.[28,54] HAV and HEV superinfection have no chronic sequalae as HCV and HDV superinfection do.

Hepatitic C Virus

Acute HCV infection in HBeAg-positive HBsAg carriers usually results in only transient HCV infection. In contrast, most acute HCV superinfection in HBeAg-negative HBsAg carriers progress to persistent HCV infection.[69] These findings suggest that the presence of underlying active HBV replication may interfere with HCV replication and thereby inhibit the persistence of HCV infection. Worldwide, concurrent chronic HCV infection is present in 10–15% of chronic HBsAg carriers and is more common in injecting drug users.[70] HCV superinfection in HBsAg carriers may lead to a decrease in serum and liver HBV DNA levels and to increase the rate of HBsAg seroclearance. Most patients with dual HBV and HCV infection have detectable serum HCV RNA but not HBeAg or HBV DNA, suggesting that HCV is the predominant cause of liver disease in such cases.[70] In addition, HCV superinfection is associated with relatively more severe liver disease during the acute phase, with earlier and more frequent progression to cirrhosis than HDV superinfection or HBV infection alone.[54]

Hepatitis D Virus

HDV is a satellite virus, which is dependent on HBV for the production of envelope proteins. HDV superinfection in HBsAg carriers almost always results in chronic HDV infection. HDV superinfection is highly prevalent in the Mediterranean countries, but its prevalence decreases recently, probably due to the use of disposable needles, safe sex, and improvement in socioeconomic standards. HDV superinfection can also suppress HBV replication. Acute HDV superinfection is usually associated with relatively more severe liver disease.[54] A higher proportion of HBsAg carriers with HDV superinfection develops cirrhosis, hepatic decompensation, and HCC, as compared with those with chronic HBV infection alone.[54,71] In Taiwan, HDV superinfection was evident in around 10% of patients with chronic HBV infection, and does not seem to increase the incidence of cirrhosis, but may accelerate the progression to cirrhosis.[54] HDV superinfection is associated with relatively milder disease in Asia than in Europe, probably related to the different geographic distribution of HDV genotypes, with genotype II being dominant in the former and genotype I in the latter.[72]

Human Immunodeficiency Virus

Due to shared routes of transmission, human immunodeficiency virus (HIV) infection is prevalent (10–15%) in patients with chronic HBV infection in the West but rare in the East, possibly because of different prevalence of HIV infection. Many long-term follow-up studies have shown that individuals with HBV and HIV coinfection tend to have higher levels of HBV DNA, lower serum ALT levels and lower rates of spontaneous HBeAg seroconversion. Despite laboratory evidence of apparently quiescent disease, there appears to be a higher rate of progression to cirrhosis in HBV and HIV coinfection.[73] HBV-HIV coinfection is associated with a higher rate of liver-related mortality than infection with either HBV or HIV alone.[74] The development of effective antiretroviral regimen has led to immune reconstitution in many HIV-infected patients. The so-called immune reconstitution flares of hepatitis B have been observed in HBV-HIV coinfected patients following the initiation of HARRT.[75]

Sequelae and Mortality

Cirrhosis and Contributing Factors

It is estimated that cirrhosis develops in approximately 20% of patients with chronic HBV infection.[1] Whether this rate is higher in perinatally acquired infection because of the longer duration of infection than adult-acquired infection remains unknown. In one clinicopathologic study from Taiwan, liver cirrhosis was noted in 21% of asymptomatic HBsAg carriers who were over age 40.[76]

The annual incidence and cumulative probability of cirrhosis in patients with chronic HBV infection varied considerably in different reported series, possibly due to inclusion of patients in different phases of infection as well as the variable severity of liver injury in each phase. Studies have identified factors contributing to cirrhosis development: HBeAg positivity at recruitment[77,78]; persistence of serum HBV DNA during follow up[79]; persistent HBeAg seropositivity[41–43]; HBeAg reversion[29,52]; delayed HBeAg seroconversion over age 40[25,37,40]; reactivation of hepatitis B[25,29] especially if reactivation at the age over 40 years[68]; and ALT flares with hepatic decompensation or recurrent severe ALT flares with high AFP or bridging hepatic necrosis.[80] HBV genotype is also a contributing factor.[32,33] In other words, many patients have developed cirrhosis during the HBeAg positive phase, as shown in a recent study that 28 (30%) of 93 patients were HBeAg positive at the onset of their cirrhosis.[81] These data implies that the ultimate outcome of chronic HBV infection appears to depend on the duration of the immune clearance phase and reactivation of hepatitis B following HBeAg seroconversion, as well as on the severity of liver damage during these phases. Other factors significantly correlated with progression to cirrhosis included advanced age, hence longer duration of infection, male gender, and concurrent HCV, HDV or HIV superinfection, alcoholism and superimposed non-alcoholic fatty liver disease.

A recent population-based cohort study (the REVEAL-HBV) from Taiwan revealed that the risk of cirrhosis increased proportionally with increasing HBV DNA levels at recruitment.[82] Serum HBV DNA levels ≥2,000 IU/mL were significantly associated with cirrhosis. Patients included in the study were mostly HBeAg negative (85%), had normal ALT levels (94%) and of a median age of 45 (30–65) years at recruitment. Consistent with the clinical finding that patients with serum HBV-DNA ≥2,000 IU/mL have an increased risk of hepatitis B reactivation,[64,67] it seems likely that the patients with these HBV DNA levels at baseline in this cohort study are prone to have hepatitis B reactivation and progress to HBeAg-negative chronic hepatitis prior to cirrhosis development.[83]

Hepatic Decompensation and Mortality

HBV replication and necroinflammaory activity may have subsided at the onset of cirrhosis. However, at least one third to one half of patients with HBV cirrhosis still have a high level of HBV replication at presentation (positive HBeAg or HBV DNA by non-PCR assays).[81,84]

About 3–5% of patients with compensated HBV cirrhosis developed jaundice, ascites, hepatic encephalopathy or esophageal variceal bleeding each year.[81] The risk of hepatic decompensation is four-fold higher in HBeAg or HBV DNA-positive patients (4% per year) than in HBeAg and HBV DNA-negative patients (1% per year).[84]

One form of hepatic decompensation in HBV cirrhosis is secondary to hepatitis flares. In two studies from Taiwan, the annual incidence of hepatitis flare was 15–25% in HBeAg-positive patients and 5–10% in HBeAg-negative patients. Some 10–15% of acute flares were complicated with jaundice and 3–5% with ascites.[85,86] The annual incidence of hepatic decompensation was 1.5% and the cumulative incidence was 30.8% during a mean duration of 102 ± 60 (12–246) months.[81]

HCC and Contributing Factors

The incidence of HCC in chronic HBV infection correlates closely with the severity of the underlying liver diseases, as summarized by Fattovich *et al.*[87] In the East Asian countries, the summary annual rate of HCC ranges from 0.2% among inactive carriers to 0.8% in patients with chronic hepatitis B without cirrhosis and 3.7% in subjects with compensated cirrhosis; the corresponding five-year cumulative incidences are 1%, 3%, and 17%, respectively. In the Western countries, the summary annual rate of HCC is 0.02% in inactive carriers, 0.3% in patients with chronic hepatitis B without cirrhosis, and 2.2% in patients with compensated cirrhosis; the corresponding five-year cumulative incidences are 0.1%, 1%, and 10%, respectively.[87] These data confirm that liver cirrhosis is a well documented risk factor of HCC development and also suggest that perinatally acquired chronic HBV infection is associated with a greater risk of HCC than infection acquired in adults because of the longer duration of infection.

Most cases of HCC are likely to have concomitant cirrhosis. Factors significantly predictive for progression to cirrhosis therefore also contribute to HCC development. Other factors significantly associated with HCC in chronic HBV infection include race (Asians and Africans); a family history of HCC; HBV genotype; basal core promoter mutations and pre-S deletion mutations[88]; aflatoxin exposure; and alcohol drinking.

In the REVEAL-HBV study, the risk of HCC also increased proportionally with increasing HBV DNA levels.[89] A nested case control study further showed that HBV DNA levels and genotype C are synergistic in increasing the risk of HCC.[91] Unfortunately, these studies did not report the prevalence of cirrhosis among HCC patients, so it is difficult to determine whether increased viral replication, known to encourage the development of cirrhosis, may have any additional impact on HCC development.

Survival of Cirrhosis

The five-year survival rate of compensated, Child'A HBV cirrhosis is approximately 80–85%, which correlates closely with the status of HBV replication. Survival probability is > 95% in patients negative for HBeAg and HBV DNA but only 60–72% in HBeAg or HBV DNA positive patients.[84,92] Among the latter, HBeAg seroclearance is associated with a 2.2-fold decrease in mortality,[93] and ALT normalization is a better predictor of improved survival than HBeAg seroclearance.[94]

Once hepatic decompensation has developed, survival probability decreases remarkably. The reported five-year survival rate of decompensated, Child'B or Child'C HBV cirrhosis varies considerably among different reports, ranging from 14 to 88% (average 30–50%).[84] In one study from Hong Kong, the five-year survival rate was significantly lower in patients with positive HBeAg at presentation (57%) than in patients seronegative for HBeAg at presentation 88%.[95] In contrast, in another study from the Netherlands, the five-year survival rate was extremely low (14%), independent of serum HBeAg at enrollment.[93] These data may suggest that survival probability correlates significantly with the status of HBV replication in patients with less severe hepatic decompensation such as Child A and B cirrhosis, but not in patients with more severe hepatic decompensation such as Child'C cirrhosis.

Conclusion

The natural history of perinatally acquired chronic HBV infection and the annual rate of each event are summarized in Fig. 1. The natural course of perinatally acquired chronic HBV infection can be divided

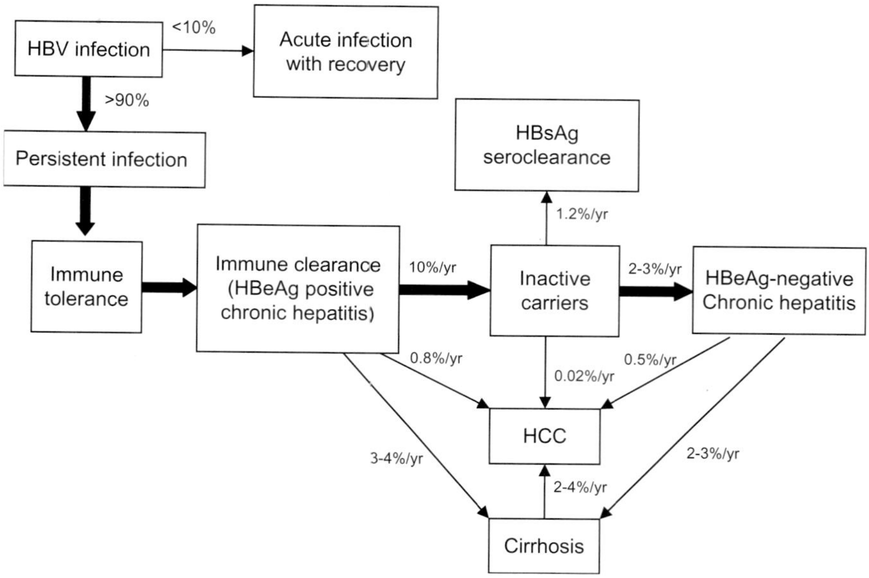

Fig. 1. Natural history of perinatally acquired chronic hepatitis B viru infection.

into three chronological phases: (1) immune tolerant phase; (2) immune clearance phase; and (3) inactive/residual phase. In adult-acquired chronic infection, however, no obvious initial immune tolerant phase is evident. HBeAg seroconversion usually marks the transition from active to inactive HBV infection with ALT normalization and resolution of hepatitis activity. HBV may reactivate and trigger immune-mediated liver injuries (HBeAg-negative chronic hepatitis) in as high as 25% of inactive carriers. This reactive phase can be viewed as a variant of immune clearance phase. Recent studies using HBsAg quantitation have shown that there is an overall correlation between serum HBsAg and HBV DNA levels, especially in HBeAg-positive patients, and the HBsAg levels decrease slowly and gradually, being the highest in the immune tolerant phase, followed by HBeAg-positive chronic hepatitis, HBeAg-negative chronic hepatitis, and the lowest in the inactive residual phase,[96–98] as shown in Table 1. However, there are more to learn before HBsAg quantitation is applied in clinical use.

HBV replication with subsequent interactions between HBV, hepatocytes and immune cells during the immune clearance phase or reactive phase may lead to hepatitis activity and disease progression. High HBV DNA levels and disease activity at enrollment and during follow-up are the best predictors of adverse clinical outcomes (hepatic decompensation, cirrhosis, HCC and liver related mortality). Sustained reduction of HBV replication before the onset of cirrhosis confers a favorable outcome, with similar survival compared to uninfected individuals. Sustained reduction of viral replication in cirrhotic patients also reduces the risk of hepatic decompensation, HCC development as well as improves survival. In other words, active HBV replication is the key driving force of disease progression, and therefore is the primary target of therapy. Recent improvements in the knowledge of the natural history of chronic hepatitis B and a detailed understanding of predictors of disease progression will help in the management of patients with chronic HBV infection.

References

1. Liaw YF, Chu CM. (2009) Hepatitis B virus infection. *Lancet* **373**: 582–592.
2. Perz JF, Armstrong GL, Farrington LA, *et al.* (2006) The contributions of hepatitis B virus and hepatitis C virus infections to cirrhosis and primary liver cancer worldwide. *J Hepatol* **45**: 529–538.
3. Iloeje UH, Yang HI, Jen CL, *et al.* (2007) Risk and predictors of mortality associated with chronic hepatitis B infection. *Clin Gastroenterol Hepatol* **5**: 921–931.
4. Milich D, Liang TJ. (2003) Exploring the biological basis of hepatitis B e antigen in hepatitis B virus infection. *Hepatology* **38**: 1075–1086.
5. Botha JF, Ritchie MJ, Dusheiko GM, *et al.* (1984) Hepatitis B virus carrier state in black children in Ovamboland: role of perinatal and horizontal infection. *Lancet* **1**: 1210–1212.
6. Chu CM, Sheen IS, Lin SM, Liaw YF. (1993) Sex difference in chronic hepatitis B virus infection: studies of serum HBeAg and alanine aminotransferase levels in 10,431 asymptomatic Chinese HBsAg carriers. *Clin Infect Dis* **16**: 709–713.
7. Shiraki K. (2000) Perinatal transmission of hepatitis B virus and its prevention. *J Gastroenterol Hepatol* **15**: E11–E15.
8. Ni YH, Huang LM, Chang MH, *et al.* (2007) Two decades of universal hepatitis B vaccination in Taiwan: impact and implication for future strategies. *Gastroenterology* 132: 1287–1293.

9. Chen CH, Yang PM, Huang GT, *et al.* (2007) Estimation of seroprevalence of hepatitis B virus and hepatitis C virus in Taiwan from a large-scale survey of free hepatitis screening participants. *J Formos Med Assoc* **106**: 148–155.
10. Chu CM, Liaw YF, Pao CC, Huang MJ. (1989) The etiology of acute hepatitis superimposed upon previously unrecognized asymptomatic HBsAg carriers. *Hepatology* **9**: 452–456.
11. Chu CM, Karayiannis P, Fowler MJ, *et al.* (1985) Natural history of chronic hepatitis B virus infection in Taiwan: studies of hepatitis B virus DNA in serum. *Hepatology* **5**: 431–434.
12. Chu CM. (2000) Natural history of chronic hepatitis B virus infection in adults with emphasis on the occurrence of cirrhosis and hepatocellular carcinoma. *J Gastroenterol Hepatol* **15** Suppl: E25–E30.
13. Chu CM, Liaw YF. (2004) Natural history differences in perinatally versus adult-acquired disease. *Current Hepatitis Reports* **3**: 123–131.
14. Liaw YF, Leung N, Kao JH, *et al.* (2008) Asian-Pacific consensus statement on the management of chronic hepatitis B: a 2008 update. *Hepatl Int* **2**: 263–283.
15. Chu CM, Liaw YF. (1987) Intrahepatic distribution of hepatitis B surface and core antigens in chronic hepatitis B virus infection. Hepatocyte with cytoplasmic/membranous hepatitis B core antigen as a possible target for immune hepatocytolysis. *Gastroenterology* **92**: 220–225.
16. Hui CK, Leung N, Yuen ST, *et al.* (2007) Natural history and disease progression in Chinese chronic hepatitis B patients in immune-tolerant phase. *Hepatology* **46**: 395–401.
17. Chu CM, Yeh CT, Lee CS, *et al.* (2002) Precore stop mutant in HBeAg-positive patients with chronic hepatitis B: clinical characteristics and correlation with the course of HBeAg-to-anti-HBe seroconversion. *J Clin Microbiol* **40**: 16–21.
18. Chu CM, Yeh CT, Sheen IS, Liaw YF. (1995) Subcellular localization of hepatitis B core antigen in relation to hepatocyte regeneration in chronic hepatitis B. *Gastroenterology* **109**: 1926–1932.
19. Liaw YF. (2003) Hepatitis flares and hepatitis B e antigen seroconversion: implication in anti-hepatitis B virus therapy. *J Gastroenterol Hepatol* **18**: 246–252.
20. Liaw YF, Pao CC, Chu CM. (1988) Changes of serum HBV-DNA in relation to serum transaminase level during acute exacerbation in patients with chronic type B hepatitis. *Liver* **8**: 231–235.
21. Tsai SL, Chen PJ, Lai MY, *et al.* (1992) Acute exacerbations of chronic type B hepatitis are accompanied by increased T cell responses to hepatitis B core and e antigens. Implications for hepatitis B e antigen seroconversion. *J Clin Invest* **89**: 87–96.
22. Liaw YF, Chu CM, Su IJ, *et al.* (1983) Clinical and histological events preceding hepatitis B e antigen seroconversion in chronic type B hepatitis. *Gastroenterology* **84**: 216–219.

23. Liaw YF, Yang SS, Chen TJ, Chu CM. (1985) Acute exacerbation in hepatitis B e antigen positive chronic type B hepatitis. A clinicopathological study. *J Hepatol* **1**: 227–233.
24. Liaw YF, Tai DI, Chu CM, *et al.* (1987) Acute exacerbation in chronic type B hepatitis: comparison between HBeAg and antibody-positive patients. *Hepatology* 7: 20–23.
25. Chu CM, Hung SJ, Lin J, *et al.* (2004) Natural history of hepatitis B e antigen to antibody seroconversion in patients with normal serum aminotransferase levels. *Am J Med* **116**: 829–834.
26. Sheen IS, Liaw YF, Tai DI, Chu CM. (1985) Hepatic decompensation associated with hepatitis B e antigen clearance in chronic type B hepatitis. *Gastroenterology* **89**: 732–735.
27. Chu CM, Lin SM, Hsieh SY, *et al.* (1999) Etiology of sporadic acute viral hepatitis in Taiwan: the role of hepatitis C virus, hepatitis E virus and GB virus-C/hepatitis G virus in an endemic area of hepatitis A and B. *J Med Virol* **58**: 154–159.
28. Chu CM, Liaw YF. (2005) Increased incidence of fulminant hepatic failure in previously unrecognized HBsAg carriers with acute hepatitis independent of etiology. *Infection* **33**: 136–139.
29. Hsu YS, Chien RN, Yeh CT, *et al.* (2002) Long-term outcome after spontaneous HBeAg seroconversion in patients with chronic hepatitis B. *Hepatology* **35**: 1522–1527.
30. Liaw YF, Chu CM, Huang MJ, *et al.* (1984) Determinants for hepatitis B e antigen clearance in chronic type B hepatitis. *Liver* **4**: 301–306.
31. Chu CJ, Hussain M, Lok AS. (2002) Hepatitis B virus genotype B is associated with earlier HBeAg seroconversion compared with hepatitis B virus genotype C. *Gastroenterology* **122**: 1756–1762.
32. Kao JH. (2007) Role of viral factors in the natural course and therapy of chronic hepatitis B. *Hepatol Int* **1**: 415–430.
33. Chu CM, Liaw YF. (2005) Genotype C hepatitis B virus infection is associated with a higher risk of reactivation of hepatitis B and progression to cirrhosis than genotype B: a longitudinal study of hepatitis B e antigen-positive patients with normal aminotransferase levels at baseline. *J Hepatl* **43**: 411–417.
34. Livingston SE, Simonetti JP, Bulkow LR, *et al.* (2007) Clearance of hepatitis B e antigen in patients with chronic hepatitis B and genotypes A, B, C, D, and F. *Gastroenterology* **133**: 1452–1457.
35. Marx G, Martin SR, Chicoine JF, Alvarez F. (2002) Long-term follow-up of chronic hepatitis B virus infection in children of different ethnic origins. *J Infect Dis* **186**: 295–301.
36. Sánchez-Tapias JM, Costa J, Mas A, *et al.* (2002) Influence of hepatitis B virus genotype on the long-term outcome of chronic hepatitis B in western patients. *Gastroenterology* **123**: 1848–1856.

37. Chu CM, Liaw YF. (2007) Chronic hepatitis B virus infection acquired in childhood: special emphasis on prognostic and therapeutic implication of delayed HBeAg seroconversion. *J Viral Hepat* **14**: 147–152.
38. Chang MH, Hsu HY, Hsu HC, *et al.* (1995) The significance of spontaneous hepatitis B e antigen seroconversion in childhood: with special emphasis on the clearance of hepatitis B e antigen before 3 years of age. *Hepatology* **22**: 1387–1392.
39. Chu CM, Sheen IS, Lin SM, Liaw YF. (1993) Sex difference in chronic hepatitis B virus infection: studies of serum HBeAg and alanine aminotransferase levels in 10,431 asymptomatic Chinese HBsAg carriers. *Clin Infect Dis* **16**: 709–713.
40. Chen YC, Chu CM, Liaw YF. (2010) Age-specific prognosis following spontaneous hepatitis B e antigen seroconversion in chronic hepatitis B. *Hepatology* **51**: 435–444.
41. Niederau C, Niederau C, Heintges T, *et al.* (1996) Long-term follow-up of HBeAg-positive patients treated with interferon alfa for chronic hepatitis B. *N Engl J Med* **334**: 1422–1427.
42. Lin SM, Lin SM, Yu ML, *et al.* (2007) Interferon therapy in HBeAg positive chronic hepatitis reduces progression to cirrhosis and hepatocellular carcinoma. *J Hepatol* **46**: 45–52.
43. Fattovich G, Fattovich G, Olivari N, *et al.* (2008) Long-term outcome of chronic hepatitis B in Caucasian patients: mortality after 25 years. *Gut* **57**: 84–90.
44. Keeffe EB, Dieterich DT, Han SH, *et al.* (2008) A treatment algorithm for the management of chronic hepatitis B virus infection in the United States: 2008 update. *Clin Gastroenterol Hepatol* **6**: 1315–1341.
45. Chu CM, Chen YC, Tai DI, Liaw YF. (2010) Level of hepatitis B virus DNA in inactive carriers with persistently normal levels of alanine aminotransferase. *Clin Gastroenterol Hepatol* **8**: 535–540.
46. Villeneuve JP, Desrochers M, Infante-Rivard C, *et al.* (1994) A long-term follow-up study of asymptomatic hepatitis B surface antigen-positive carriers in Montreal. *Gastroenterology* **106**: 1000–1005.
47. Manno M, Cammà C, Schepis F, *et al.* (2004) Natural history of chronic HBV carriers in northern Italy: morbidity and mortality after 30 years. *Gastroenterology* **127**: 756–763.
48. Liaw YF, Sheen IS, Chen TJ, *et al.* (1991) Incidence, determinants and significance of delayed clearance of serum HBsAg in chronic hepatitis B virus infection: a prospective study. *Hepatology* **13**: 627–631.
49. Yuen MF, Wong DK, Sablon E, *et al.* (2004) HBsAg seroclearance in chronic hepatitis B in the Chinese: virological, histological, and clinical aspects. *Hepatology* **39**: 1694–1701.
50. Chu CM, Law YF. (2007) HBsAg seroclearance in asymptomatic carriers of high endemic areas: appreciably high rates during a long-term follow-up. *Hepatology* **45**: 1187–1192.

51. Kim JH, Lee JH, Park SJ, *et al.* (2008) Factors associated with natural seroclearance of hepatitis B surface antigen and prognosis after seroclearance: a prospective follow-up study. *Hepatogastroenterology* **55**: 578–581.
52. McMahon BJ, Holck P, Bulkow L, Snowball M. (2001) Serologic and clinical outcomes of 1536 Alaska Natives chronically infected with hepatitis B virus. *Ann Intern Med* **135**: 759–768.
53. Liu J, Yang HI, Lee MH, *et al.* (2010) REVEAL-HBV Study Group. Incidence and determinants of spontaneous hepatitis B surface antigen seroclearance: a community-based follow-up study. *Gastroenterology* **139**: 474–482.
54. Liaw YF, Chen YC, Sheen IS, *et al.* (2004) Impact of acute hepatitis C virus superinfection in patients with chronic hepatitis B virus infection. *Gastroenterology* **126**: 1024–1029.
55. Chu CM, Liaw YF. (2010) Hepatitis B surface antigen seroclearance during chronic HBV infection. *Antivir Ther* **15**: 133–143.
56. Chu CM, Lin DY, Liaw YF. (2007) Does increased body mass index with hepatic steatosis contribute to seroclearance of hepatitis B virus (HBV) surface antigen in chronic HBV infection? *Int J Obes (Lond)* **31**: 871–875.
57. Tai DI, Lin SM, Sheen IS, *et al.* (2009) Long-term outcome of hepatitis B e antigen-negative hepatitis B surface antigen carriers in relation to changes of alanine aminotransferase levels over time. *Hepatology* **49**: 1859–1867.
58. Chen YC, Sheen IS, Chu CM, Liaw YF. (2002) Prognosis following spontaneous HBsAg seroclearance in chronic hepatitis B patients with or without concurrent infection. *Gastroenterology* **123**: 1084–1089.
59. Perrillo RP. (2001) Acute flares in chronic hepatitis B: the natural and unnatural history of an immunologically mediated liver disease. *Gastroenterology* **120**: 1009–1022.
60. Funk ML, Rosenberg DM, Lok AS. (2002) World-wide epidemiology of HBeAg-negative chronic hepatitis B and associated precore and core promoter variants. *J Viral Hepat* **9**: 52–61.
61. Bortolotti F, Guido M, Bartolacci S, *et al.* (2006) Chronic hepatitis B in children after e antigen seroclearance: final report of a 29-year longitudinal study. *Hepatology* **43**: 556–562.
62. Chu CM, Liaw YF. (2007) Spontaneous relapse of hepatitis in inactive HBsAg carrier. *Hepatol Int* **1**: 311–315.
63. de Franchis R, Meucci G, Vecchi M, *et al.* (1993) The natural history of asymptomatic hepatitis B surface antigen carriers. *Ann Intern Med* **1**: 118: 191–194.
64. Zacharakis GH, Koskinas J, Kotsiou S, *et al.* (2005) Natural history of chronic HBV infection: a cohort study with up to 12 years follow-up in North Greece (part of the Interreg I-II/EC-project). *J Med Virol* 77: 173–179.
65. Papatheodoridis GV, Chrysanthos N, Hadziyannis E, *et al.* (2008) Longitudinal changes in serum HBV DNA levels and predictors of progression during the

natural course of HBeAg-negative chronic hepatitis B virus infection. *J Viral Hepat* **15**: 434–441.

66. Chu CM, Liaw YF. (2007) Predictive factors for reactivation of hepatitis B following hepatitis B e antigen seroconversion in chronic hepatitis B. *Gastroenterology* **133**: 1458–1465.
67. Feld JJ, Ayers M, El-Ashry D, *et al.* (2007) Hepatitis B virus DNA prediction rules for hepatitis B e antigen-negative chronic hepatitis B. *Hepatology* **46**: 1057–1070.
68. Chu CM, Liaw YF. (2009) Incidence and risk factors of progression to cirrhosis in inactive carriers of hepatitis B virus. *Am J Gastroenterol* **104**: 1693–1699.
69. Chu CM, Yeh CT, Sheen IS, Liaw YF. (2002) Acute hepatitis C virus (HCV) infection in chronic carriers of hepatitis B virus (HBV): the impact of underlying active HBV replication on persistence of HCV infection and antibody responses to HCV. *Gut* **51**: 95–99.
70. Liaw YF. (1995) Role of hepatitis C virus in dual and triple hepatitis virus infection. *Hepatology* **22**: 1101–1108.
71. Fattovich G, Boscaro S, Noventa F, *et al.* (1987) Influence of hepatitis delta virus infection on progression to cirrhosis in chronic hepatitis type B. *J Infect Dis* **155**: 931–935.
72. Wu JC, Choo KB, Chen CM, *et al.* (1995) Genotyping of hepatitis D virus by restriction-fragment length polymorphism and relation to outcome of hepatitis D. *Lancet* **346**: 939–941.
73. Colin JF, Cazals-Hatem D, Loriot MA, *et al.* (1999) Influence of human immunodeficiency virus infection on chronic hepatitis B in homosexual men. *Hepatology* **29**: 1306–1310.
74. Thio CL, Seaberg EC, Skolasky R Jr, *et al.* (2002) HIV-1, hepatitis B virus, and risk of liver-related mortality in the Multicenter Cohort Study (MACS). *Lancet* **360**: 1921–1926.
75. Drake A, Mijch A, Sasadeusz J. (2004) Immune reconstitution hepatitis in HIV and hepatitis B coinfection, despite lamivudine therapy as part of HAART. *Clin Infect Dis* **39**: 129–132.
76. Yang PM, Chen DS, Lai MY, *et al.* (1987) Clinicopathologic studies of asymptomatic HBsAg carriers: with special emphasis on carriers older than 40 years. *Hepatogastroenterology* **34**: 251–254.
77. Yu MW, Hsu FC, Sheen IS, *et al.* (1997) Prospective study of hepatocellular carcinoma and liver cirrhosis in asymptomatic chronic hepatitis virus carriers. *Am J Epidemiol* **145**: 1039–1047.
78. Yang HI, Lu SN, Liaw YF, *et al.* (2002) Hepatitis B e antigen and the risk of hepatocellular carcinoma. *N Engl J Med* **347**: 168–174.
79. Brunetto MR, Oliveri F, Coco B, *et al.* (2002) Outcome of anti-HBe positive chronic hepatitis B in alpha-interferon treated and untreated patients: a long term cohort study. *J Hepatol* **36**: 263–270.

80. Liaw YF, Tai DI, Chu CM, Chen TJ. (1988) The development of cirrhosis in patients with chronic type B hepatitis: a prospective study. *Hepatology* **8**: 493–496.
81. Chen YC, Chu CM, Yeh CT, Liaw YF. (2007) Natural course following the onset of cirrhosis in patients with chronic hepatitis B: a long-term follow-up study. *Hepatol Intl* **1**: 267–273.
82. Iloeje UH, Yang HI, Su J, *et al.* (2006) Predicting cirrhosis risk based on the level of circulating hepatitis B viral load. *Gastroenterology* **130**: 678–686.
83. Liaw YF. (2006) Hepatitis B virus replication and liver disease progression: The impact of antiviral therapy. *Antivir Ther* **11**: 669–679.
84. Chu CM, Liaw YF. (2006) Hepatitis B virus-related cirrhosis: natural history and treatment. *Semin Liver Dis* **26**: 142–152.
85. Liaw YF, Lin DY, Chen TJ, Chu CM. (1989) Natural course after the development of cirrhosis in patients with chronic type B hepatitis: a prospective study. *Liver* **9**: 235–241.
86. Liaw YF, Chen JJ, Chen TJ. (1990) Acute exacerbation in patients with liver cirrhosis: a clinicopathological study. *Liver* **10**: 177–184.
87. Fattovich G, Bortolotti F, Donato F. (2008) Natural history of chronic hepatitis B: special emphasis on disease progression and prognostic factors. *J Hepatol* **48**: 335–352.
88. Liaw YF. (2010) Does chemotherapy prevent HBV-related hepatocellular carcinoma? Pros. *Dig Liver Dis* **42**(Suppl 3): S293–S297.
89. Chen CJ, Yang HI, Su J, *et al.* (2006) Risk of hepatocellular carcinoma across a biological gradient of serum hepatitis B virus DNA level. *JAMA* **295**: 65–73.
90. Yang HI, Yeh SH, Chen PJ, *et al.* (2008) REVEAL-HBV study group. Associations between hepatitis B virus genotype and mutants and the risk of hepatocellular carcinoma. *J Natl Cancer Inst* **100**: 1134–1143.
91. Yu MW, Yeh SH, Chen PJ, *et al.* (2005) Hepatitis B virus genotype and DNA level and hepatocellular carcinoma: a prospective study in men. *J Natl Cancer Inst* **97**: 265–272.
92. Liaw YF, Sung JJY, Chow WC, *et al.* (2004) Lamivudine for patients with chronic hepatitis B and advanced liver disease. *N Engl J Med* **351**: 1521–1531.
93. de Jongh FE, Janssen HL, de Man RA, *et al.* (1992) Survival and prognostic indicators in hepatitis B surface antigen-positive cirrhosis of the liver. *Gastroenterology* **103**: 1630–1635.
94. Fattovich G, Giustina G, Realdi G, *et al.* (1997) Long-term outcome of hepatitis B e antigen-positive patients with compensated cirrhosis treated with interferon alfa. European Concerted Action on Viral Hepatitis (EUROHEP). *Hepatology* **26**: 1338–1342.
95. Chung HT, Lai CL, Lok AS. (1995) Pathogenic role of hepatitis B virus in hepatitis B surface antigen-negative decompensated cirrhosis. *Hepatology* **22**: 25–29.

96. Nguyen T, Thompson AJ, Bowden S, *et al.* (2010) Hepatitis B surface antigen levels during the natural history of chronic hepatitis B: a perspective on Asia. *J Hepatol* **52**: 508–513.
97. Jaroszewicz J, Calle Serrano B, Wursthorn K, *et al.* (2010) Hepatitis B surface antigen (HBsAg) levels in the natural history of hepatitis B virus (HBV)-infection: a European perspective. *J Hepatol* **52**: 514–522.
98. Chan HL, Wong VW, Wong GL, *et al.* (2010) A longitudinal study on the natural history of serum hepatitis B surface antigen changes in chronic hepatitis B. *Hepatology* **52**: 1232–1241.

Chapter 2

Immune Response and Viral Hepatitis B

David R. Milich*

Introduction

The hepatitis B virus (HBV) is a major cause of infectious liver disease throughout the world. An estimated one million people die annually as a result of complications of HBV infection. According to the CDC, there are 1.2 million carriers of HBV in the U.S. and approximately 400 million worldwide.

The clinical consequences of HBV infection are extremely variable. After clinically apparent acute hepatitis B infection in adults, approximately 90% of affected adults recover without sequelae, and develop immunity to the virus; however, the clinical course can be quite variable. In another 0.1–0.5% of cases, a fulminant hepatitis results in which cell death in the liver is so extensive that fewer than 20–30% of the patients survive. Finally, 5–10% of adults exposed to HBV develop a chronic infection. The nature of the chronic infection is also variable, ranging from continuous hepatocellular necrosis and inflammation (chronic active hepatitis, CAH) to an asymptomatic carrier (ASC) state. In contrast to adult infection, neonatal HBV

* Vaccine Research Institute of San Diego, 3030 Bunker Hill Street, Suite 205, San Diego, CA 92109, Tel: (858) 581-3960, Fax: (858) 581-3970, E-mail: dmilich@vrisd.org

infection is rarely cleared and as many as 90% of perinatally infected children become chronically infected.[1] Vertical transmission of HBV is the major source of chronic infection in endemic areas in Asia. Chronic HBV infection is associated with hepatocellular carcinoma (HCC). In fact, the risk of HCC is 100-fold in patients with chronic HBV infection.[2] Effective HBV vaccine programs have interrupted the vicious cycle of perinatal infection, chronicity, and late-term complications including HCC which plagues endemic areas.

Asymptomatic HBV carriers can have relatively normal liver morphology and function despite the presence of high levels of viral replication within hepatocytes. This and a number of other observations have led investigators to the conclusion that liver injury and the subsequent clearance of virus is mediated by the immune response, and that the HBV is not directly cytopathic.[3] An implication of this conclusion is that an inadequate immune response leads to chronicity. Although the success of HBV immunization programs attests to the role of antibody in protective immunity, evidence supports the role of the cellular immune response in the pathogenesis of HBV-associated liver disease, although both arms of the immune response are involved in viral clearance. The lack of inbred animal models of HBV infection and the difficulty of cultivating the virus *in vitro* have hampered immunopathogenic studies. At least four general approaches have been employed to examine immune responses to HBV proteins: clinical and serologic studies; human *in vitro* cellular assays; murine *in vivo* (i.e., transgenic) and *in vitro* model systems; and immunologic studies in the animal models of hepadnavirus infection. The purpose of this review is to attempt to interpret the findings from these varied approaches in the context of the central issues in HBV immunology and immunopathogenesis that are yet to be fully resolved. These issues include (a) mechanisms of persistence; (b) variability of chronic disease; (c) role of virus variants; and (d) mechanisms of liver injury and viral clearance.

Serology of Acute and Chronic HBV Infections

The serology of acute hepatitis B (AH-B) infection has been extensively studied. The HBsAg, including the preS region, is an early marker of HBV infection. Although the anti-HBs response is often delayed in appearance, recent data suggest that anti-preS responses may occur quite early in infection.[4] Nevertheless, the humoral responses to the envelope antigens can show great variation from patient to patient during infection and after vaccination, ranging from high to non-responder phenotypes. With respect to the HBcAg, high-titered antibodies are regularly produced by virtually 100% of HBV-infected patients. The high frequency of anti-HBc production is due, at least in part, to the fact that HBcAg can function as a T cell-dependent antigen.[5] IgM anti-HBc appears early in acute hepatitis B, and IgM and IgG anti-HBc can persist with slowly decreasing titers for many years. In contrast, seroconversion to anti-HBe status occurs later than anti-HBc, can be quite variable in terms of onset, and correlates with viral clearance.[6]

The serology of chronic hepatitis B (CH-B) infection has been reported to be remarkably homogeneous considering the diversity of clinical classifications and degrees of liver injury. For example, the sera of CH-B patients contain HBV virions, subviral particles (HBsAg and preS), the HBeAg, and antibody to the nucleocapsid (anti-HBc), but remain seronegative for anti-envelope and anti-HBe antibodies for many years until HBV DNA clearance and loss of serum HBeAg.[7] Serum HBsAg remains positive in many resolved as well as unresolved CH-B infections, and correspondingly anti-HBs often remains undetectable even after HBV clearance.

The PreS Envelope Antigens

The immune responses to the preS antigens of the HBV envelope have been less well studied. Anti-preS antibodies have been shown to be virus neutralizing in chimpanzee protection studies.[8–12] Several groups have examined the humoral immune response to these antigens during HBV infection. Neurath *et al.*[13] demonstrated the presence of preS(1)-specific antibodies in the sera of individuals recovering from acute HBV infection, and noted that anti-preS(1) antibodies can be detected prior to anti-HBs or anti-HBc seroconversion but decline rapidly after production of anti-HBs and anti-HBc. The human anti-preS(1) response can include multiple specificities since peptide sequences represented by 1–21, 12–32, 32–53, and 94–117 are variably recognized by sera from different HBV-infected patients.[14,15] Analysis of the temporal occurrence of anti-preS(1) in an HBV-infected chimpanzee also illustrated the early appearance of this antibody, which was detected shortly after preS(1) antigen appeared during acute disease.[16] Several studies have examined preS(1) antibody production in acute and chronic HBV infection and found that antibodies to preS(1) appear at an early stage of acute resolving HBV infection, but the antibodies are absent[17] or of differing specificity[14] in the sera of patients with an HBV infection entering a chronic course.

PreS(2)-specific antibodies appear early during infection and decline in titer rather rapidly.[13] Several studies have suggested a transient nature for antibodies to preS(2) during HBV infection.[18,19] However, in chronic HBV infection, preS(2)-specific antibodies were detected in only 1% of 358 symptomatic carriers seropositive for anti-HBe and in none of 67 sera that were seropositive for HBeAg.[20] Further temporal analysis of anti-preS production during acute and chronic HBV infections also indicated a correlation between HBV clearance and anti-preS production. No anti-preS(2) was detected in acute patients who progressed to chronicity or in patients with chronic hepatitis with persistence of HBsAg and preS antigens regardless of HBeAg/anti-HBe status.[21,22] These results indicate that preS(2) antibody is not produced by asymptomatic carriers regardless of HBeAg/anti-HBe status, and in this respect anti-preS(2) parallels anti-HBs region antibody production.

One caveat to the inability to detect anti-preS as well as anti-HBs and anti-HBe antibodies in chronic patients may be the simultaneous presence of excess amounts of the respective antigens in sera.

Increased Complexity of the Serology of Chronic HBV Infection

If cellular immune responses are responsible for liver cell injury and HBV clearance, there should be serological evidence of these underlying cellular processes. However, the presence of HBsAg and HBeAg in the serum of chronic patients may mask the ability to detect circulating antibodies and may obscure the onset of seroconversion. The available commercial assays usually detect anti-HBs and anti-HBe antibodies only after the antigens have been cleared from the serum. To address the possibility that antibodies to the structural HBV antigens are produced but not detected in CH-B infection, several groups have developed sensitive immunoassays for the detection of anti-HBe and anti-envelope antibodies, regardless of the presence of the antigens in serum.[23–25] Using these more sensitive immunoassays, a series of serum samples from 200 HBsAg- and HBeAg-positive CH-B patients with various degrees of liver disease were analyzed.[23] All patients were seronegative for antibodies specific for the envelope antigens or the HBeAg when the current commercial assays were utilized. In contrast, virtually all chronically infected patients with liver disease, and approximately 50% of chronic patients without liver disease, demonstrated anti-HBe and anti-envelope antibodies when sera were tested in the more sensitive immunoassays. Furthermore, the asymptomatic patients could be serologically distinguished from the symptomatic patients based on antibody fine specificity, titer, and IgG subclass. For example, the production of anti-HBc and anti-HBe antibodies of the IgG_3 isotype correlated with symptomatic infection. This and other studies reporting the presence of circulating HBsAg- and HBeAg-containing immune complexes (IC)[23,26–31] and free anti-HBe antibodies[23,25] in the sera of CH-B patients demonstrate that not all chronically infected HBV patients are immunologically unresponsive, as suggested by the current commercial assays. In fact, only 44.4% of even asymptomatic chronic carriers were characterized by the absence of anti-HBe and anti-envelope antibody production. Anti-HBe seroconversion can occur many years prior to the loss of HBeAg or liver injury. Similarly, anti-envelope antibodies may co-exist with virions and subviral HBsAg particles for many years before viral clearance and loss of HBsAg.

The specificity and source of Th cell function for anti-HBc, anti-HBe, and anti-envelope antibody production in the context of CH-B infection may also be surmised from a complete serologic analysis. We proposed that low titer and IgG_3 subclass-deficient anti-HBc and anti-HBe responses in asymptomatic versus chronic active hepatitis patients indicate reduced Th cell function in asymptomatic carriers, and may also reflect differential engagement of Th cell subsets among these patient groups. The serologic profile and absence of liver disease in asymptomatic carriers is consistent with an exclusive Th_2-like response. Alternatively, the serology and evidence of immune-mediated cytotoxic responses in chronic active hepatitis patients suggest a Th_1-like or combined Th_1/Th_2-like response. In addition to the diagnostic value of more sensitive immunoassays, these results suggest possible therapeutic approaches for terminating the chronic carrier state. For example, it may be beneficial to convert a predominantly Th_2-type HBeAg-specific Th cell response into a Th_1-predominant response that may mediate viral clearance more efficiently. In fact, the administration of recombinant IL-12 to HBeAg transgenic (Tg) mice can shift a predominant Th_2-like response to a predominant Th_1-like response[32] and inhibit HBV replication in the liver of HBV-Tg mice.[33] Studies in chronic HBV carriers have demonstrated that $_r$IL-12 enhanced HBV-specific T cell responses *in vitro*.[34–36] In two studies, $_r$IL-12 administered to chronic HBV patients increased HBV-specific T cell reactivity and exerted significant antiviral activity.[37,38] The antiviral effects were dose-dependent, but severe toxicity at high $_r$IL-12 doses compromises this therapy. In an attempt to limit systemic IL-12 levels and possible toxicity, IL-12-mediated gene therapy using an adenovirus vector encoding murine IL-12 was recently tested in woodchucks chronically infected with WHV.[39] All woodchucks with viral loads below 10^{10} viral genomes (vg/mL) demonstrated significant and sustained reduction in viremia, accompanied by a reduction in hepatic WHV DNA, a loss of WHeAg and WHsAg, and improved liver histology, which correlated with an induction of T cell immunity.[39] Interestingly, none of the woodchucks with higher viremia levels responded to therapy. This result is consistent with the hypothesis that serum subviral envelope particles and the HBeAg function as immune tolerogens and the degree of tolerance is antigen dose dependent (see below).

Human T Cell Responses in Acute and Chronic HBV Infection

CD4⁺ Th Cell Response

The T cell responses that occur during HBV infection appear to differ depending on the viral antigen, the T cell function (i.e., Th vs. CTL), and the clinical outcome. For example, during acute self-limited HBV infection a vigorous Th cell ($CD4^+$) response (determined by proliferation or cytokine production) specific for the HBc/HBeAgs is observed.[40–42] The HBc/HBeAg-specific Th cell response is quite heterogeneous and involves recognition of many T cell epitopes. In contrast, the Th cell responses elicited by the envelope antigens during acute HBV infection are considerably less vigorous.[41,43,44] This is somewhat surprising, given the fact that the acute infection is rapidly cleared, and given the ability to detect HBsAg-specific Th cell sensitization after vaccination.[45–47] At least two studies suggests that HBsAg-specific Th cell activation may occur quite early in the preclinical phase of infection, and may therefore elude detection.[48,49] Similarly, during chronic HBV infection the HBc/HBeAg-specific Th cell responses appear superior to the HBsAg-specific responses[41,42,50–55] however, all T cell responses are significantly more difficult to detect in CH-B patients than in AH-B patients.[41,42,56] These studies are consistent with the murine data in a non-infectious system, which indicated that HBcAg is approximately 100-fold more immunogenic than HBsAg at both the Th cell and B cell levels. This suggests that the differential immunogenicity may be due to intrinsic characteristics of these two particulate proteins, at least in terms of the endocytic, MHC Class II-restricted antigen presentation pathway.[5,57] Consistent with the murine observations, molecular analyses of human HLA Class II molecules demonstrated that high affinity ligands for a variety of Class II alleles were more prevalent for the HBc/eAgs than for the envelope antigens, especially the HBsAg, which was extremely limited.[58]

A number of findings suggest that Th cells specific for HBc/HBeAg play an important role in viral clearance. For example, HBc/HBeAg-specific Th cell sensitization appears to correlate with active liver disease,[51,59] and with the clearance of HBsAg and HBV-DNA from the serum.[41,42,56]

How might HBc/HBeAg-specific Th cells affect HBV clearance?

(a) The induction or maintenance of HBcAg/HBeAg-specific CTL may require Th cell function, which is consistent with the correlation between the HBc/HBeAg-specific Th cell and CTL responses.[60,61] We have recently observed in mice that the $CD4^+$ Th cell dependence of HBcAg-specific $CD8^+$ CTL varies depending on the efficiency of the methods used to induce the HBcAg-specific $CD8^+$ CTL response. Less than optimal induction requires the presence of HBcAg-specific $CD4^+$ Th cells.[62] CTL function may be dependent on Th cells for a continuous supply of IL-2 or for the IFNγ necessary to induce MHC Class I expression on hepatocytes.
(b) $CD4^+$ Th cells may be directly cytotoxic for infected hepatocytes via secretion of inflammatory cytokines (i.e., TNFα).
(c) Cytokines produced by $CD4^+$ Th cells (i.e., IL-2, IFNs, TNFα) may down-regulate HBV replication by virtue of their effects on HBV gene expression, as shown in HBsAg-Tg murine model systems[63–67] (see below).
(d) Lastly, HBc/HBeAg-specific $CD4^+$ Th cells can help HBV envelope-specific B cells produce neutralizing antibodies by the process of "intermolecular/intrastructural" T cell help.[68] In this process, envelope-specific B cells that bind virions process the complete virus particle, including the HBcAg, and present all viral peptides in the context of MHC Class II molecules on the B cell surface. Therefore, HBcAg peptides as well as envelope-derived peptides can serve as ligands for Th cell recognition, and anti-HBs antibody production can result from HBcAg-specific Th cell function. Although antibodies to HBcAg do not prevent infection, the fact that HBc/HBeAg-specific Th cells can elicit anti-envelope antibodies that are virus-neutralizing may explain the reported ability of HBcAg vaccination to protect against HBV liver disease.[69,70] The protective potential of nucleocapsid-specific Th cells has been extended in the woodchuck model

system to include immunization with synthetic Th cell sites derived from the WHcAg.[71] Immunization with WHcAg residues 97–110 protected woodchucks from experimental infection.

CD8⁺ CTL Response

Virus-specific CTL are believed to play a major role in viral clearance by virtue of their ability to directly kill infected cells that express the appropriate peptide presented by MHC Class I molecules,[72] and/or to secrete antiviral factors such as IFNγ and TNFα upon recognition of MHC class/peptide complexes.[73] Studies of patients with self-limited acute hepatitis demonstrated that CTL specific for multiple sites within the envelope, polymerase, and HBc/HBeAg proteins are detectable.[61,74,75] Although these studies have concentrated primarily on HLA-A2-restricted CTL responses, the heterogeneity of CTL sites recognized on the various proteins is interesting. Only one HLA-A2-restricted site has been identified on HBc/HBeAg (i.e., 18–27), whereas nine distinct sites within envelope proteins and five sites within the polymerase protein were identified.[61,74,75] Therefore, the dominance of the HBc/HBeAg in terms of $CD4^+$ Th cell function does not extend to CTL recognition, which appears to be directed more extensively toward the envelope and polymerase proteins. This may be at least partially explained by the rather unique ability of exogenous HBsAg to enter the MHC Class I processing pathway and induce $CD8^+$, Class I-restricted CTL as efficiently as endogenously synthesized HBsAg.[47,76–78] The HBcAg does not share this ability.[79] Induction of a CTL response to HBcAg in mice requires genetic immunization with either plasmid DNA[62,79] or recombinant retroviral or *Vaccinia* infection.[80,81] It is also of interest that HBcAg and the HBeAg are cross-reactive at the level of CTL recognition in man[82] and in mice.[80,81]

Given the efficiency of HBsAg in eliciting CTL during acute infection[74] and in the mouse model after immunization,[77] it is remarkable that envelope-specific CTL are undetectable or only marginally detectable[83] during chronic HBV infection. Similarly, CTL specific for all the HBV proteins are difficult to detect during persistent HBV infection, and are most efficiently detected among liver-derived lymphocytes. These observations suggest that HBV-specific CTL play an important role in viral clearance.[60,61,74] It has been suggested that the weakness of the CTL response is most likely responsible for the indolent necro-inflammatory liver disease characteristic of chronic infection, and that the most important factor determining whether a given infection will be transient or persistent is the strength of the CTL response.[84] In view of the current human T cell and serologic data, it is tempting to speculate that HBsAg-specific CTL may be quite important early in resolving an acute infection, but may be less important after a persistent infection has become established, possibly because of long-term exposure to circulating HBsAg. During chronic HBV infection, HBc/HBeAg-specific T cell responses may become more relevant to viral clearance. Although it is likely that the HBc/HBeAg-specific Th cell and CTL repertoires are limited by persistent infection, they may be less "eroded" than the envelope-specific T cell repertoires. It is also possible that the primary T cell defect in chronic HBV infection resides at the level of insufficient $CD4^+$ Th cell function and CTL precursors are present but quiescent. In fact, a study of chronically infected patients who experienced a spontaneous or IFNα-induced remission suggested that these patients develop a CTL response to HBV proteins similar in strength and specificity to AH-B patients.[85] This suggests that long-term exposure to HBV antigens has virtually no adverse effect (i.e., clonal deletion or anergy) on the $CD8^+$ CTL repertoire. This is certainly not true for the $CD4^+$ Th cell repertoire exposed to secretory HBeAg[86,87] and HBsAg[88] in Tg mouse models (see below). The relative efficiency of tolerance induction in $CD4^+$ versus $CD8^+$ T cells in general or during a perinatal or adult viral infection is a complex issue involving questions of central (i.e., thymic) versus peripheral tolerance, access of the antigens in question to the thymus and/or peripheral lymphoid tissue, endogenous versus exogenous antigen presentation, T cell receptor affinity, etc. However, knowledge of the functional state of the HBV-specific $CD4^+$ and $CD8^+$ T cell repertoires in chronically infected patients will be required before we can expect to fully understand viral persistence.

Innate Immunity to HBV

The role of the innate immune system in HBV infection has not been as extensively studied as adaptive immunity. In general, innate immune responses are critical for early viral containment and for efficient transition to virus-specific adaptive immunity required for complete eradication of infection. However, micro-array studies in HBV acutely infected chimpanzees suggest that HBV is a "stealth" virus that does not induce innate immune response-related genes within the liver.[89] However, it is not clear whether this also applies to natural human infection. Several observations argue against the "stealth" nature of HBV in human infection. For example, circulating NK cells peak at the earliest stage of the HBV incubation phase and decline concurrently with the decline of Niremia preceding ALT elevations and the peak in HBV-specific CD8$^+$ CTLs, suggesting that IFNγ produced by intra-hepatic innate immune cells inhibits HBV replication without cytotoxicity.[90] A recent study followed two HBV seronegative blood donors who subsequently became HBsAg- and HBV DNA-positive. Both innate and adaptive immune responses were observed early in infection. However, NK and NKT cells demonstrated a faster kinetics than HBV-specific CD4$^+$ and CD8$^+$ T cells.[48]

We recently demonstrated that the HBcAg interacts with the murine innate immune system in several unique ways. First, the primary antigen presenting cells (APC) for the HBcAg are B cells and not dendritic cells (DC); however, splenic B1a and B1b cells (members of the innate immune system) more efficiently present soluble HBcAg to naïve CD4$^+$ T cells than splenic B2 cells.[91] Secondly, TLR activation, another innate immune response, was also examined. The HBc/HBeAg proteins were not ligands for human TLRs; however, the ssRNA contained within full-length HBcAg particles does function as a TLR7 ligand and enhances both HBcAg-specific CD4$^+$ T cell and B cell activation.[91] Because bacterial, yeast, and mammalian ssRNA encapsidated within HBcAg particles all function as TLR 7 ligands, it is possible that the pregenomic RNA and/or partially double-stranded HBV DNA present in HBcAg particles during a natural infection may serve as TLR ligands either within hepatocytes or more likely in APCs.

Mechanisms of Liver Injury and Clearance

Transgenic Models of Immune-mediated Liver Injury

Several groups have produced Tg mice that produce all of the HBV gene products and replicate the viral genome in hepatocytes and in other tissues as well.[92–94] Significantly, the HBV gene products and the process of replication in these Tg mice are not directly cytopathic for hepatocytes, further emphasizing the role of immune-mediated mechanisms in liver injury.[92–94] Studies employing the adoptive transfer of $CD8^+$, HBsAg-specific CTL into HBV envelope-expressing Tg mice have demonstrated that CTL can induce an acute necro-inflammatory liver disease similar to natural acute HBV infection.[95] The investigators describe a three-step process through which the liver disease progresses. The first step involves the attachment of the donor CTL to HBsAg-positive hepatocytes that are triggered to undergo apoptosis.[96] Thereafter, between 4 and 12 h after injection, the CTL recruit host-derived antigen non-specific inflammatory cells (i.e., polymorphonuclear cells) that amplify the effects of the CTL (step 2). This process results in necro-inflammatory foci in which hepatocellular necrosis extends well beyond the location of CTL, suggesting that most hepatocytes are killed by cells other than the donor CTL.[97] In these studies the liver injury in most Tg lineages is transient and is confined to no more than 5% of hepatocytes. However, in recipient Tg mice that over-express and accumulate HBsAg filaments the disease process proceeds to step 3, in which approximately half of the mice die of liver failure within 24–72 h of CTL transfer.[97] The investigators suggest that this process resembles the histopathologic features of HBV-induced fulminant hepatitis in humans characterized by widespread necrosis of HBsAg-laden hepatocytes and diffuse lymphomononuclear inflammatory cell infiltrate and Kupffer cell hyperplasia.[97] Fulminant hepatitis has been reported to occur after perinatal or adult infection with the HBeAg-negative HBV variant,[98–101] suggesting that the HBeAg may moderate an otherwise overly aggressive inflammatory response. We have suggested that the HBeAg may accomplish this by preferentially eliciting HBe/HBcAg-specific Th_2-like cells and by depleting inflammatory Th_1-like cells.[102,103]

In addition to the direct hepatocyte injury triggered by HBsAg-specific CTL, a second non-cytolytic mechanism has been described in which the cytokines secreted by the HBsAg-specific CTL profoundly suppressed hepatocellular HBV gene expression and HBV replication in Tg mice.[64,104] The cytokines responsible for these non-cytolytic antiviral effects were CTL-derived IFNγ and CTL-induced TNFα.[104] The antiviral regulatory potential of inflammatory cytokines was confirmed by the fact that administration of recombinant TNFα,[63] IL-2, and to a lesser extent IFNα and IFNβ[65] also inhibited HBV gene expression in HBV envelope Tg mice. Furthermore, these cytokine effects were mediated by a post-transcriptional mechanism involving the degradation of cytoplasmic HBV mRNA.[66] In an *in vitro* system, TNFα, IFNγ, and IFNα have been shown to inhibit the HBV core/pregenomic (C/P) promoter by transcriptional mechanisms.[105] Another study reported that IFNα but not IFNγ inhibited the C/P promoter.[106] Interestingly, systemic treatment of HBsAg-Tg mice *in vivo* with IFNγ did not affect HBV gene expression, whereas CTL-delivered IFNγ did.[65] This suggested that systemic IFNγ may have multiple effects, some of which may be cross-regulatory in terms of inhibition of HBV gene expression. Using these Tg models it has been shown that all of the viral gene products, including the viral replicative intermediates, are susceptible to the effects of these CTL-derived inflammatory cytokines, suggesting that HBV is exquisitely sensitive to this effect. Recently, the source of the inflammatory cytokines responsible for inhibition of HBV gene expression has been extended to include $CD4^+$ Th_1 cells.[67] Furthermore, it was demonstrated that the HBsAg-specific $CD4^+$ Th cells could be elicited by DNA immunization in HBsAg-Tg mice. These studies suggest that non-cytolytic mechanisms contribute substantially to viral clearance during acute HBV infection. Similarly, massive acute infections observed in both the woodchuck and duck hepadna virus systems are cleared with relatively mild disease, suggesting that non-cytolytic clearance mechanisms may be important in natural infection.[107,108] In contrast, *in vivo* transfection of a replication-competent HBV DNA construct produced a transient infection in a rat model, and viral clearance was accompanied by significant liver injury.[109] Similarly, clearance of a chronic duck hepadna virus infection during antiviral therapy appeared to require hepatocyte turnover.[110] Lastly, the role of neutralizing antibody production in resolving woodchuck and duck acute infections should not be underestimated. Neutralizing antibody production is not relevant in the Tg mouse replication models because no cell-to-cell virus infection occurs.

Recent studies also indicate that HBV-nonspecific inflammatory cells and platelets[111] play major roles in augmenting HBV-specific $CD8^+$ CTL-mediated hepatocytosis and in the production of inhibitory cytokines. The recruitment of HBV-nonspecific cells into the liver is mediated by the IFNγ-inducible chemokines CXCL9 (Mig) and CXCL10 (IFNα-inducible protein-10),[112] IL-8,[113] and neutrophil-derived matrix metallo-proteinases (MMP-8, MMP-9).[114,115]

HBV Persistence

Tolerogenicity of HBc/HBeAgs

Infants born to HBeAg-positive HBV carrier mothers invariably become persistently infected. To investigate the role of immunologic tolerance mechanisms in the chronic infection of the newborn, we have generated HBeAg-expressing Tg mice. HBeAg-Tg mice represent a model system to examine the consequences of *in utero* and/or perinatal exposure to HBeAg on HBc/HBeAg-specific immune responses. Characterization of tolerance in HBeAg-Tg mice and mice rendered neonatally tolerant indicated that (a) T cells, but not B cells, were rendered tolerant by HBeAg present in the serum at a concentration of 10–100 ng/mL; (b) T cell tolerance elicited by HBeAg also extends to HBcAg-specific T cells; (c) Tg mice produced anti-HBc but not anti-HBe antibodies upon immunization; (d) the immunoglobulin G (IgG) but not the IgM anti-HBc response was diminished in HBeAg-Tg mice; and (e) the T cell tolerance induced by a single neonatal exposure to HBeAg was reversible and persisted for 12–16 weeks.[116] It is of interest that many characteristics of immune tolerance found in HBeAg-Tg mice parallel the long-term immunologic status of neonates born to HBeAg-positive HBV carrier mothers. For example, infants infected perinatally often remain HBeAg positive and produce anti-HBc, but not anti-HBe antibodies. The correlation of human serological observations with murine experimental data suggests the aberrant immunological responses of neonates born to carrier mothers may also result from *in utero* exposure to HBeAg, as occurs in the transgenic model. More recent studies in HBeAg-Tg mice, which do not express HBeAg until after birth, indicate that perinatal exposure to HBeAg is also sufficient to elicit HBe/HBcAg-specific Th cell tolerance without the requirement for HBeAg to cross the placenta.[117,118]

A study of HBcAg-Tg and HBeAg-Tg mice bred to T cell receptor (TCR)-Tg mice expressing receptors specific for different HBc/HBeAg epitopes revealed three phenotypes of HBe/HBcAg-specific $CD4^+$ T cell tolerance: (a) profound T cell tolerance mediated by clonal deletion; (b) non-deletional T cell tolerance dependent on the structure, location, and/or concentration of the tolerogen mediated by *in vivo* anergy; and (c) T cell clonal ignorance. Importantly, the HBeAg appears more efficient at eliciting T cell tolerance than the HBcAg, and this split immune tolerance may have significant implications during a natural HBV infection.[117] This study indicated the clonal nature of T cell tolerance. High affinity HBc/HBeAg-specific Th cells are very susceptible to clonal deletion in the thymus or the periphery, whereas intermediate and low affinity Th cells are prone to anergy or remain "ignorant" and escape tolerance, respectively. Therefore, T cell tolerance toward HBcAg is clonal and heterogeneous in terms of mechanisms of induction, and this heterogeneity is likely to be evident even within an individual HBV-infected patient. It is important to note that the HBe and HBc antigens are cross-reactive at the T cell level. Therefore, HBc/HBe-specific T cell tolerance would preclude Th cell function necessary for anti-HBe and maximal IgG anti-HBc antibody production, and the ability of HBc/HBe-specific Th cells to elicit antibodies to envelope proteins[68] would also be impaired. Moreover, HBc/HBe-specific Th cell tolerance may diminish the strength of a CTL response ($CD4^+$- and $CD8^+$-mediated), which may be required for the elimination of virus-infected hepatocytes. Because of the central regulatory role of HBe/HBc-specific Th cells, the functional absence of this Th cell specificity may impair viral clearance and result in chronic infection.

For example, TCR-Tg mice, expressing an intermediate affinity TCR specific for the HBc/HBeAgs, bred with HBcAg-Tg mice spontaneously produce IgG anti-HBc antibody between three to six weeks of age. However, the presence of HBeAg in the serum of TCR × HBcAg × HBeAg triple-Tg mice prevents anti-HBc seroconversion. Serum HBeAg mediates this immunoregulatory effect by eliciting non-deletional tolerance in HBc/HBeAg-specific Th cells. This result suggests that hepadnaviruses have retained a secretory form of the nucleoprotein because it functions as a T cell tolerogen and regulates the immune response to the intracellular, structural nucleocapsid.[118] Therefore, HBeAg-mediated immune regulation may predispose to chronicity during perinatal infections and prevent severe liver injury during adult infections (see Table 1).

Table 1. HBV Persistence

Cause:	T cell tolerance (deletional/non-deletional) (Note: T cell tolerance is CLONAL)
Tolerogens:	Serum HBeAg 1. Deletion of HI affinity HBc/HBeAg-sp. T cells 2. Anergy in LO affinity HBc/HBeAg-sp. T cells 3. Shift to Th_2-like response
	Serum HBsAg 1. Deletion/anergy of HBsAg-sp. T cells 2. Absorption of anti-HBs 3. MHC blockade 4. CTL exhaustion
Modifying Factors:	*in utero* infection Neonatal vs. adult infection Immunocompetence Immunogenetics Immune escape variants Innate immunity
Manifestation of T cell tolerance:	lo affinity HBV-sp. $CD4^+$/$CD8^+$ T cell repertoire ↑ PD-1/PD-1L ↑ Tregs ↑ CTLA-4

Greater amounts of excess envelope proteins than HBeAg are secreted into the serum during chronic HBV infection and serve as tolerogens in Tg mouse models.[88] Another effect of this excess HBsAg is to "soak up" neutralizing anti-HBs antibody present in the serum of CH-B patients, thereby inhibiting virus neutralization. Another possible effect of large amounts of circulating envelope proteins is to saturate MHC molecules sufficiently to compete with other possibly more relevant antigens for MHC binding.[86] Although not directly examined, circulating HBsAg may deplete $CD4^+$ and $CD8^+$ T cells by clonal exhaustion due to high viral load[119] or by Fas-mediated apoptosis as demonstrated for circulating HBeAg.[102] The unique ability of exogenous HBsAg to enter the MHC Class I processing pathway[78] also suggests mechanisms whereby secretory HBsAg may lead to virus-specific immunosuppression during chronic HBV infection. For example, HBsAg-specific B cells that bind serum HBsAg may be targets of HBsAg-specific CTL[76] or T cells that take up HBsAg via the transferrin receptor[120] may become targets of MHC Class I or Class II-restricted CTL.

Immune Escape Variants

One of the most efficient mechanisms used by viruses to evade immune recognition by the host is the production of variants not "seen" by either neutralizing antibody or by the cellular arm of the immune response. Several studies have reported the occurrence of vaccine-induced or transplant-related humoral escape mutants involving residue 319 within the S region.[121–123] An important group-specific or "a" determinant of HBsAg is believed to be at least partially composed of the hydrophilic region from residues 298 to 321. This region is highly conserved and believed to form two loops by intramolecular disulfide bridging between the five cysteines within this sequence.[124] Mutation of the conserved glycine at residue 319 has been shown to cause the loss of neutralizing antibody binding.[125] The rare detection of this S region mutation during natural HBV infection suggests the polyclonal nature of neutralizing antibody production.

A similar argument has been made that the occurrence of CTL escape mutants during acute HBV infection is unlikely due to the strength and polyclonal nature of the CTL response.[84] However, the much weaker and possibly monoclonal or oligoclonal CTL response of chronically infected patients may be much more susceptible to point mutations that reduce MHC molecule binding or T cell receptor (TCR) recognition. In fact, such mutations within the HBcAg CTL epitope 18–27 have been reported in two patients with chronic HBV infection.[126] Similarly, a number of mutations within the HBcAg have been described that correspond to $CD4^+$ Th cell recognition sites in chronic HBV patients in which liver injury has occurred.[127–129] These results suggest that $CD4^+$ Th cells as well as $CD8^+$ CTL and antibody may exert selective pressure for the emergence of immune escape mutants. However, one recent study suggested that selection of viral variants by CTL is not a common event in the majority of CH-B patients.[130] A novel type of CTL escape mutant has been described in the HBV system. An escape mutation within the HBcAg CTL epitope 18–27 that also interferes with recognition of the wild-type epitope and acts as a TCR antagonist was demonstrated in two patients.[131] Cells infected exclusively with the variant or simultaneously with both the variant and the wild-type virus would be protected from elimination by CTL specific for the wild-type epitope. Another putative target for escape may be the viral response elements that regulate sensitivity to the effects of inflammatory cytokines produced by Th_1 cells and CTL. This is an attractive hypothesis because such a mutation would have a much more profound effect than a point mutation in a single T cell recognition site.[84]

By far the most common HBV variant occurs as a result of a point mutation within the pre-core region (codon 28) that generates a stop codon, preventing translation of the pre-core/core P25 protein from which HBeAg is derived.[132,133] As discussed previously, secretory HBeAg may play an important role in eliciting T cell tolerance perinatally and may down-regulate inflammatory clearance mechanisms by depleting HBeAg-specific Th_1 cells during adult infections. These characteristics are clearly advantageous for viral persistence. Therefore, why would HBeAg-negative variants be selected during the course of a chronic infection? One proposal has been that the pre-core mutation may confer a growth advantage on the variant virus independent of the immune response.[134,135] However, transfection of mutant WHV-DNA (i.e., eAg-negative) into the livers of adult woodchucks resulted in equivalent levels of viral replication as compared to wild-type WHV.[136] We have suggested that while secretory HBeAg is advantageous for the virus during the "tolerance" phase of chronic infection, during the "injury" phase of chronic infection or during an acute infection the HBeAg-specific Th_1/Th_2 cell balance may shift in favor of the inflammatory Th_1 cell subset. The HBeAg can elicit Th_1 cells under some circumstances. For example, treatment of HBeAg-Tg with either IFNγ or IFNα shifts HBeAg-specific Th cells toward the Th_1 cell phenotype.[32] Therefore, depending on the context of exposure, HBeAg may become disadvantageous for the virus and HBeAg-negative variants may emerge.[102] A mechanism to explain the selection of the HBeAg-negative variant during chronic HBV infection has recently been proposed. When HBc/HBeAg-specific CTLs were adoptively transferred into HBcAg-Tg, HBeAg-Tg, or HBc/HBeAg-dbl-Tg mice, the duration of liver injury and inflammation were significantly greater in HBeAg-Tg recipient mice than in HBcAg-Tg recipients.[62] Importantly, liver injury in HBc/HBeAg-dbl-Tg recipients was similar to the injury observed in HBeAg-Tg mice. The finding that hepatocytes expressing wild-type HBV (containing both HBcAg and HBeAg) are more susceptible to CTL-mediated clearance than hepatocytes expressing only HBcAg suggest that the HBeAg-negative variant may have a selective advantage over wild-type HBV within the livers of patients with chronic infection during an immune response and may represent a CTL escape mutant.[62] Historically, it has been problematic to understand the selective advantage of an HBeAg-negative variant at the level of CTL recognition because the infected hepatocyte expresses the CTL-cross-reactive HBcAg regardless of HBeAg co-expression. The observation that differential CTL recognition of the HBc/HBeAgs expressed in hepatocytes *in vivo* may result in preferential clearance of wild-type HBV could resolve this dilemma. Because hepatocytes do not serve as effective antigen-presenting cells for exogenous antigens and exogenous HBeAg has not been shown to elicit CTLs, it is unlikely that secreted HBeAg is a target of the immune selection process. Indeed, secreted HBeAg likely serves as a counterbalance that moderates HBc/HBeAg-specific liver injury due to its immunoregulatory properties.[102,116–118,137,138] It may seem counterintuitive that HBeAg functions both as a tolerogen/immunoregulatory protein and as a viral target for CTL recognition and the mediator of subsequent liver injury. How can a immunoregulatory protein also be the target of immune-mediated selection? In this regard, it may be helpful to discriminate between secreted HBeAg and cytosolic HBeAg. The secreted HBeAg can tolerize HBc/HBeAg-specific $CD4^+$ T cells, and this function is most apparent during the asymptomatic tolerance phase of chronic HBV infection. However, once HBeAg-specific T cell tolerance is at least partially broken, usually in young adulthood in perinatally infected chronic carriers, cytosolic HBeAg not only may serve as a CTL target but may actually be a more efficient CTL target than intracellular HBcAg in the liver. While secreted HBeAg is advantageous for the virus during the early phase of infection, cytosolic HBeAg becomes a liability for the virus as HBeAg-specific T cell tolerance wanes. Production of HBeAg-negative CTL escape mutants and their selection by the host immune response may represent another mechanism for viral persistence in later stages of chronic infection. In this way the HBeAg-negative virus escapes the CTL response targeted at cytosolic HBeAg, but the escape mutant is not entirely successful because the virus also loses the function of its secreted immunoregulatory protein, serum HBeAg. Although the emergence of an HBeAg-negative variant may sometimes correlate with reduced liver injury and viral clearance,[139–141] because HBcAg remains as a CTL target and because the immunoregulatory function of secreted HBeAg is lost, the HBeAg-negative immune escape variant may actually become more pathogenic in some circumstances, most likely depending on the viral load and stage of infection.[137,142–144]

Manifestations of HBV-Specific T Cell Tolerance

As previously discussed, secretion of viral HBeAg and HBsAg elicits $CD4^+/CD8^+$ T cell tolerance mediated by deletional and non-deletional mechanisms. The HBV-specific T cell repertoire that survives clonal deletion is by definition composed of low to intermediate affinity T cells. As T cell tolerance wanes, usually in early adulthood of perinatally infected chronic carriers, the surviving low affinity HBV-specific T cells continue to be exposed to circulating HBeAg and HBsAg, which may severely affect the T cell effector function. For example, HBV-specific $CD8^+$ T cells during chronic HBV infection express elevated levels of both PD-1 and its ligand, PD-1L, and are hyporesponsive to activation *in vitro* in HBV-specific recall cultures.[145–147] Furthermore, the functional responses improve when PD-1 signaling is blocked, indicating that paracrine PD-1/PD-1L signaling suppresses HBV-specific CTL activity in chronic infection.[147] Recent studies suggest that $CD4^+$ Tregs contribute to the defective immune responses characteristic of chronic HBV infection. Chronic HBV patients have higher percentages of circulating $CD4^+$, $CD25^+$, and CTLA-4^+ T cells than patients recovered from acute infection or healthy controls.[148,149] Increased expression of $FOXP_3$ in PBMCs from chronic HBV patients further suggest the presence of Treg cells.[149] We have recently identified a novel HBeAg-specific $CD4^-/CD8^-$ [double negative (DN)] Treg cell in HBeAg × TCR-dbl-Tg mice and, interestingly, this DN Treg cell population is not present in HBcAg × TCR-dbl-Tg mice.[162]

Immunogenetics

The immunogenetics of the murine response to HBV envelope antigens has been extensively studied. In summary, the murine studies indicate that at least two MHC Class II genes[150–152] and one MHC Class III gene (S region, C4)[152,153] regulate the anti-HBs immune response. Additional MHC-linked genes influence immune responses to the envelope preS regions. For example, immunization with preS(2) region-containing HBsAg particles can bypass non-response to the HBsAg and immunization with preS(1) + preS(2) region-containing HBsAg particles can bypass non-response to the HBsAg and preS(2) regions.[15,154,155] The envelope polypeptides of the HBV present an array of T cell determinants to the host's immune system, the specificity of T cell recognition is determined by MHC-linked genes, and T helper cell recognition of one region can circumvent non-response to another region. This level of complexity in inbred mice strains predicted a greater complexity in the outbred human population.

The suggestion that MHC-linked genes may also control the human immune response to HBsAg was first made by Walker *et al.*,[156] who observed a significant excess of HLA-DR7 and a total absence of HLA-DR1 in HBsAg-vaccinated low or non-responders. This observation has been confirmed and extended by numerous investigators.[157,158] Alleles that are strongly associated with high response to HBsAg vaccination are *DRB1*01 (DR1), DRB1*11 (DR5), DRB1*15 (DR2), DQB1*0501*, and *DPB1*0401*, whereas alleles strongly associated with non/poor response are *DRB1*03, DRB1*07, DQB1*02* and *DPB1*1101.*[159,160] In addition to the associations between single MHC Class II alleles and anti-HBs response, several associations with extended haplotypes have been reported, and most associations with MHC Class I alleles have been observed within the context of the extended haplotypes. Moreover, HBsAg vaccine non-responders have a significantly increased prevalence of *C4A* gene deletions and of the non-expressed *C4AQO* alleles compared to responders.[161]

In terms of MHC associations in relation to HBV susceptibility, persistence, and treatment, a number of studies have been published. For example, MHC Class II *DR13* is consistently associated with HBV clearance globally and MHC Class II *DRB1*11/*12* alleles are associated with HBV persistence in major populations.[160] Although MHC associations with susceptibility or protection from persistent HBV infection are robust, such associations are also inconsistent from one global region to another and even within the same populations. This may not be surprising given the racial diversity, variations in study design, methodology, and most importantly the numerous viral proteins and complex immune response and regulatory mechanisms involved in HBV clearance and persistence. It is, therefore, unlikely that MHC genotype is a determinative factor in resistance or susceptibility to chronic HBV infection but may play a modifying role.

Summary

The HBV has evolved as a successful human pathogen not because of its ability to spread easily among adults (i.e., blood-borne infection) or because of extraordinary immune evasion mechanisms. In fact, HBV infection in adults usually results in subacute or acute resolution. The HBV appears relatively sensitive to combined innate/adaptive immune responses mediated by polyclonal immune cells producing soluble factors (neutralizing antibodies, cytokines) in concert with a variety of cellular effector mechanisms. Rather HBV is a successful human pathogen due to its ability to establish persistence primarily following vertical transmission from mother to child either *in utero* or more frequently perinatally. In this circumstance, relative immune system immaturity and the secreted viral proteins combine to shift the viral-host balance toward immune T cell tolerance as opposed to immunogenicity. We suggest that the determinative cause of HBV persistence is clonal T cell tolerance to the structural nucleocapsid and envelope antigens. As depicted in Table 1, a number of factors may modify the quantitative and qualitative characteristics of T cell tolerance and there may be a number of manifestations of T cell tolerance during a chronic infection. However, a review of the literature indicates that cause and effect relationships are often confused.

Acknowledgment

The author is grateful for the contribution of all the co-author/collaborators who participated in the studies cited from his laboratory, and to Melissa Grace for editorial assistance. This work was supported by NIH grants R01 AI049730 and R01 AI20720.

References

1. Stevens CE, Beasley RP, Tsui J, Lee WC. (1975) Vertical transmission of hepatitis B antigen in Taiwan. *N Engl J Med* **292**: 771–774.
2. Beasley RP, Hwang LY, Lin CC, Chien CS. (1981) Hepatocellular carcinoma and hepatitis B virus. A prospective study of 22 707 men in Taiwan. *Lancet* **2**: 1129–1133.
3. Alberti A, Trevisan A, Fattovich G, *et al.* (1984) The role of hepatitis B virus replication and hepatocyte membrane expression in the pathogenesis of HBV-related hepatic damage. In: FV Chisari, *Advances in Hepatitis Research.* Masson Publishing, New York, pp. 134–143.
4. Neurath AR, Kent SB, Parker K, *et al.* (1986) Antibodies to a synthetic peptide from the preS 120–145 region of the hepatitis B virus envelope are virus neutralizing. *Vaccine* **4**: 35–37.
5. Milich DR, McLachlan A. (1986) The nucleocapsid of hepatitis B virus is both a T-cell-independent and a T-cell-dependent antigen. *Science* **234**: 1398–1401.
6. Hoofnagle JH, Seef LB, Bates ZB, *et al.* (1978) Serologic responses in hepatitis B. In: GN Vyas, SN Cohen, R Schmid (eds), *Viral Hepatitis.* Franklin Institute Press, Philadelphia, pp. 219–228.
7. Hoofnagle JH, Dusheiko GM, Seeff LB, *et al.* (1981) Seroconversion from hepatitis B e antigen to antibody in chronic type B hepatitis. *Ann Intern Med* **94**: 744–748.
8. Emini EA, Larson V, Eichberg J, *et al.* (1989) Protective effect of a synthetic peptide comprising the complete preS2 region of the hepatitis B virus surface protein. *J Med Virol* **28**: 7–12.
9. Itoh Y, Takai E, Ohnuma H, *et al.* (1986) A synthetic peptide vaccine involving the product of the pre-S(2) region of hepatitis B virus DNA: protective efficacy in chimpanzees. *Proc Natl Acad Sci USA* **83**: 9174–9178.
10. Neurath AR, Seto B, Strick N. (1989) Antibodies to synthetic peptides from the preS1 region of the hepatitis B virus (HBV) envelope (env) protein are virus-neutralizing and protective. *Vaccine* **7**: 234–236.
11. Thornton GB, Milich D, Chisari FV, *et al.* (1987) Immune response in primates to the pre-S(2) region of hepatitis B surface antigen: identification of a protective determinant. In: RM Chanock, RA Lerner, F Brown (eds), *Vaccines '87.* Cold Spring Harbor Laboratory, Cold Spring Harbor, NY, pp. 77–80.
12. Thornton GB, Moriarty A, Milich D, *et al.* (1989) Protection of chimpanzees from hepatitis B infection after immunization with synthetic peptides: identification of protective epitopes in the pre-S region. In: F Brown (ed.), *Modern Approaches to New Vaccines Including Prevention of Aids.* Cold Spring Harbor Laboratory, Cold Spring Harbor, NY, pp. 467–71.
13. Neurath AR, Kent SB, Strick N, *et al.* (1985) Hepatitis B virus contains pre-S gene-encoded domains. *Nature* **315**: 154–156.

14. Alberti A, Cavalletto D, Chemello L, *et al.* (1990) Fine specificity of human antibody response to the PreS1 domain of hepatitis B virus. *Hepatology* **12**: 199–203.
15. Milich DR, McLachlan A, Chisari FV, *et al.* (1986) Immune response to the pre-S(1) region of the hepatitis B surface antigen (HBsAg): a pre-S(1)-specific T cell response can bypass nonresponsiveness to the pre-S(2) and S regions of HBsAg. *J Immunol* **137**: 315–322.
16. Klinkert MQ, Theilmann L, Pfaff E, Schaller H. (1986) Pre-S1 antigens and antibodies early in the course of acute hepatitis B virus infection. *J Virol* **58**: 522–525.
17. Theilmann L, Klinkert MQ, Gmelin K, *et al.* (1987) Detection of antibodies against pre-S1 proteins in sera of patients with hepatitis B virus (HBV) infection. *J Hepatol* **4**: 22–28.
18. Hellstrom U, Sylvan S, Kuhns M, Sarin V. (1986) Absence of pre-S2 antibodies in natural hepatitis B virus infection. *Lancet* **2**: 889–893.
19. Petit MA, Maillard P, Capel F, Pillot J. (1986) Immunochemical structure of the hepatitis B surface antigen vaccine-II. Analysis of antibody responses in human sera against the envelope proteins. *Mol Immunol* **23**: 511–523.
20. Okamoto H, Usuda S, Imai M, *et al.* (1986) Antibody to the receptor for polymerized human serum albumin in acute and persistent infection with hepatitis B virus. *Hepatology* **6**: 354–359.
21. Budkowska A, Dubreuil P, Capel F, Pillot J. (1986) Hepatitis B virus pre-S gene-encoded antigenic specificity and anti-pre-S antibody: relationship between anti-pre-S response and recovery. *Hepatology* **6**: 360–368.
22. Coursaget P, Adamowicz P, Bourdil C, *et al.* (1988) Anti-pre-S2 antibodies in natural hepatitis B virus infection and after immunization. *Vaccine* **6**: 357–361.
23. Maruyama T, McLachlan A, Iino S, *et al.* (1993) The serology of chronic hepatitis B infection revisited. *J Clin Invest* **91**: 2586–2595.
24. Maruyama T, Thornton GB, Iino S, *et al.* (1992) Use of anti-peptide antibodies for the design of antigen-specific immune complex assays. *J Immunol Methods* **155**: 65–75.
25. Sallberg M, Norder H, Magnius LO. (1990) Comparison of class and subclass distribution of antibodies to the hepatitis B core and B e antigens in chronic hepatitis B. *J Med Virol* **30**: 1–6.
26. Anh-Tuan N, Novak E. (1981) Hepatitis B surface antigen circulating immune complexes (HBsAg-CICs) in patients with hepatitis B and asymptomatic HBsAg carriers. *Clin Exp Immunol* **43**: 246–253.
27. Brown SE, Steward MW, Viola L, *et al.* (1983) Chronic liver disease: the detection and characterization of circulating immune complexes. *Immunology* **49**: 673–683.
28. Pernice W, Sodomann CP, Luben G, *et al.* (1979) Antigen-specific detection of HBsAG-containing immune complexes in the course of hepatitis B virus infection. *Clin Exp Immunol* **37**: 376–380.

29. Rath S, Devey ME. (1988) IgG subclass composition of antibodies to HBsAg in circulating immune complexes from patients with hepatitis B virus infections. *Clin Exp Immunol* **72**: 164–167.
30. Sallberg M, Norder H, Lindh G, Magnius LO. (1991) IgG subclasses in circulating immune complexes with hepatitis B e antigen in chronic hepatitis B. *Clin Exp Immunol* **84**: 116–121.
31. Thomas HC, De Villiers D, Potter B, *et al.* (1978) Immune complexes in acute and chronic liver disease. *Clin Exp Immunol* **31**: 150–157.
32. Milich DR, Wolf SF, Hughes JL, Jones JE. (1995) Interleukin 12 suppresses autoantibody production by reversing helper T-cell phenotype in hepatitis B e antigen transgenic mice. *Proc Natl Acad Sci USA* **92**: 6847–6851.
33. Cavanaugh VJ, Guidotti LG, Chisari FV. (1997) Interleukin-12 inhibits hepatitis B virus replication in transgenic mice. *J Virol* **71**: 3236–3243.
34. Lohr HF, Pingel S, Bocher WO, *et al.* (2002) Reduced virus specific T helper cell induction by autologous dendritic cells in patients with chronic hepatitis B — restoration by exogenous interleukin-12. *Clin Exp Immunol* **130**: 107–114.
35. Rossol S, Marinos G, Carucci P, *et al.* (1997) Interleukin-12 induction of Th1 cytokines is important for viral clearance in chronic hepatitis B. *J Clin Invest* **99**: 3025–3033.
36. Xiong SQ, Lin BL, Gao X, *et al.* (2007) IL-12 promotes HBV-specific central memory CD8+ T cell responses by PBMCs from chronic hepatitis B virus carriers. *Int Immunopharmacol* **7**: 578–587.
37. Carreno V, Zeuzem S, Hopf U, *et al.* (2000) A phase I/II study of recombinant human interleukin-12 in patients with chronic hepatitis B. *J Hepatol* **32**: 317–324.
38. Rigopoulou EI, Suri D, Chokshi S, *et al.* (2005) Lamivudine plus interleukin-12 combination therapy in chronic hepatitis B: antiviral and immunological activity. *Hepatology* **42**: 1028–1036.
39. Crettaz J, Otano I, Ochoa L, *et al.* (2009) Treatment of chronic viral hepatitis in woodchucks by prolonged intrahepatic expression of interleukin-12. *J Virol* **83**: 2663–2674.
40. Ferrari C, Bertoletti A, Penna A, *et al.* (1991) Identification of immunodominant T cell epitopes of the hepatitis B virus nucleocapsid antigen. *J Clin Invest* **88**: 214–222.
41. Ferrari C, Penna A, Bertoletti A, *et al.* (1990) Cellular immune response to hepatitis B virus-encoded antigens in acute and chronic hepatitis B virus infection. *J Immunol* **145**: 3442–3449.
42. Jung MC, Diepolder HM, Spengler U, *et al.* (1995) Activation of a heterogeneous hepatitis B (HB) core and e antigen-specific CD4+ T-cell population during seroconversion to anti-HBe and anti-HBs in hepatitis B virus infection. *J Virol* **69**: 3358–3368.

43. Cupps TR, Hoofnagle JH, Ellis RW, *et al.* (1990) *In vitro* immune responses to hepatitis B surface antigens S and preS2 during acute infection by hepatitis B virus in humans. *J Infect Dis* **161**: 412–419.
44. Jung MC, Spengler U, Schraut W, *et al.* (1991) Hepatitis B virus antigen-specific T-cell activation in patients with acute and chronic hepatitis B. *J Hepatol* **13**: 310–317.
45. Celis E, Kung PC, Chang TW. (1984) Hepatitis B virus-reactive human T lymphocyte clones: antigen specificity and helper function for antibody synthesis. *J Immunol* **132**: 1511–1516.
46. Ferrari C, Penna A, Bertoletti A, *et al.* (1989) The preS1 antigen of hepatitis B virus is highly immunogenic at the T cell level in man. *J Clin Invest* **84**: 1314–1319.
47. Jin Y, Shih WK, Berkower I. (1988) Human T cell response to the surface antigen of hepatitis B virus (HBsAg). Endosomal and nonendosomal processing pathways are accessible to both endogenous and exogenous antigen. *J Exp Med* **168**: 293–306.
48. Fisicaro P, Valdatta C, Boni C, *et al.* (2009) Early kinetics of innate and adaptive immune responses during hepatitis B virus infection. *Gut* **58**: 974–982.
49. Vento S, Rondanelli EG, Ranieri S, *et al.* (1987) Prospective study of cellular immunity to hepatitis-B-virus antigens from the early incubation phase of acute hepatitis B. *Lancet* **2**: 119–122.
50. Ferrari C, Penna A, Sansoni P, *et al.* (1986) Selective sensitization of peripheral blood T lymphocytes to hepatitis B core antigen in patients with chronic active hepatitis type B. *Clin Exp Immunol* **66**: 497–506.
51. Inoue M, Kakumu S, Yoshioka K, *et al.* (1989) Hepatitis B core antigen-specific IFN-gamma production of peripheral blood mononuclear cells in patients with chronic hepatitis B virus infection. *J Immunol* **142**: 4006–4011.
52. Lohr HF, Gerken G, Schlicht HJ, *et al.* (1993) Low frequency of cytotoxic liver-infiltrating T lymphocytes specific for endogenous processed surface and core proteins in chronic hepatitis B. *J Infect Dis* **168**: 1133–1139.
53. Sylvan SP, Hellstrom UB, Flehmig B. (1987) Characterization of cell mediated immune responses to the hepatitis B core protein in man. *Clin Exp Immunol* **68**: 233–241.
54. Vento S, Chen SH, Giuliani-Piccari G, *et al.* (1987) Cellular immunity to nucleocapsid and pre-S determinants in asymptomatic carriers of hepatitis B virus. *Immunology* **62**: 593–598.
55. Vento S, Hegarty JE, Alberti A, *et al.* (1985) T lymphocyte sensitization to HBcAg and T cell-mediated unresponsiveness to HBsAg in hepatitis B virus-related chronic liver disease. *Hepatology* **5**: 192–197.
56. Lohr HF, Weber W, Schlaak J, *et al.* (1995) Proliferative response of CD4+ T cells and hepatitis B virus clearance in chronic hepatitis with or without hepatitis B e-minus hepatitis B virus mutants. *Hepatology* **22**: 61–68.

57. Milich DR, McLachlan A, Moriarty A, Thornton GB. (1987) Immune response to hepatitis B virus core antigen (HBcAg): localization of T cell recognition sites within HBcAg/HBeAg. *J Immunol* **139**: 1223–1231.
58. Godkin A, Davenport M, Hill AV. (2005) Molecular analysis of HLA class II associations with hepatitis B virus clearance and vaccine nonresponsiveness. *Hepatology* **41**: 1383–1390.
59. Tsai SL, Chen PJ, Lai MY, *et al.* (1992) Acute exacerbations of chronic type B hepatitis are accompanied by increased T cell responses to hepatitis B core and e antigens. Implications for hepatitis B e antigen seroconversion. *J Clin Invest* **89**: 87–96.
60. Missale G, Redeker A, Person J, *et al.* (1993) HLA-A31- and HLA-Aw68-restricted cytotoxic T cell responses to a single hepatitis B virus nucleocapsid epitope during acute viral hepatitis. *J Exp Med* **177**: 751–762.
61. Penna A, Chisari FV, Bertoletti A, *et al.* (1991) Cytotoxic T lymphocytes recognize an HLA-A2-restricted epitope within the hepatitis B virus nucleocapsid antigen. *J Exp Med* **174**: 1565–1570.
62. Frelin L, Wahlstrom T, Tucker AE, *et al.* (2009) A mechanism to explain the selection of the hepatitis e antigen-negative mutant during chronic hepatitis B virus infection. *J Virol* **83**: 1379–1392.
63. Gilles PN, Fey G, Chisari FV. 1992. Tumor necrosis factor alpha negatively regulates hepatitis B virus gene expression in transgenic mice. *J Virol* **66**: 3955–3960.
64. Guidotti LG, o K, Hobbs MV, *et al.* (1994) Cytotoxic T lymphocytes inhibit hepatitis B virus gene expression by a noncytolytic mechanism in transgenic mice. *Proc Natl Acad Sci USA* **91**: 3764–3768.
65. Guidotti LG, Guilhot S, Chisari FV. (1994) Interleukin-2 and alpha/beta interferon down-regulate hepatitis B virus gene expression *in vivo* by tumor necrosis factor-dependent and -independent pathways. *J Virol* **68**: 1265–1270.
66. Guilhot S, Guidotti LG, Chisari FV. (1993) Interleukin-2 downregulates hepatitis B virus gene expression in transgenic mice by a posttranscriptional mechanism. *J Virol* **67**: 7444–7449.
67. Mancini M, Hadchouel M, Davis HL, *et al.* (1996) DNA-mediated immunization in a transgenic mouse model of the hepatitis B surface antigen chronic carrier state. *Proc Natl Acad Sci U S A* **93**: 12496–12501.
68. Milich DR, McLachlan A, Thornton GB, Hughes JL. (1987) Antibody production to the nucleocapsid and envelope of the hepatitis B virus primed by a single synthetic T cell site. *Nature* **329**: 547–549.
69. Murray K, Bruce SA, Hinnen A, *et al.* (1984) Hepatitis B virus antigens made in microbial cells immunise against viral infection. *Embo J* **3**: 645–650.
70. Tabor E, Gerety RJ. (1984) Possible role of immune responses to hepatitis B core antigen in protection against hepatitis B infections. *Lancet* **1**: 172.

71. Menne S, Maschke J, Tolle TK, *et al.* (1997) Characterization of T-cell response to woodchuck hepatitis virus core protein and protection of woodchucks from infection by immunization with peptides containing a T-cell epitope. *J Virol* **71**: 65–74.
72. Yewdell JW, Bennink JR. (1992) Cell biology of antigen processing and presentation to major histocompatibility complex class I molecule-restricted T lymphocytes. *Adv Immunol* **52**: 1–123.
73. Ramsay AJ, Ruby J, Ramshaw IA. (1993) A case for cytokines as effector molecules in the resolution of virus infection. *Immunol Today* **14**: 155–157.
74. Nayersina R, Fowler P, Guilhot S, *et al.* (1993) HLA A2 restricted cytotoxic T lymphocyte responses to multiple hepatitis B surface antigen epitopes during hepatitis B virus infection. *J Immunol* **150**: 4659–4671.
75. Rehermann B, Fowler P, Sidney J, *et al.* (1995) The cytotoxic T lymphocyte response to multiple hepatitis B virus polymerase epitopes during and after acute viral hepatitis. *J Exp Med* **181**: 1047–1058.
76. Barnaba V, Franco A, Alberti A, *et al.* (1990) Selective killing of hepatitis B envelope antigen-specific B cells by class I-restricted, exogenous antigen-specific T lymphocytes. *Nature* **345**: 258–260.
77. Schirmbeck R, Melber K, Kuhrober A, *et al.* (1994) Immunization with soluble hepatitis B virus surface protein elicits murine H-2 class I-restricted CD8+ cytotoxic T lymphocyte responses *in vivo*. *J Immunol* **152**: 1110–1119.
78. Schirmbeck R, Melber K, Mertens T, Reimann J. (1994) Antibody and cytotoxic T-cell responses to soluble hepatitis B virus (HBV) S antigen in mice: implication for the pathogenesis of HBV-induced hepatitis. *J Virol* **68**: 1418–1425.
79. Kuhober A, Pudollek HP, Reifenberg K, *et al.* (1996) DNA immunization induces antibody and cytotoxic T cell responses to hepatitis B core antigen in H-2b mice. *J Immunol* **156**: 3687–3695.
80. Ishikawa T, Kakumu S, Yoshioka K, *et al.* (1993) Relative immunogenicity of hepatitis B virus-encoded antigens as targets for cytotoxic T-cell response. *Immunology* **80**: 313–318.
81. Townsend K, Sallberg M, O'Dea J, *et al.* (1997) Characterization of CD8+ cytotoxic T-lymphocyte responses after genetic immunization with retrovirus vectors expressing different forms of the hepatitis B virus core and e antigens. *J Virol* **71**: 3365–3374.
82. Bertoletti A, Ferrari C, Fiaccadori F, *et al.* (1991) HLA class I-restricted human cytotoxic T cells recognize endogenously synthesized hepatitis B virus nucleocapsid antigen. *Proc Natl Acad Sci USA* **88**: 10445–10449.
83. Barnaba V, Franco A, Alberti A, *et al.* (1989) Recognition of hepatitis B virus envelope proteins by liver-infiltrating T lymphocytes in chronic HBV infection. *J Immunol* **143**: 2650–2655.

84. Chisari FV, Ferrari C. (1995) Hepatitis B virus immunopathogenesis. *Annu Rev Immunol* **13**: 29–60.
85. Rehermann B, Lau D, Hoofnagle JH, Chisari FV. (1996) Cytotoxic T lymphocyte responsiveness after resolution of chronic hepatitis B virus infection. *J Clin Invest* **97**: 1655–1665.
86. Milich DR, McLachlan A, Raney AK, *et al.* (1991) Autoantibody production in hepatitis B e antigen transgenic mice elicited with a self T-cell peptide and inhibited with nonself peptides. *Proc Natl Acad Sci USA* **88**: 4348–4352.
87. Milich DR, Schodel F, Peterson DL, *et al.* (1995) Characterization of self-reactive T cells that evade tolerance in hepatitis B e antigen transgenic mice. *Eur J Immunol* **25**: 1663–1672.
88. Wirth S, Guidotti LG, o K, *et al.* (1995) Breaking tolerance leads to autoantibody production but not autoimmune liver disease in hepatitis B virus envelope transgenic mice. *J Immunol* **154**: 2504–2515.
89. Wieland S, Thimme R, Purcell RH, Chisari FV. (2004) Genomic analysis of the host response to hepatitis B virus infection. *Proc Natl Acad Sci USA* **101**: 6669–6674.
90. Webster GJ, Reignat S, Maini MK, *et al.* (2000) Incubation phase of acute hepatitis B in man: dynamic of cellular immune mechanisms. *Hepatology* **32**: 1117–1124.
91. Lee BO, Tucker A, Frelin L, *et al.* (2009) Interaction of the hepatitis B core antigen and the innate immune system. *J Immunol* **182**: 6670–6681.
92. Araki K, Miyazaki J, Hino O, *et al.* (1989) Expression and replication of hepatitis B virus genome in transgenic mice. *Proc Natl Acad Sci USA* **86**: 207–211.
93. Farza H, Hadchouel M, Scotto J, *et al.* (1988) Replication and gene expression of hepatitis B virus in a transgenic mouse that contains the complete viral genome. *J Virol* **62**: 4144–4152.
94. Guidotti LG, Matzke B, Schaller H, Chisari FV. (1995) High-level hepatitis B virus replication in transgenic mice. *J Virol* **69**: 6158–6169.
95. Moriyama T, Guilhot S, Klopchin K, *et al.* (1990) Immunobiology and pathogenesis of hepatocellular injury in hepatitis B virus transgenic mice. *Science* **248**: 361–364.
96. Ando K, Guidotti LG, Wirth S, *et al.* (1994) Class I-restricted cytotoxic T lymphocytes are directly cytopathic for their target cells *in vivo. J Immunol* **152**: 3245–3253.
97. Ando K, Moriyama T, Guidotti LG, *et al.* (1993) Mechanisms of class I restricted immunopathology. A transgenic mouse model of fulminant hepatitis. *J Exp Med* **178**: 1541–1554.
98. Brunetto MR, Giarin MM, Oliveri F, *et al.* (1991) Wild-type and e antigen-minus hepatitis B viruses and course of chronic hepatitis. *Proc Natl Acad Sci USA* **88**: 4186–4190.

99. Carman WF, Fagan EA, Hadziyannis S, *et al.* (1991) Association of a precore genomic variant of hepatitis B virus with fulminant hepatitis. *Hepatology* **14**: 219–222.
100. Liang TJ, Hasegawa K, Rimon N, *et al.* (1991) A hepatitis B virus mutant associated with an epidemic of fulminant hepatitis. *N Engl J Med* **324**: 1705–1709.
101. Terazawa S, Kojima M, Yamanaka T, *et al.* (1991) Hepatitis B virus mutants with precore-region defects in two babies with fulminant hepatitis and their mothers positive for antibody to hepatitis B e antigen. *Pediatr Res* **29**: 5–9.
102. Milich DR, Chen MK, Hughes JL, Jones JE. (1998) The secreted hepatitis B precore antigen can modulate the immune response to the nucleocapsid: a mechanism for persistence. *J Immunol* **160**: 2013–2021.
103. Milich DR, Schodel F, Hughes JL, *et al.* (1997) The hepatitis B virus core and e antigens elicit different Th cell subsets: antigen structure can affect Th cell phenotype. *J Virol* **71**: 2192–2201.
104. Guidotti LG, Ishikawa T, Hobbs MV, *et al.* (1996) Intracellular inactivation of the hepatitis B virus by cytotoxic T lymphocytes. *Immunity* **4**: 25–36.
105. Romero R, Lavine JE. (1996) Cytokine inhibition of the hepatitis B virus core promoter. *Hepatology* **23**: 17–23.
106. Tur-Kaspa R, Teicher L, Laub O, *et al.* (1990) Alpha interferon suppresses hepatitis B virus enhancer activity and reduces viral gene transcription. *J Virol* **64**: 1821–1824.
107. Jilbert AR, Wu TT, England JM, *et al.* (1992) Rapid resolution of duck hepatitis B virus infections occurs after massive hepatocellular involvement. *J Virol* **66**: 1377–1388.
108. Kajino K, Jilbert AR, Saputelli J, *et al.* (1994) Woodchuck hepatitis virus infections: very rapid recovery after a prolonged viremia and infection of virtually every hepatocyte. *J Virol* **68**: 5792–5803.
109. Takahashi H, Fujimoto J, Hanada S, Isselbacher KJ. (1995) Acute hepatitis in rats expressing human hepatitis B virus transgenes. *Proc Natl Acad Sci USA* **92**: 1470–1474.
110. Fourel I, Cullen JM, Saputelli J, *et al.* (1994) Evidence that hepatocyte turnover is required for rapid clearance of duck hepatitis B virus during antiviral therapy of chronically infected ducks. *J Virol* **68**: 8321–8330.
111. Iannacone M, Sitia G, Isogawa M, *et al.* (2005) Platelets mediate cytotoxic T lymphocyte-induced liver damage. *Nat Med* **11**: 1167–1169.
112. Kakimi K, Lane TE, Wieland S, *et al.* (2001) Blocking chemokine responsive to gamma-2/interferon (IFN)-gamma inducible protein and monokine induced by IFN-gamma activity *in vivo* reduces the pathogenetic but not the antiviral potential of hepatitis B virus-specific cytotoxic T lymphocytes. *J Exp Med* **194**: 1755–1766.
113. Mahe Y, Mukaida N, Kuno K, *et al.* (1991) Hepatitis B virus X protein transactivates human interleukin-8 gene through acting on nuclear factor kB and

CCAAT/enhancer-binding protein-like cis-elements. *J Biol Chem* **266**: 13759–13763.

114. Sitia G, Isogawa M, Iannacone M, *et al.* (2004) MMPs are required for recruitment of antigen-nonspecific mononuclear cells into the liver by CTLs. *J Clin Invest* **113**: 1158–1167.
115. Sitia G, Isogawa M, Kakimi K, *et al.* (2002) Depletion of neutrophils blocks the recruitment of antigen-nonspecific cells into the liver without affecting the antiviral activity of hepatitis B virus-specific cytotoxic T lymphocytes. *Proc Natl Acad Sci USA* **99**: 13717–13722.
116. Milich DR, Jones JE, Hughes JL, *et al.* (1990) Is a function of the secreted hepatitis B e antigen to induce immunologic tolerance in utero? *Proc Natl Acad Sci USA* **87**: 6599–6603.
117. Chen M, Sallberg M, Hughes J, *et al.* (2005) Immune tolerance split between hepatitis B virus precore and core proteins. *J Virol* **79**: 3016–3027.
118. Chen MT, Billaud JN, Sallberg M, *et al.* (2004) A function of the hepatitis B virus precore protein is to regulate the immune response to the core antigen. *Proc Natl Acad Sci USA* **101**: 14913–14918.
119. Moskophidis D, Lechner F, Pircher H, Zinkernagel RM. (1993) Virus persistence in acutely infected immunocompetent mice by exhaustion of antiviral cytotoxic effector T cells. *Nature* **362**: 758–761.
120. Franco A, Paroli M, Testa U, *et al.* (1992) Transferrin receptor mediates uptake and presentation of hepatitis B envelope antigen by T lymphocytes. *J Exp Med* **175**: 1195–1205.
121. Carman WF, Zanetti AR, Karayiannis P, *et al.* (1990) Vaccine-induced escape mutant of hepatitis B virus. *Lancet* **336**: 325–329.
122. McMahon G, McCarthy LA, Dottavia D, *et al.* (1991) Surface antigen and polymerase gene variation in hepatitis B virus isolates from a monoclonal antibody treated liver transplant patient. In: FB Hollinger, SM Lemon, H Margolis (eds), *Viral Hepatitis and Liver Disease.* Williams & Wilkins, pp. 216–219.
123. Moriyama K, Nakajima E, Hohjoh H, *et al.* (1986) Identification and transmission of hepatitis B virus-related variants. *Proc Natl Acad Sci USA* **83**: 6608–6612.
124. Howard CR, Brown SE, Hogben DN, *et al.* (1984) Analysis of antibody responses to hepatitis B surface antigen. In: GN Vyas, J Dienstag, JH Hoofnagle (eds), *Viral Hepatitis and Liver Disease.* Grune & Stratton, Philadelphia, pp. 561–572.
125. Waters JA, Kennedy M, Voet P, *et al.* (1992) Loss of the common "A" determinant of hepatitis B surface antigen by a vaccine-induced escape mutant. *J Clin Invest* **90**: 2543–2547.
126. Bertoletti A, Costanzo A, Chisari FV, *et al.* (1994) Cytotoxic T lymphocyte response to a wild type hepatitis B virus epitope in patients chronically infected by variant viruses carrying substitutions within the epitope. *J Exp Med* **180**: 933–943.

127. Carman WF, Thursz M, Hadziyannis S, *et al.* (1995) Hepatitis B e antigen negative chronic active hepatitis: hepatitis B virus core mutations occur predominantly in known antigenic determinants. *J Viral Hepat* **2**: 77–84.
128. Ehata T, Omata M, Yokosuka O, *et al.* (1992) Variations in codons 84–101 in the core nucleotide sequence correlate with hepatocellular injury in chronic hepatitis B virus infection. *J Clin Invest* **89**: 332–338.
129. Hosono S, Tai PC, Wang W, *et al.* (1995) Core antigen mutations of human hepatitis B virus in hepatomas accumulate in MHC class II-restricted T cell epitopes. *Virology* **212**: 151–162.
130. Rehermann B, Pasquinelli C, Mosier SM, Chisari FV. (1995) Hepatitis B virus (HBV) sequence variation of cytotoxic T lymphocyte epitopes is not common in patients with chronic HBV infection. *J Clin Invest* **96**: 1527–1534.
131. Bertoletti A, Sette A, Chisari FV, *et al.* (1994) Natural variants of cytotoxic epitopes are T-cell receptor antagonists for antiviral cytotoxic T cells. *Nature* **369**: 407–410.
132. Brunetto M, Stemler M, Schodel F, *et al.* (1989) Identification of HBV variants which cannot produce precore derived HBeAg and may be responsible for severe hepatitis. *Ital G Gastroenterol* **21**: 151–154.
133. Carman WF, Jacyna MR, Hadziyannis S, *et al.* (1989) Mutation preventing formation of hepatitis B e antigen in patients with chronic hepatitis B infection. *Lancet* **2**: 588–591.
134. Guidotti LG, Matzke B, Pasquinelli C, *et al.* (1996) The hepatitis B virus (HBV) precore protein inhibits HBV replication in transgenic mice. *J Virol* **70**: 7056–7061.
135. Lamberts C, Nassal M, Velhagen I, *et al.* (1993) Precore-mediated inhibition of hepatitis B virus progeny DNA synthesis. *J Virol* **67**: 3756–3762.
136. Chen HS, Kew MC, Hornbuckle WE, *et al.* (1992) The precore gene of the woodchuck hepatitis virus genome is not essential for viral replication in the natural host. *J Virol* **66**: 5682–5684.
137. Milich D, Liang TJ. (2003) Exploring the biological basis of hepatitis B e antigen in hepatitis B virus infection. *Hepatology* **38**: 1075–1086.
138. Milich DR. (1997) Influence of T-helper cell subsets and crossregulation in hepatitis B virus infection. *J Viral Hepat* **4**: 48–59.
139. Akarca US, Greene S, Lok AS. (1994) Detection of precore hepatitis B virus mutants in asymptomatic HBsAg-positive family members. *Hepatology* **19**: 1366–1370.
140. Okamoto H, Yotsumoto S, Akahane Y, *et al.* (1990) Hepatitis B viruses with precore region defects prevail in persistently infected hosts along with seroconversion to the antibody against e antigen. *J Virol* **64**: 1298–1303.
141. Takeda K, Akahane Y, Suzuki H, *et al.* (1990) Defects in the precore region of the HBV genome in patients with chronic hepatitis B after sustained seroconversion from HBeAg to anti-HBe induced spontaneously or with interferon therapy. *Hepatology* **12**: 1284–1289.

142. Angus PW, Locarnini SA, McCaughan GW, *et al.* (1995) Hepatitis B virus precore mutant infection is associated with severe recurrent disease after liver transplantation. *Hepatology* **21**: 14–18.
143. Bonino F, Rosina F, Rizzetto M, *et al.* (1986) Chronic hepatitis in HBsAg carriers with serum HBV-DNA and anti-HBe. *Gastroenterology* **90**: 1268–1273.
144. Chu CM, Yeh CT, Lee CS, *et al.* (2002) Precore stop mutant in HBeAg-positive patients with chronic hepatitis B: clinical characteristics and correlation with the course of HBeAg-to-anti-HBe seroconversion. *J Clin Microbiol* **40**: 16–21.
145. Boni C, Fisicaro P, Valdatta C, *et al.* (2007) Characterization of hepatitis B virus (HBV)-specific T-cell dysfunction in chronic HBV infection. *J Virol* **81**: 4215–4225.
146. Evans A, Riva A, Cooksley H, *et al.* (2008) Programmed death 1 expression during antiviral treatment of chronic hepatitis B: impact of hepatitis B e-antigen seroconversion. *Hepatology* **48**: 759–769.
147. Maier H, Isogawa M, Freeman GJ, Chisari FV. (2007) PD-1:PD-L1 interactions contribute to the functional suppression of virus-specific CD8+ T lymphocytes in the liver. *J Immunol* **178**: 2714–2720.
148. Stoop JN, van der Molen RG, Baan CC, *et al.* (2005) Regulatory T cells contribute to the impaired immune response in patients with chronic hepatitis B virus infection. *Hepatology* **41**: 771–778.
149. Xu D, Fu J, Jin L, *et al.* (2006) Circulating and liver resident CD4+CD25+ regulatory T cells actively influence the antiviral immune response and disease progression in patients with hepatitis B. *J Immunol* **177**: 739–747.
150. Milich DR, Chisari FV. (1982) Genetic regulation of the immune response to hepatitis B surface antigen (HBsAg). I. H-2 restriction of the murine humoral immune response to the a and d determinants of HBsAg. *J Immunol* **129**: 320–325.
151. Milich DR, Leroux-Roels GG, Chisari FV. (1983) Genetic regulation of the immune response to hepatitis B surface antigen (HBsAg). II. Qualitative characteristics of the humoral immune response to the a, d, y determinants of HBsAg. *J Immunol* **130**: 1395–1400.
152. Milich DR, Leroux-Roels GG, Louie RE, Chisari FV. (1984) Genetic regulation of the immune response to hepatitis B surface antigen (HBsAg). IV. Distinct H-2-linked Ir genes control antibody responses to different HBsAg determinants on the same molecule and map to the I-A and I-C subregions. *J Exp Med* **159**: 41–56.
153. Milich, D. (2002) Influence of C4A deficiency on nonresponse to HBsAg vaccination: a new immune response gene. *J Hepatol* **37**: 396.
154. Milich DR, McNamara MK, McLachlan A, *et al.* (1985) Distinct H-2-linked regulation of T-cell responses to the pre-S and S regions of the same hepatitis

B surface antigen polypeptide allows circumvention of nonresponsiveness to the S region. *Proc Natl Acad Sci USA* **82**: 8168–8172.

155. Milich DR, Thornton GB, Neurath AR, *et al.* (1985) Enhanced immunogenicity of the pre-S region of hepatitis B surface antigen. *Science* **228**: 1195–1199.
156. Walker M, Szmuness W, Stevens CE, Rubinstein P. (1981) Genetics of anti-HBs responsiveness. *Transfusion* **21**: 601.
157. Alper CA. (1995) The human immune response to hepatitis B surface antigen. *Exp Clin Immunogenet* **12**: 171–181.
158. Craven DE, Awdeh ZL, Kunches LM, *et al.* (1986) Nonresponsiveness to hepatitis B vaccine in health care workers. Results of revaccination and genetic typings. *Ann Intern Med* **105**: 356–360.
159. Milich DR, Leroux-Roels GG. (2003) Immunogenetics of the response to HBsAg vaccination. *Autoimmun Rev* **2**: 248–257.
160. Singh R, Kaul R, Kaul A, Khan K. (2007) A comparative review of HLA associations with hepatitis B and C viral infections across global populations. *World J Gastroenterol* **13**: 1770–1787.
161. Hohler T, Stradmann-Bellinghausen B, Starke R, *et al.* (2002) C4A deficiency and nonresponse to hepatitis B vaccination. *J Hepatol* **37**: 387–392.
162. Lee BO, Jones JE, Peter CJ, *et al.* (2011) Identification of a unique double-negative regulatory T cell population. *Immunol* **134**: 434–447.

Chapter 3

Epidemiology of Viral Hepatitis B-related Hepatocellular Carcinoma

Katherine A. McGlynn, Alison A. Evans and W. Thomas London*

Introduction

Chronic infection with hepatitis B virus (HBV), defined as the presence of HBsAg in the serum for at least six months, is recognized both clinically and epidemiologically as a common cause of death from liver failure, cirrhosis, and hepatocellular carcinoma (HCC). Globally, 350–400 million people are chronically infected with HBV.[1] Among these individuals, 500,000–700,000 will die each year from liver diseases, including 35–87 million who will die of HCC. For the year 2000, 620,000 persons were estimated to have died worldwide from HBV-related causes: 580,000 (94%) from cirrhosis and HCC, and 40,000 (6%) from acute hepatitis B infection.[2]

History

Several pathologists first proposed in the 1950s the possibility that chronic viral infections of the liver could lead to liver cancer.[3] They

* Correspondence Author: Thomas London, Fox Chase Cancer Center 333 Cottman AVe. Philadelphia, PA 19111. Tel: 215-728-2204, Fax: 215-214-4053. E-mail: wt_london@fccc.edu

noted that at least 80% of liver cancers occurred in cirrhotic livers. Edmondson[3] pointed out that in the United States and Western Europe, 3–10% of male patients with cirrhosis developed HCC, whereas in Asia and Africa 15–50% of men with cirrhosis developed liver cancer. They proposed that cirrhosis in Western countries was mainly due to alcohol abuse, but in Asia and Africa, where alcoholism was uncommon, some other agent caused cirrhosis and HCC. They included nutritional deficiencies, exposure to toxins, and chronic viral infections among the possible candidates.

We now know that chronic infection with HBV or HCV is causally associated with 80–95% of all HCCs in the world. In the regions of the world where HCC is most common, HBV infection is associated with most cases of cirrhosis and 70–80% of HCC cases. Areas of the world with high incidence and mortality rates for HCC have high prevalences of chronic HBV infection. The reverse is also true. Countries with prevalences of chronic HBV infection of greater than 2% have increased incidence and mortality rates from HCC.[4,5] Initially, case reports, case series, and cross-sectional studies from various Asian, African, and Western countries showed associations of HBsAg seropositivity with HCC.[6] Later, case-control studies in all regions of the world consistently showed that chronic HBV infection was much more common among cases than controls. Odds ratios ranged from 5:1 to 65:1, thus leading the International Agency for Research on Cancer (IARC) in 1994 to classify HBV as carcinogenic to humans.[6]

Beginning in 1981, prospective studies of persons chronically infected with HBV have demonstrated very high relative risks for HCC, ranging from 5 to 103.[6–8] In Beasley's study of male government workers in Taiwan, the age-adjusted annual incidence of HCC was 474 per 100,000 in HBsAg-positive men, compared with 6 per 100,000 in HBsAg-negative men.[7] The study by Evans *et al.*[8] shows the magnitude of the effects of chronic HBV infection. The study enrolled an unselected general population of 58,545 men (15.0% HBsAg-positive) and 25,340 women (10.7% HBsAg-positive) in Haimen City, China, in 1992–1993. After eight years of follow-up, HCC was the major cause of death in the cohort, accounting for 977 of the 3,487 deaths in the population. Among men, the cumulative risks for death from HCC were 0.5% for HBV uninfected (HBsAg-negative at entry) and 8% for

HBsAg-positive; among women, the cumulative risks were 0.1% for HBV uninfected and 2.0% for HBsAg-positive.

In areas of the world with high incidences of HCC and high prevalences of chronic HBV infection, about 70% of such infections are acquired in the perinatal period or in early childhood.[5] Thus, among HBV carriers in endemic areas, those born to HBV-infected mothers are likely to have been infected longest and are at higher risk of HCC than HBV carriers with HBsAg-negative mothers.[9] Among all HBV carriers, the lifetime risk of HCC is estimated to be 10–25%.

Global Prevalence of Chronic HBV Infection

The World Health Organization estimates that there are approximately 350 million chronic HBV carriers worldwide (Fig. 1). The prevalence of chronic infection, however, varies greatly among different regions.

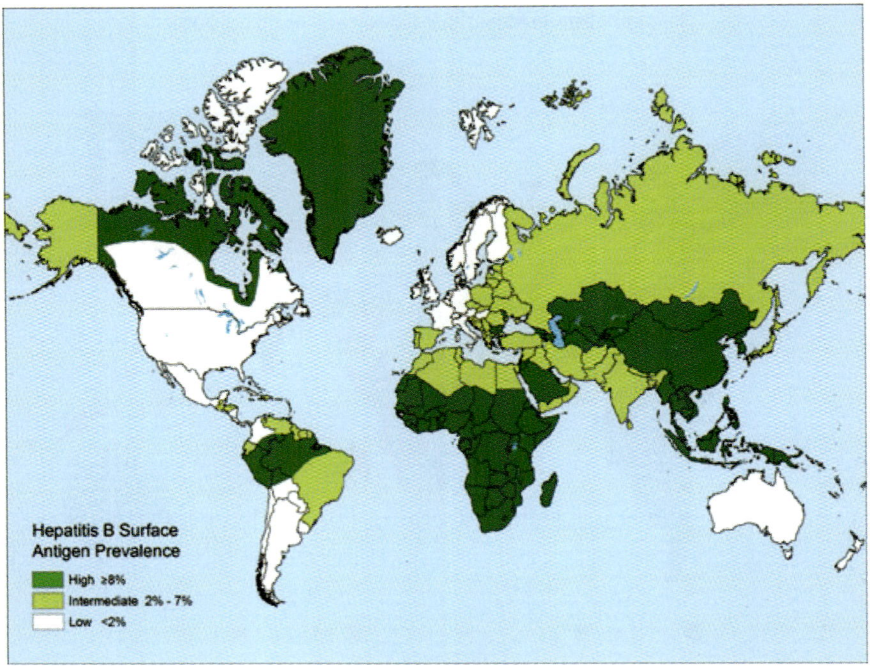

Fig. 1. Prevalence of chronic HBV infection, 2006. (From: Centers for Disease Control and Prevention, Atlanta, Georgia, 2008.)

HBV infection is endemic in eastern Asia and sub-Saharan Africa, in particular in the western countries of sub-Saharan Africa. In these HBV high-rate areas, 8–10% of the populations are chronically infected. In HBV intermediate-rate areas, such as the Middle East, India, and southern Europe, 2–7% of the populations are infected. In contrast, among the populations of North America, Australia, and Northern and Western Europe, the prevalence of chronic infection is less than 2%.

Liver Cancer Incidence, Mortality, and Survival

Primary liver cancer is the sixth most common cancer in the world and the third most common cause of cancer mortality.[10] There is wide geographic variability in liver cancer incidence, however, which correlates closely with the prevalence of chronic HBV infection (Fig. 2). The great majority of liver cancer cases (>80%) occurs in either sub-Saharan Africa or in Eastern Asia, with one country alone, China, accounting

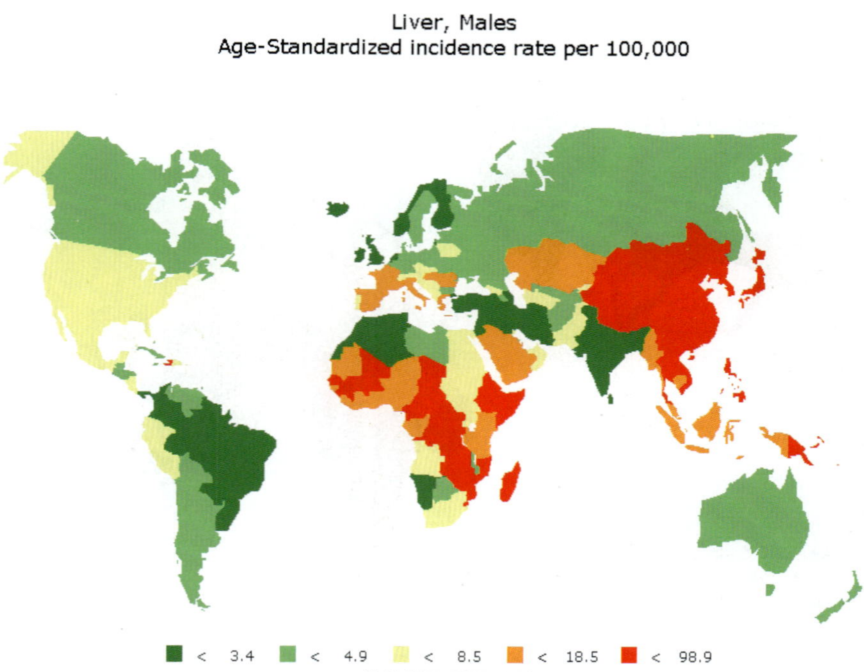

Fig. 2. Global incidence of primary liver cancer among males, 2002. (From: Globocan 2002, International Agency for Research on Cancer, 2008.)

for over 50% of all cases. Countries in North America, South America, Northern Europe, and Australia tend to have low incidence rates, while countries in Central and Southern Europe tend have intermediate rates. Some of the highest incidence rates in the world have been reported by cancer registries in Korea. For example, the registry in Busan, Korea, reported that male incidence was 49.8/100,000 and female incidence was 14.9/100,000 for the period 1998–2002.[11] During that same period, some of the lowest incidence rates in the world were reported by the Netherlands, where male incidence was 1.8/100,000 and female incidence was 0.7/100,000 (Fig. 3).

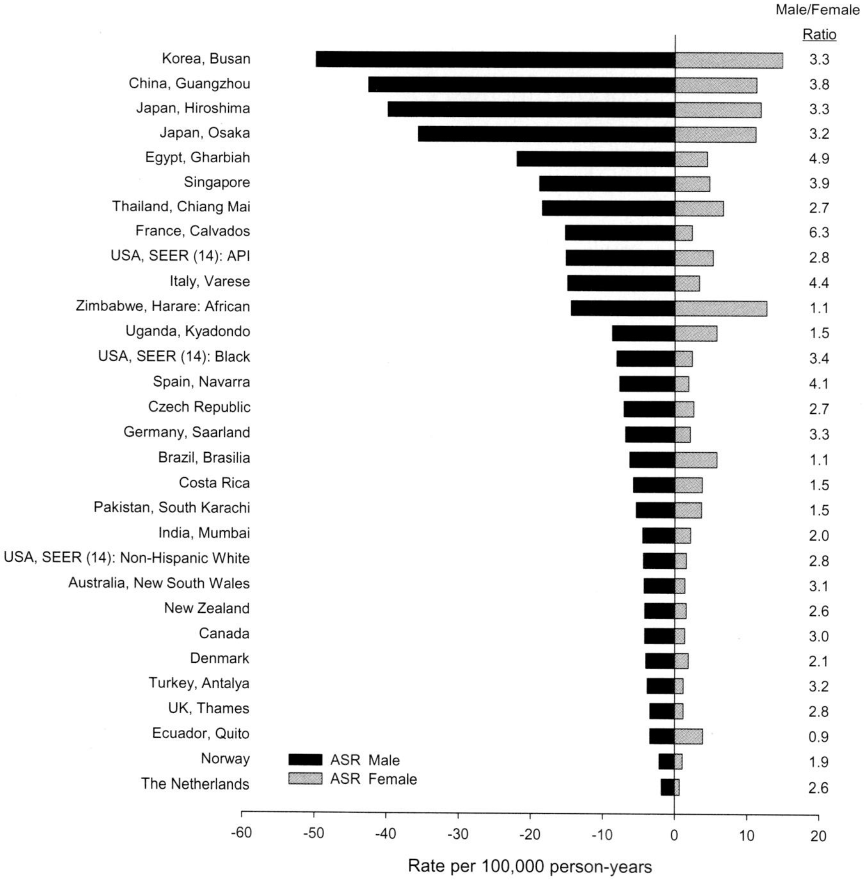

Fig. 3. Age-standardized incidence rates (ASR) of primary liver cancer by gender in selected registries (Curado *et al.*, 2007).

Between 1978–1982 and 1998–2002, trends in liver cancer incidence varied by geographical location.[11,12] Some high-rate Asian areas, such as Hong Kong, Shanghai, and Singapore experienced declines in incidence among both males and females. Male incidence rates in Japan also experienced a decline, while female incidence rates in Japan continued to increase. Trends in high-rate African countries are less clear, as few registries in Africa have long-term data. The declining incidence rates in Hong Kong, Shanghai, and Singapore may be related to hepatitis B vaccination programs and improvements in public health. Declining incidence rates in Japan, however, are more likely to be related to the decline in the population prevalence of hepatitis C virus infection (Fig. 4).

In contrast to trends in some high-risk areas, registries in a number of low-rate areas reported increases in incidence between 1978–1982 and 1998–2002.[11,12] Included among these registries are those in Australia, the United States, Canada, and the United Kingdom. Reasons for the increased incidence in low-rate areas are not certain. It is likely, however, that the increased prevalence of chronic infection with hepatitis C virus is a key determinant. Increased incidence of diabetes and obesity may also be affecting the liver cancer rates.

Survival rates of primary liver cancer are uniformly poor in high- and low-rate areas. As a result, mortality rates are very similar to incidence rates. Overall, the IARC estimates that the age-standardized worldwide incidence rates of primary liver cancer are 15.7/100,000 among males and 5.8/100,000 among females. The comparable mortality rates are 14.9/100,000 and 5.7/100,000.[10] While improvements in survival have recently been reported in the United States, the one-year survival rate still remains below 50%.[13]

Demographic Factors

Age

The global age distribution of HCC varies by incidence, gender, and possibly also by etiology. In general, incidence rates tend to peak at younger ages in high-risk areas than in low-risk areas, but in almost

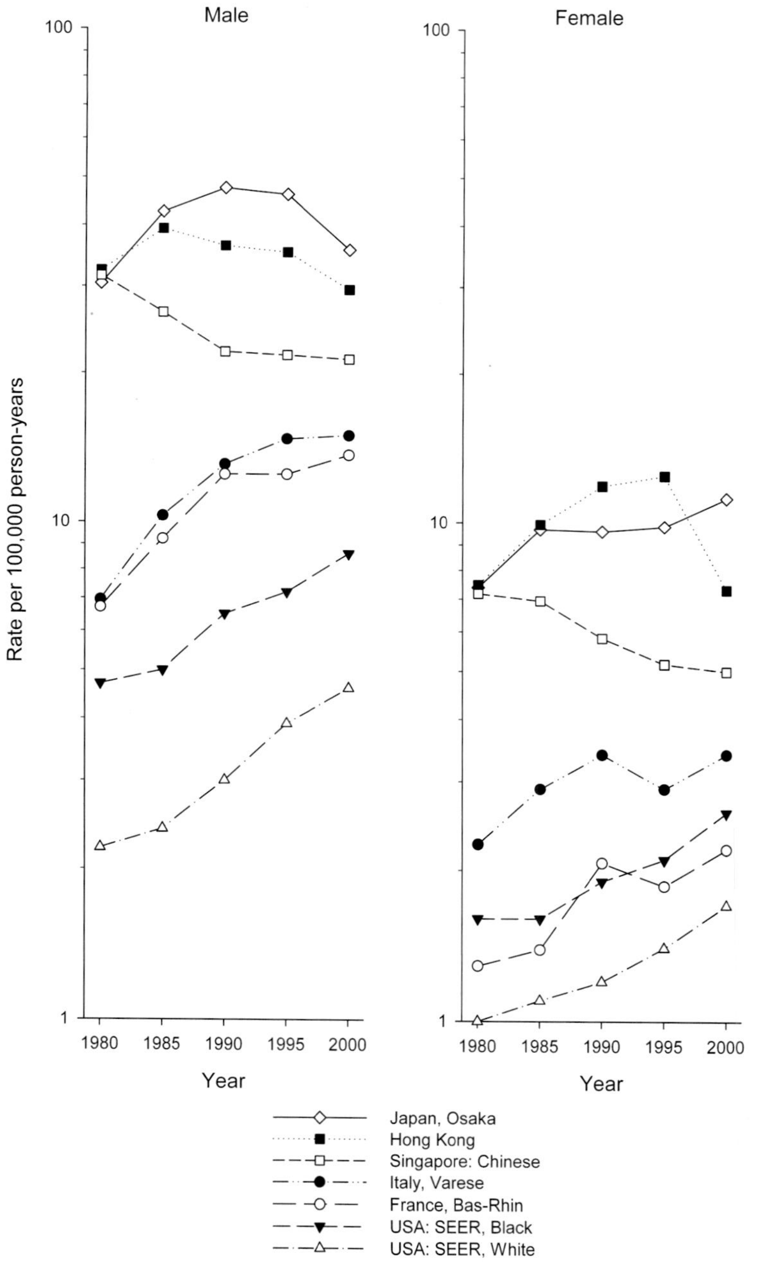

Fig. 4. Trends in the incidence of primary liver cancer by gender, 1978–2002 (Curado *et al.*, 2007; Parkin *et al.*, 2005.)

all regions, female incidence peaks at later ages than male incidence. In low-risk populations, such as the United States, male incidence peaks at ages 65–69 years, while female incidence peaks five years later, at ages 70–74 years. In comparison, in high-risk areas of sub-Saharan Africa, such as Zimbabwe, male incidence peaks at ages 50–54 years, while female incidence peaks at ages 55–59 years. In high-risk Asian areas, such as Korea, male incidence peaks at ages 55–59 years, while female incidence peaks at ages 60–64 years.

Sex

In almost all areas of the world, the incidence of liver cancer is higher among men than women. The greatest differences between male and female rates no longer occur among the highest-risk populations, however,[14] but among the populations of Central and Southern Europe.[12] Typical among these male:female ratios are the ones reported from Calvados, France (6.3); Geneva, Switzerland (5.6); and Varese, Italy (4.4). In comparison, typical M:F ratios reported in high-risk Asian registries are those of Hong Kong (4.0); Busan, Korea (3.3); and Singapore (3.9). The only registries in the world that report ratios at or near 1 are in South America. Typical ratios there are those of Cali, Colombia (1.1); Quito, Ecuador (0.9); and Brasilia, Brazil (1.1). The difference in rates between males and females in high-risk areas may have become less pronounced because male incidence is declining faster than female incidence in some populations.

The reasons that males have higher rates of liver cancer than females are not completely understood, but may be partly explained by the sex-specific prevalence of risk factors. Males are more likely to be chronically infected with HBV, consume alcohol, smoke cigarettes, and have increased iron stores. Androgenic hormones and increased genetic susceptibility may also be related to the male excess in liver cancer.

Ethnicity

Liver cancer incidence rates can vary greatly among persons of different ethnicities living in the same region. For example, in Los Angeles,

Korean men have incidence rates approximately five and a half times higher than do white men (Table 1). Similarly, in Singapore, Chinese men have rates 2.6 times greater than rates among Indian men, while in San Francisco, Chinese men have rates 3.8 times the rate of white men. Within the SEER registries of the United States, Asians and Pacific Islanders have the highest incidence rates (11.7/100,000), while whites have the lowest rates (3.9/100,000). Intermediate to

Table 1. Primary Liver Cancer Incidence Rates by Ethnicity Within Selected Cancer Registries, 1998–2002.[1]

	Male		Female	
Registry	**Rate**[2]	**Rate/base**	**Rate**[2]	**Rate/base**
Los Angeles				
White[3]	5.0	1.00	2.1	1.00
Black	9.0	1.80	2.7	1.29
Japanese	7.9	1.58	7.3	3.48
Hispanic	10.2	2.04	4.3	2.05
Filipino	12.5	2.50	2.6	1.24
Chinese	17.2	3.44	6.1	2.90
Korean	27.5	5.50	10.2	4.86
San Francisco				
White[3]	5.0	1.00	1.7	1.00
Japanese	7.0	1.40	3.7	2.18
Hispanic	11.1	2.22	4.5	2.65
Black	10.9	2.18	3.3	1.94
Filipino	10.4	2.08	3.7	2.18
Chinese	18.9	3.78	4.8	2.82
Singapore				
Indian[3]	8.1	1.00	3.2	1.00
Malay	12.3	1.52	3.6	1.13
Chinese	21.3	2.63	5.0	1.56

[1] *Source*: Cancer Incidence in Five Continents, Vol. IX, 2007.
[2] All rates are age-adjusted to the world standard population and are calculated per 100,000 person-years.
[3] Base ethnicity for rate comparison in each group.

these groups are the rates of Hispanics (8.0/100,000), African-Americans (7.0/100,000), and American Indians and Alaska Natives (6.6/100,000). This variation in incidence almost certainly reflects differences in the likelihood of infection with HBV or HCV, although genetic susceptibility and different patterns of exposure to other risk factors may also play a role.

Risk Factors

Aflatoxin

Aflatoxin B_1 (AFB_1) is a mycotoxin elaborated by fungi of the *Aspergillus* species. The fungi grow readily on foodstuffs, such as corn and peanuts, stored in warm, damp conditions. Although there are four principal aflatoxins — B_1, B_2, G_1, and G_2 — AFB_1 is the most potent in animal studies.[15] Since 1987, IARC has classified aflatoxin as a human carcinogen.[15] Once ingested, AFB_1 is metabolized to an active intermediate, AFB_1-*exo*-8,9-epoxide, which is later detoxified through a variety of metabolic processes. The intermediate epoxide has been shown to bind and damage DNA, primarily at the N7 position of guanine.[16] The characteristic genetic change associated with AFB_1 is a G to T transversion in the third base of codon 249 of the *p53* gene.[17] The *p53* 249^{ser} mutation has been observed in 30–60% of tumors arising in persons living in high aflatoxin areas.[18,19]

Many ecological studies of AFB_1 contamination of food conducted in the 1970s and 1980s were compatible with a role for the carcinogen in HCC. Stronger evidence of an AFB_1–HCC link was supplied subsequently by studies based on the detection of AFB_1 markers in biosamples. Further, an interaction of AFB_1 and HBV infection on HCC risk was revealed in a short-term prospective study in Shanghai, China,[20] that estimated AFB_1 increased the risk of HCC fourfold, HBV increased the risk sevenfold, and the combination of AFB_1 and HBV increased the risk 60-fold.

In most areas of the world where AFB_1 exposure is widespread, chronic HBV infection is also highly prevalent. Though HBV vaccination in these areas should be the major preventive tactic, persons already chronically infected with HBV will not benefit from vaccination. HBV carriers could benefit, however, by eliminating AFB_1 exposure. In addition to primary prevention, secondary prevention strategies involving medications (e.g., Oltipraz) or hot tea infusions (e.g., sulforaphane, green tea polyphenols) have also been considered.[21]

Alcohol

The evidence in support of a positive association between alcohol consumption and HCC led IARC to conclude in 1988 that there was a causal relationship.[22] The mechanism by which alcohol increases risk is not well understood, however. Other unresolved issues include the comparative risk in men versus women and the combined effect of alcohol and HBV versus the effect of alcohol and HCV.

The increased risk associated with alcohol is difficult to quantify precisely as most HCC occurs on a background of cirrhosis, and persons with cirrhosis often reduce their alcohol consumption.[23,24] As a result, estimates of risk from case-control studies are likely to be underestimates.[25] Nevertheless, earlier studies of HCC found alcohol to be a more significant risk factor in low-incidence areas than in high-incidence areas. This may have been due to lower mean alcohol consumption in high-risk populations and/or due to the dominant effect of chronic HBV infection in high-risk areas masking any additional risk due to alcohol consumption. While some studies in HBV-endemic populations have shown a positive association between alcohol consumption and HCC,[26] other studies have not.[27] In comparison, most studies in populations where HCV is the dominant virus have found alcohol to be a significant risk factor.[28] For example, a recently published meta-analysis of 46 Japanese studies concluded that there was convincing evidence that alcohol increases the risk of HCC in Japan.[29] A direct comparison of the effect of alcohol with each virus, however, reported little difference in the risk.[30] Data also suggest that alcohol is associated with HCC in the absence of either HBV or HCV infection, though higher levels of consumption are probably required for HCC in the absence of viral infection.[30] Very high levels of consumption, however, are not likely to be associated with increased HCC risk because persons who consume at extreme levels rarely survive long enough to develop HCC.[23]

Whether alcohol is more strongly associated with HCC in women than in men has been difficult to study, given that women are less likely to be heavy drinkers and less likely to develop HCC than men. A greater effect of alcohol on women has been hypothesized based on differences in alcohol dehydrogenase activity[31] and evidence of a greater association between alcohol and cirrhosis among women.[32] In the Brescia HCC study, no substantial sex difference in risk of HCC with alcohol consumption was reported.[30] A meta-analysis published in 2001, however, found that risks associated with alcohol were higher among women than men at all levels of consumption.[33]

The mechanism by which alcohol increases HCC risk is not well understood. Animal and human studies provide little evidence that ethanol is a carcinogen.[34] Some of the mechanisms by which alcohol might increase risk include the production of acetaldehyde and free radicals during alcohol metabolism, cytochrome p4502E1 induction, modulation of cell regeneration, promotion or exacerbation of nutritional deficiencies, and alterations of the immune system.[35] It is certain that alcohol induces cirrhosis, and cirrhosis is a factor in 60–90% of HCCs. Whether alcohol is related to HCC independent of cirrhosis is less clear, although at least one study[36] reported that the standardized incidence rate (SIR) of HCC among alcoholics without cirrhosis was 3.1 (95% CI = 1.6–5.3). In comparison, the same study reported that the risk of HCC among alcoholics with cirrhosis was over 10 times greater, with an SIR = 34.3 (95% CI = 17.1–61.3).

Tobacco

The effect of cigarette smoking on the risk of HCC has been examined in numerous cohort and case-controls studies since the early 1980s.[37] The accumulated evidence is compatible with a weak to modest association between smoking and HCC that has been classified as causal by the IARC.[38] A recent meta-analysis of 31 studies reported a pooled relative risk (RR) of 1.56 (95% CI = 1.29–1.87) in a comparison of current smokers to never-smokers and a pooled

RR = 1.49 (95% CI = 1.06–2.10) in a comparison of former smokers to never-smokers.[39] Though the meta-analysis did not find statistically significant heterogeneity across the studies with regard to gender, country, alcohol, or HBV/HCV infection, interactions of smoking with these factors remain uncertain. For example, higher risks of smoking have been reported in some studies to be restricted to HBV-positive persons[40] and in other studies to HBV-negative persons.[41,42] Similarly, interactions between smoking and HCV infections have been reported.[40] In addition, the association between alcohol consumption and cigarette smoking may influence the risk associated with smoking, as indicated by significant interactions between alcohol and smoking on the risk of HCC in Greece[41] and in the United States.[43]

Pesticides

Studies in experimental animals have suggested that some pesticides may be associated with the development of HCC.[44] In particular, experimental animals exposed to 1,1,1-trichloro-2,2-bis(p-chlorophenyl) ethane (DDT) show dose-related increases in liver tumors.[44] Humans occupationally exposed to DDT as pesticide applicators have been reported to have increased rates of HCC in some[45] but not all[46] studies. An ecologic study in the United States, however, reported a significant correlation between adipose levels of 1,1,dichloro-2,2, bis(p-chlorophenyl) ethylene (DDE, the persistent metabolite of DDT) and liver cancer mortality rates among Caucasians.[47] To date, only one direct examination of DDT and HCC in humans has been reported. A nested case-control study conducted among a cohort of persons in Linxian, China, found that the risk of developing HCC increased with increasing serum DDT concentration.[48] The study found no association with DDE, suggesting that recent exposure to DDT might be necessary to increase risk.

Pesticides other than DDT have also been suggested to be associated with HCC. A case-control study of an Egyptian population found that rural men with HCC had significantly higher levels of organophosphorus compounds and carbamates than did rural men without HCC.[49] No association was evident among urban males or females, however.

Though studies of pesticides and HCC are few, the suggestive evidence of an association argues that more studies should be done. The use of pesticides is widespread globally and exposure to them is very heavy in some populations.

Iron-related Conditions

Hereditary hemochromatosis (HH) is an autosomal recessive disorder that is characterized by excessive dietary iron absorption and subsequent deposition in the parenchymal cells of the liver, pancreas, heart, joints, and pituitary gland.[50] Most HH is associated with mutations in the *HFE* gene on chromosome 6. Prior to the cloning of the *HFE* gene in 1996,[51] studies had estimated the carrier rate at 10% among populations of Northern European ancestry. Simultaneous with the identification of the *HFE* gene was the demonstration that two missense mutations, *C282Y* and *H63D*, accounted for the great majority of HH.[51,52] Homozygosity for the *C282Y* mutation is responsible for between 50% and 80% of HH, consistent with the observation that the *C282Y* mutation causes a greater loss of *HFE* protein function than does the *H63D* mutation.[51] Approximately 5% of HH occurs among *H63D* heterozygotes, another 5% among *C292Y/H63D* compound heterozygotes, and 7% among individuals who carry neither mutation.[53]

The relative risk of liver cancer in persons with HH has been estimated to range between 93 and 200.[54,55] These estimates are likely to be high, however, as most studies have included only patients being seen at clinics, not asymptomatic persons. In any case, the risk of HCC in HH is not uniformly distributed, but is increased in the presence of a variety of co-factors including male gender, age greater than 50 years, drinking, smoking, and HBV and HCV infections.[56] Though the vast majority of HCC in persons with HH develops in cirrhotic livers, HCC in non-cirrhotic livers has also been reported.[57] Despite these observations, the majority of HH patients who are diagnosed and treated prior to suffering irreversible liver damage have been reported to have a normal life expectancy.[55]

Individuals with another inherited metabolic disorder, porphyria cutanea tarda, also are subject to hepatic iron overload and are also at increased risk of HCC, though the risk is not as high as in hemochromatosis.[58]

African iron overload, first described by Strachan in 1929,[59] has also been associated with an increased risk of HCC.[60] The consumption of traditional iron-rich beer is a major risk factor for African iron overload, but evidence suggests that there is also a genetic component to risk that is unrelated to mutations in the *HFE* gene.[61]

Endogenous and Exogenous Hormones

Studies conducted in the 1970s and 1980s established that oral contraceptive (OC) use is associated with increased risk of benign hepatic adenomas.[62] Evidence from case-control studies suggests that there may also be a relationship between long-term OC use (>5 years) and risk of HCC.[63] The increased risk appears to be greater in the absence of viral infections,[63] although it should be noted that OC use is substantively more common in areas with a low prevalence of hepatic viral infections. Short-term OC use (<5 years), has not been associated with risk.[64] In contrast to the evidence for OC use, neither injectable progestogen use nor post-menopausal hormone replacement use has been associated with increased risk of HCC.[63] There is some evidence, in fact, that hormone replacement therapy may be associated with decreased risk.[65]

Evidence from case reports suggests that anabolic steroids may cause HCC,[66] but no epidemiologic study of anabolic steroids and HCC has yet been reported. Due to the scarcity of data, IARC has classified the evidence for carcinogenicity in humans as "limited."[15]

A role for endogenous hormones in the etiology of HCC has been proposed to explain the male excess of HCC in almost all countries. Experimental animal studies have supported a role for androgens in hepatocarcinogenesis,[67] but whether androgens are also a risk factor in humans is not clear. In studies conducted in Taiwan, Yu *et al.* reported a positive association between increased testosterone levels and HCC among men chronically infected with HBV.[68] In a study that examined both HBV-infected and uninfected men, Yuan *et al.*

found that infected men had significantly higher testosterone levels than uninfected men and suggested that the difference might explain the testosterone-HCC relationship.[69]

Among women, HCC risk has been associated with increasing parity.[70] The mechanism by which increased parity would increase HCC risk is not clear, but may be related to the altered estrogen profile in pregnancy,[71] or to the promotional effect of estrogens on a liver that is chronically infected with HBV. In support of a role for estrogen are recent data showing an increased risk of HCC among women with younger ages at menarche and older ages at menopause.[72]

Host Susceptibility

Susceptibility to HBV-related HCC arises both from susceptibility to chronic HBV infection and from susceptibility factors in progression to HCC. It has not always been possible to distinguish between these factors in observational studies.

Susceptibility to Chronic HBV Infection

Individuals infected with HBV early in life, particularly perinatally, have long been known to have the highest risk of development of chronic infection. Studies of genetic polymorphisms in chronic infection have often revealed associations with alterations of genes involved in immune response. Studies in West Africa have revealed that resistance to persistent HBV infection is associated with the MHC class II allele *HLA-DRB1*1302*.[73] Moreover, Thursz *et al.* have also shown that Gambians who were heterozygous for MHC class II region genes (*HLA-DR* and *HLA-DQ*) were significantly less likely to be chronic HBV carriers compared to homozygotes. No such association was found for MHC class I alleles. Class II alleles have also been shown to be significantly associated with HBV persistence in populations where infection occurs more commonly in adulthood.[74] In particular, Thio *et al.* identified the three-locus haplotype of *DQA1*0501*, *DQB1*0301*, and *DRB1*1102* as strongly associated with persistent infection in African-Americans in a prospective cohort.

Studies of genetic variation in cytokine-related genes have frequently found association of persistent infection with the TNF-α promoter, which is in strong linkage disequilibrium with the *HLA-DR* locus. Hohler *et al.*[75] first found that the –238 (G/A) polymorphism was associated with maintenance of chronic infection in German adults. Other studies have examined a larger number of polymorphisms in the TNF-α promoter region in other populations. While the specific polymorphisms found to be involved have varied somewhat in different studies, TNF-α promoter polymorphisms, in general, have frequently been found to have positive associations.[76,77]

HIV infection increases the probability of development of chronicity in HBV infection as well, with cumulative incidence rates around 20% in studies of homosexual men infected with HIV-1, compared to the expected 1% in healthy adults and ~5% in HIV-positive adults prior to the advent of highly active antiretroviral therapy (HAART).[78,79]

Susceptibility to HCC Once Chronic HBV Infection is Established

Factors thought to be involved in the susceptibility to development of HCC in persons chronically infected with HBV include genetics of both the host and the virus, coinfection with other viruses, and presence of certain medical conditions such as diabetes. Yu *et al.* have recently reviewed the evidence for involvement of genes regulating methylation in HCC development in several epidemiologic studies. Their findings suggest that lower activity variants of methylenetetrahydrofolate reductase (MTHFR) in combination with high-activity variants of thymidylate synthase (TYMS) result in reduced risk of HBV-associated HCC.[80]

Viral Factors

Viral Load

While measurement of HBV viral load has emerged as a useful clinical marker of response to antiviral therapy, its value as a prognostic marker has only recently been understood. In a prospective cohort study in Haimen City, China, Chen *et al.* measured HBV viral load at study entry in 2,763 HBsAg-positive adults.[81] In 11 years of follow-up, there were 231 deaths from HCC in this subgroup. There was a significant increase in mortality as viral load increased. Those in the highest viral load category ($>10^5$ particle/mL) had a RR of 11.2 (95% CI 3.6–35.0) compared to those who had undetectable HBV DNA at baseline. Significantly, the RR associated with high viral load did not change with increased follow-up time. Results from a similar study of HBV-infected adults in Taiwan revealed the same relationship,[82] and subsequent studies have supported these findings in other populations.

Viral Genotype

Eight major genotype groups (A through H) have been identified for HBV by examination of whole genome sequence variation,[83] and each type has a different geographic distribution. A number of studies in Asia, where types B and C dominate, have identified type C as associated with later HBeAg seroconversion and a higher risk of HCC.[84,85] This may be the result of the higher frequencies of nucleotide 1762 A >T mutations seen in genotype C. Type C infections may also result in higher long-term viral load, which is in itself a risk factor for HCC. The roles of the other HBV genotypes in HCC development have not been as exhaustively studied as types B and C, but associations exist there as well. Livingston *et al.* have found that genotype F is found in the majority of HBV-infected persons with HCC among Alaskan Natives. Genotypes A and F, common in the West, appear to be associated with a lower risk of HCC and other serious sequelae.[86]

Viral Mutations

Three variants of HBV have received substantial attention in studies of viral pathogenicity. The HBV genome contains areas of overlapping open reading frames. The pre-core mutation is a G to A change at nucleotide 1896 and is associated with HBeAg seroconversion with continued high to moderate levels of HBV DNA in serum. The Basal Core Promoter (BCP) mutation is a double mutation at nucleotide 1762 (A>T) and nucleotide 1764 (G>A) also associated with HBeAg seroconversion with continued viremia. BCP in particular has been associated with adverse outcomes in both prospective and cross-sectional studies.[87,88] A particularly important question in this area of epidemiologic research is whether sequence variants of HBV associated with adverse outcomes are latecomers in the pathway to disease or whether they persist over the long time of the latent period. Chou *et al.*[87] have shown that most common variants have their strongest effect in subjects who develop HCC within three years of their determination, but BCP and genotype C in particular have lesser but still statistically significant effects up to nine years from entry. For BCP, the strongest effects are seen in those in whom the presence of the double mutation persists over several years. In contrast, pre-core mutation was shown to have a protective effect in the short term and no long-term effect of persistence. Yuan *et al.*[89] recently estimated the effect of BCP double mutation load on HCC risk in a prospective cohort study of HBsAg positive adults in China. Presence of the mutation conferred a 2.47-fold increased risk (95% CI 1.04–5.85) overall, and this relationship was not altered by length of follow-up. Moreover, subjects in the highest mutation load category (≥ 500 copies/mL) were at 14.6-fold higher risk compared to those with no evidence of BCP mutation (95% CI 2.4–88.0).

HBV and Other Cancers

Chronic HBV infection has been hypothesized to increase risk of cancers other than HCC, largely because of evidence that HBV can infect and possibly replicate in cells other than hepatocytes. Detection of HBV DNA in the human pancreas was reported as early as 1984,[90] and there are several lines of evidence suggesting that HBV replication and/or viral antigen expression may occur in pancreatic cells.[91,92] Epidemiologic evidence of an association with pancreatic cancer has been sparse. In one recent publication, however, Hassan *et al.* (2008) found a RR of 2.5 (95% CI 1.5–4.2) for anti-HBc positive versus negative in a hospital-based case-control study at a tertiary referral center in the United States. Interestingly, there was no association of pancreatic cancer with anti-HCV positivity in the same study, and there was evidence of risk modification resulting in an increased risk to diabetics who were also anti-HBc positive.[93]

Epidemiologic studies of the relationship of HBV infection to bile duct cancers have also been revealing. Hsing *et al.*, in a population-based case-control study in Shanghai, China, found a 2.4-fold increased risk (95% CI 1.2–4.5) of extra-hepatic bile duct cancer in subjects who were HBsAg-positive versus negatives, but no increased risk of cancers of the gallbladder or the ampulla of Vater.[94] Subjects with evidence of past infection (anti-HBc positive, HBsAg negative) did not exhibit increased risk compared to those without evidence of past or current infection, and no association was found for anti-HCV positivity for cancer at any of the sites studied.

Future Directions and Unresolved Issues

In this chapter we have described factors associated with an increased risk of individuals chronically infected with HBV to develop HCC. Some factors, like host genes, coinfection with HCV, and exposure to alcohol can be distinguished from hepatitis B viral infection, but for others this may be difficult. For example, mutations in the basal core promoter (BCP, $1762^{T}/1764^{A}$) of the virus are associated with an increased risk of HCC,[89,95] but why? Persons with inactive HBV infections have lower frequencies of these mutations than either patients with chronic hepatitis or HCC. A study from Taiwan[95] showed higher rates of $\mathit{1762^{T}/1764^{A}}$ mutations associated with clinical stages of disease (inactive disease 3%, chronic hepatitis 11%, cirrhosis, 32%, HCC 64%). Does this mean that BCP mutations drive the liver disease to HCC, or are the mutations passengers and simply occur more frequently in the setting of more severe liver disease, or are they markers of duration of liver disease?

These questions are becoming more important with the development of numerous biomarkers for HCC.[96] As serological biomarkers like fucosylated hemopexin, fucosylated kininogen,[97] RASSF1A hypermethylation,[98] and others are developed, and as their specificity in detecting HCC increases, the question of when in the course of development of HCC they become detectable becomes interesting epidemiologically. Do these markers parallel the occurrence of BCP mutations? Are they biologically related to such mutations? Newer biomarkers may be able to detect different stages of fibrosis (Fibrotest, glyccirrhotest),[99] making for more complex but ultimately more informative epidemiological studies.

References

1. WHO. Hepatitis B Fact Sheet (#204). 2008 [cited 2009 January]; Available from: http: //www.who.int/mediacentre/factsheets/fs204/en/
2. Goldstein ST, Zhou F, Hadler SC, *et al.* (2005) A mathematical model to estimate global hepatitis B disease burden and vaccination impact. *Int J Epidemiol* **34**: 1329–1339.
3. Edmondson HA. (1958) *Tumors of the Liver and Intrahepatic Bile Ducts.* Washington, DC, Armed Forces Institute of Pathology, pp. 32–33.
4. Parkin DM. (2001) Global cancer statistics in the year 2000. *Lancet Oncol* **2**: 533–543.
5. Centers for Disease Control. (2002) Geographic distribution of chronic HBV infection. Hepatitis B slide set (Internet). [cited January, 2009]; Available from:
6. IARC. (1994) *Hepatitis Viruses.* Lyon, International Agency for Research on Cancer.
7. Beasley RP, Hwang LY, Lin CC, Chien CS. (1981) Hepatocellular carcinoma and hepatitis B virus. A prospective study of 22 707 men in Taiwan. *Lancet* **2**(8256): 1129–1133.
8. Evans AA, Chen G, Ross EA, *et al.* (2002) Eight-year follow-up of the 90,000-person Haimen City cohort: I. Hepatocellular carcinoma mortality, risk factors, and gender differences. *Cancer Epidemiol Biomarkers Prev* **11**: 369–376.
9. Larouze B, Saimot G, Lustbader ED, *et al.* (1976) Host responses to hepatitis-B infection in patients with primary hepatic carcinoma and their families. A case/control study in Senegal, West Africa. *Lancet* **2**: 534–538.
10. Parkin DM, Bray F, Ferlay J, Pisani P. (2005) Global cancer statistics, 2002. *CA Cancer J Clin* **55**: 74–108.
11. Curado MP, Edwards B, Shin HR, *et al.* (2007) *Cancer Incidence in Five Continents.* Lyon, IARC Scientific Publications.
12. McGlynn KA, Tsao L, Hsing AW, *et al.* (2001) International trends and patterns of primary liver cancer. *Int J Cancer* **94**: 290–296.

13. Altekruse SF, McGlynn KA, Reichman ME. (2009) Hepatocellular carcinoma incidence, mortality, and survival trends in the United States from 1975 to 2005. *J Clin Oncol* **27**: 1485–1491.
14. Parkin DM, Bray FI, Devesa SS. (2001) Cancer burden in the year 2000. The global picture. *Eur J Cancer* **37**(Suppl 8): S4–66.
15. IARC. (1987) Overall evaluations of carcinogenicity: an updating of IARC Monographs volumes 1 to 42. *IARC Monogr Eval Carcinog Risks Hum Suppl* **7**: 1–440.
16. Garner RC, Miller EC, Miller JA. (1972) Liver microsomal metabolism of aflatoxin B 1 to a reactive derivative toxic to Salmonella typhimurium TA 1530. *Cancer Res* **32**: 2058–2066.
17. Foster PL, Eisenstadt E, Miller JH. (1983) Base substitution mutations induced by metabolically activated aflatoxin B1. *Proc Natl Acad Sci USA* **80**: 2695–2698.
18. Bressac B, Kew M, Wands J, Ozturk M. (1991) Selective G to T mutations of p53 gene in hepatocellular carcinoma from southern Africa. *Nature* **350**: 429–431.
19. Hsu IC, Metcalf RA, Sun T, *et al.* (1991) Mutational hotspot in the p53 gene in human hepatocellular carcinomas. *Nature* **350**: 427–428.
20. Qian GS, Ross RK, Yu MC, *et al.* (1994) A follow-up study of urinary markers of aflatoxin exposure and liver cancer risk in Shanghai, People's Republic of China. *Cancer Epidemiol Biomarkers Prev* **3**: 3–10.
21. Groopman JD, Kensler TW, Wild CP. (2008) Protective interventions to prevent aflatoxin-induced carcinogenesis in developing countries. *Annu Rev Public Health* **29**: 187–203.
22. IARC. (1988) *Alcohol Drinking*. Lyon: IARC Press, pp. 255–259.
23. Arico S, Corrao G, Torchio P, *et al.* (1994) A strong negative association between alcohol consumption and the risk of hepatocellular carcinoma in cirrhotic patients. A case-control study. *Eur J Epidemiol* **10**: 251–257.
24. La Vecchia C, Negri E, Cavalieri d'Oro L, Franceschi S. (1998) Liver cirrhosis and the risk of primary liver cancer. *Eur J Cancer Prev* **7**: 315–320.
25. La Vecchia C. (2007) Alcohol and liver cancer. *Eur J Cancer Prev* **16**: 495–497.
26. Oshima A, Tsukuma H, Hiyama T, *et al.* (1984) Follow-up study of HBs Ag-positive blood donors with special reference to effect of drinking and smoking on development of liver cancer. *Int J Cancer* **34**: 775–779.
27. Yu MW, Hsu FC, Sheen IS, *et al.* (1997) Prospective study of hepatocellular carcinoma and liver cirrhosis in asymptomatic chronic hepatitis B virus carriers. *Am J Epidemiol* **145**: 1039–1047.
28. Aizawa Y, Shibamoto Y, Takagi I, *et al.* (2000) Analysis of factors affecting the appearance of hepatocellular carcinoma in patients with chronic hepatitis C. A long term follow-up study after histologic diagnosis. *Cancer* **89**: 53–59.

29. Tanaka K, Tsuji I, Wakai K, *et al.* (2008) Alcohol drinking and liver cancer risk: an evaluation based on a systematic review of epidemiologic evidence among the Japanese population. *Jpn J Clin Oncol* **38**: 816–838.
30. Donato F, Tagger A, Gelatti U, *et al.* (2002) Alcohol and hepatocellular carcinoma: the effect of lifetime intake and hepatitis virus infections in men and women. *Am J Epidemiol* **155**: 323–331.
31. Frezza M, di Padova C, Pozzato G, *et al.* (1990) High blood alcohol levels in women. The role of decreased gastric alcohol dehydrogenase activity and first-pass metabolism. *N Engl J Med* **322**: 95–99.
32. Corrao G, Arico S, Zambon A, *et al.* (1997) Female sex and the risk of liver cirrhosis. *Scand J Gastroenterol* **32**: 1174–1180.
33. Bagnardi V, Blangiardo M, La Vecchia C, Corrao G. (2001) A meta-analysis of alcohol drinking and cancer risk. *Br J Cancer* **85**: 1700–1705.
34. McCoy GD, Hecht SS, Katayama S, Wynder EL. (1981) Differential effect of chronic ethanol consumption on the carcinogenicity of N-nitrosopyrrolidine and N′-nitrosonornicotine in male Syrian golden hamsters. *Cancer Res* **41**: 2849–2854.
35. Seitz HK, Poschl G, Simanowski UA. (1998) Alcohol and cancer. *Recent Dev Alcohol* **14**: 67–95.
36. Adami HO, Hsing AW, McLaughlin JK, *et al.* (1992) Alcoholism and liver cirrhosis in the etiology of primary liver cancer. *Int J Cancer* **51**: 898–902.
37. London WT, McGlynn KA. (2006) Liver cancer. In: Schottenfeld D, Fraumeni JF Jr, (eds). Cancer Epidemiology and Prevention. 3rd ed. New York, Oxford University Press, pp. 763–786.
38. IARC. (2004) *Tobacco Smoke and Involuntary Smoking*. Lyon, IARC Press.
39. Gandini S, Botteri E, Iodice S, *et al.* (2008) Tobacco smoking and cancer: a meta-analysis. *Int J Cancer* **122**: 155–164.
40. Mori M, Hara M, Wada I, *et al.* (2000) Prospective study of hepatitis B and C viral infections, cigarette smoking, alcohol consumption, and other factors associated with hepatocellular carcinoma risk in Japan. *Am J Epidemiol* **151**: 131–139.
41. Kuper H, Tzonou A, Kaklamani E, *et al.* (2000) Tobacco smoking, alcohol consumption and their interaction in the causation of hepatocellular carcinoma. *Int J Cancer* **85**: 498–502.
42. Wang LY, You SL, Lu SN, *et al.* (2003) Risk of hepatocellular carcinoma and habits of alcohol drinking, betel quid chewing and cigarette smoking: a cohort of 2416 HBsAg-seropositive and 9421 HBsAg-seronegative male residents in Taiwan. *Cancer Causes Control* **14**: 241–250.
43. Marrero JA, Fontana RJ, Fu S, *et al.* (2005) Alcohol, tobacco and obesity are synergistic risk factors for hepatocellular carcinoma. *J Hepatol* **42**: 218–224.
44. IARC. (1991) *Occupational Exposures in Insecticide Application, and Some Pesticides*. Lyon, IARC Press.

45. Giordano F, Dell'Orco V, Giannandrea F, *et al.* (2006) Mortality in a cohort of pesticide applicators in an urban setting: sixty years of follow-up. *Int J Immunopathol Pharmacol* **19**(4 Suppl): 61–65.
46. Cocco P, Fadda D, Billai B, *et al.* (2005) Cancer mortality among men occupationally exposed to dichlorodiphenyltrichloroethane. *Cancer Res* **65**: 9588–9594.
47. Cocco P, Kazerouni N, Zahm SH. (2000) Cancer mortality and environmental exposure to DDE in the United States. *Environ Health Perspect* **108**: 1–4.
48. McGlynn KA, Abnet CC, Zhang M, *et al.* (2006) Serum concentrations of 1,1,1-trichloro-2,2-bis(p-chlorophenyl)ethane (DDT) and 1,1-dichloro-2,2-bis (p-chlorophenyl)ethylene (DDE) and risk of primary liver cancer. *J Natl Cancer Inst* **98**: 1005–1010.
49. Ezzat S, Abdel-Hamid M, Eissa SA, *et al.* (2005) Associations of pesticides, HCV, HBV, and hepatocellular carcinoma in Egypt. *Int J Hyg Environ Health* **208**: 329–339.
50. Allen KJ, Gurrin LC, Constantine CC, *et al.* (2008) Iron-overload-related disease in HFE hereditary hemochromatosis. *N Engl J Med* **358**: 221–230.
51. Feder JN, Gnirke A, Thomas W, *et al.* (1996) A novel MHC class I-like gene is mutated in patients with hereditary haemochromatosis. *Nat Genet* **13**: 399–408.
52. Beutler E, Gelbart T, West C, *et al.* (1996) Mutation analysis in hereditary hemochromatosis. *Blood Cells Mol Dis* **22**: 187–194; discussion 194a–194b.
53. Hanson EH, Imperatore G, Burke W. (2001) HFE gene and hereditary hemochromatosis: a HuGE review. Human Genome Epidemiology. *Am J Epidemiol* **154**: 193–206.
54. Bradbear RA, Bain C, Siskind V, *et al.* (1985) Cohort study of internal malignancy in genetic hemochromatosis and other chronic nonalcoholic liver diseases. *J Natl Cancer Inst* **75**: 81–84.
55. Niederau C, Erhardt A, Haussinger D, Strohmeyer G. (1999) Haemochromatosis and the liver. *J Hepatol* **30**(Suppl 1): 6–11.
56. Deugnier Y. (2003) Iron and liver cancer. *Alcohol* **30**: 145–150.
57. Goh J, Callagy G, McEntee G, *et al.* (1999) Hepatocellular carcinoma arising in the absence of cirrhosis in genetic haemochromatosis: three case reports and review of literature. *Eur J Gastroenterol Hepatol* **11**: 915–919.
58. Linet MS, Gridley G, Nyren O, *et al.* (1999) Primary liver cancer, other malignancies, and mortality risks following porphyria: a cohort study in Denmark and Sweden. *Am J Epidemiol* **149**: 1010–1015.
59. Strachan AS. (1929) Haemosiderosis and Haemochromatosis in South African Natives with a Comment on the Htiology of haemochromatosis. Glasgow, University of Glasgow Press.
60. Gordeuk VR, McLaren CE, MacPhail AP, *et al.* (1996) Associations of iron overload in Africa with hepatocellular carcinoma and tuberculosis: strachan's 1929 thesis revisited. *Blood* **87**: 3470–3476.

61. McNamara L, MacPhail AP, Gordeuk VR, *et al.* (1998) Is there a link between African iron overload and the described mutations of the hereditary haemochromatosis gene? *Br J Haematol* **102**: 1176–1178.
62. Edmondson HA, Henderson B, Benton B. (1976) Liver-cell adenomas associated with use of oral contraceptives. *N Engl J Med* **294**: 470–472.
63. WHO. (1997) Oral contraceptives and liver cancer. Results of the Multicentre International Liver Tumor Study (MILTS). *Contraception* **56**: 275–284.
64. Maheshwari S, Sarraj A, Kramer J, El-Serag HB. (2007) Oral contraception and the risk of hepatocellular carcinoma. *J Hepatol* **47**: 506–513.
65. Fernandez E, Gallus S, Bosetti C, *et al.* (2003) Hormone replacement therapy and cancer risk: a systematic analysis from a network of case-control studies. *Int J Cancer* **105**: 408–412.
66. Soe KL, Soe M, Gluud C. (1992) Liver pathology associated with the use of anabolic-androgenic steroids. *Liver* **12**: 73–79.
67. De Maria N, Manno M, Villa E. (2002) Sex hormones and liver cancer. *Mol Cell Endocrinol* **193**: 59–63.
68. Yu MW, Chang HC, Chang SC, *et al.* (2003) Role of reproductive factors in hepatocellular carcinoma: impact on hepatitis B- and C-related risk. *Hepatology* **38**: 1393–1400.
69. Yuan JM, Ross RK, Stanczyk FZ, *et al.* (1995) A cohort study of serum testosterone and hepatocellular carcinoma in Shanghai, China. *Int J Cancer* **63**: 491–493.
70. La Vecchia C, Negri E, Franceschi S, D'Avanzo B. (1992) Reproductive factors and the risk of hepatocellular carcinoma in women. *Int J Cancer* **52**: 351–354.
71. Cole P, MacMahon B, Brown JB. (1976) Oestrogen profiles of parous and nulliparous women. *Lancet* **2**(7986): 596–599.
72. Mucci LA, Kuper HE, Tamimi R, *et al.* (2001) Age at menarche and age at menopause in relation to hepatocellular carcinoma in women. *Bjog* **108**: 291–294.
73. Thursz MR, Kwiatkowski D, Allsopp CE, *et al.* (1995) Association between an MHC class II allele and clearance of hepatitis B virus in the Gambia. *N Engl J Med* **332**: 1065–1069.
74. Thio CL, Carrington M, Marti D, *et al.* (1999) Class II HLA alleles and hepatitis B virus persistence in African Americans. *J Infect Dis* **179**: 1004–1006.
75. Hohler T, Kruger A, Gerken G, *et al.* (1998) A tumor necrosis factor-alpha (TNF-alpha) promoter polymorphism is associated with chronic hepatitis B infection. *Clin Exp Immunol* **111**: 579–582.
76. Kim YJ, Lee HS, Yoon JH, *et al.* (2003) Association of TNF-alpha promoter polymorphisms with the clearance of hepatitis B virus infection. *Hum Mol Genet* **12**: 2541–2546.
77. Kummee P, Tangkijvanich P, Poovorawan Y, Hirankarn N. (2007) Association of HLA-DRB1*13 and TNF-alpha gene polymorphisms with clearance of

chronic hepatitis B infection and risk of hepatocellular carcinoma in Thai population. *J Viral Hepat* **14**: 841–848.

78. Bodsworth NJ, Cooper DA, Donovan B. (1991) The influence of human immunodeficiency virus type 1 infection on the development of the hepatitis B virus carrier state. *J Infect Dis* **163**: 1138–1140.
79. Hadler SC, Judson FN, O'Malley PM, *et al.* (1991) Outcome of hepatitis B virus infection in homosexual men and its relation to prior human immunodeficiency virus infection. *J Infect Dis* **163**: 454–459.
80. Yu MC, Yuan JM, Lu SC. (2008) Alcohol, cofactors and the genetics of hepatocellular carcinoma. *J Gastroenterol Hepatol* **23**(Suppl 1): S92–S97.
81. Chen G, Lin W, Shen F, *et al.* (2006) Past HBV viral load as predictor of mortality and morbidity from HCC and chronic liver disease in a prospective study. *Am J Gastroenterol* **101**: 1797–1803.
82. Chen CJ, Yang HI, Su J, *et al.* (2006) Risk of hepatocellular carcinoma across a biological gradient of serum hepatitis B virus DNA level. *Jama* **295**: 65–73.
83. Kao JH. (2002) Hepatitis B viral genotypes: clinical relevance and molecular characteristics. *J Gastroenterol Hepatol* **17**: 643–650.
84. Yu MW, Yeh SH, Chen PJ, *et al.* (2005) Hepatitis B virus genotype and DNA level and hepatocellular carcinoma: a prospective study in men. *J Natl Cancer Inst* **97**: 265–272.
85. Yuen MF, Sablon E, Yuan HJ, *et al.* (2003) Significance of hepatitis B genotype in acute exacerbation, HBeAg seroconversion, cirrhosis-related complications, and hepatocellular carcinoma. *Hepatology* **37**: 562–567.
86. Livingston SE, Simonetti JP, McMahon BJ, *et al.* (2007) Hepatitis B virus genotypes in Alaska Native people with hepatocellular carcinoma: preponderance of genotype F. *J Infect Dis* **195**: 5–11.
87. Chou YC, Yu MW, Wu CF, *et al.* (2008) Temporal relationship between hepatitis B virus enhancer II/basal core promoter sequence variation and risk of hepatocellular carcinoma. *Gut* **57**: 91–97.
88. Chen CH, Lee CM, Hung CH, *et al.* (2007) Clinical significance and evolution of core promoter and precore mutations in HBeAg-positive patients with HBV genotype B and C: a longitudinal study. *Liver Int* **27**: 806–815.
89. Yuan JM, Ambinder A, Fan Y, *et al.* (2009) Prospective evaluation of hepatitis B 1762T/1764A mutations on hepatocellular carcinoma development in Shanghai, China. *Cancer Epidemiol Biomarkers Prev* **18**: 590–594.
90. Dejean A, Lugassy C, Zafrani S, *et al.* (1984) Detection of hepatitis B virus DNA in pancreas, kidney and skin of two human carriers of the virus. *J Gen Virol* **65**(Pt 3): 651–655.
91. Lanford RE, Michaels MG, Chavez D, *et al.* (1995) Persistence of extrahepatic hepatitis B virus DNA in the absence of detectable hepatic replication in patients with baboon liver transplants. *J Med Virol* **46**: 207–212.

92. Shimoda T, Shikata T, Karasawa T, *et al.* (1981) Light microscopic localization of hepatitis B virus antigens in the human pancreas. Possibility of multiplication of hepatitis B virus in the human pancreas. *Gastroenterology* **81**: 998–1005.
93. Hassan MM, Li D, El-Deeb AS, *et al.* (2008) Association between hepatitis B virus and pancreatic cancer. *J Clin Oncol* **26**: 4557–4562.
94. Hsing AW, Zhang M, Rashid A, *et al.* (2008) Hepatitis B and C virus infection and the risk of biliary tract cancer: a population-based study in China. *Int J Cancer* **122**: 1849–1853.
95. Kao JH, Chen PJ, Lai MY, Chen DS. (2003) Basal core promoter mutations of hepatitis B virus increase the risk of hepatocellular carcinoma in hepatitis B carriers. *Gastroenterology* **124**: 327–334.
96. Block TM, Marrero J, Gish RG, *et al.* (2008) The degree of readiness of selected biomarkers for the early detection of hepatocellular carcinoma: notes from a recent workshop. *Cancer Biomark* **4**: 19–33.
97. Comunale MA, Lowman M, Long RE, *et al.* (2006) Proteomic analysis of serum associated fucosylated glycoproteins in the development of primary hepatocellular carcinoma. *J Proteome Res* **5**: 308–315.
98. Chan KC, Lai PB, Mok TS, *et al.* (2008) Quantitative analysis of circulating methylated DNA as a biomarker for hepatocellular carcinoma. *Clin Chem* **54**: 1528–1536.
99. Callewaert N, Van Vlierberghe H, Van Hecke A, *et al.* (2004) Noninvasive diagnosis of liver cirrhosis using DNA sequencer-based total serum protein glycomics. *Nat Med* **10**: 429–434.

Chapter 4

Molecular Biology of HBV-related Hepatocellular Carcinoma

Amanda J. Hodgson* and Betty L. Slagle*,†

Abstract

Chronic infection with hepatitis B virus (HBV) is a risk factor for hepatocellular carcinoma (HCC). This occurs through a complex, multistep process that likely involves both viral and host factors, as well as interactions with other cancer risk factors. In 1996, a model was proposed for HBV-associated tumor formation that is now supported by evidence from many laboratories using a variety of experimental model systems. Central to this model is chronic HBV infection, in which high levels of virus are produced over decades. The resulting host immune response to virus-infected cells leads to hepatocyte damage, cell death, and compensatory regeneration to restore lost cells. The ability of the liver to regenerate also provides a process in which cells containing a mutation providing a growth advantage can be selected. During virus replication, subgenomic portions of HBV are able to integrate into the host genome. While this is not part of

* Department of Molecular Virology and Microbiology, Baylor College of Medicine, Houston, TX 77030, U.S.A.

† Corresponding author: Betty L. Slagle, Ph.D. Mailstop BCM-385, One Baylor Plaza, Houston, TX 77030, USA, Tel: 713-798-3006, Fax: 713-798-5075, E-mail: bslagle@bcm.edu.

the virus life cycle, the viral inserts contribute to chromosome instability, including in some instances the loss of important tumor suppressor genes. In addition, viral inserts may encode novel gene products with properties that can be oncogenic. HBV has developed strategies to benefit virus replication in non-dividing G0 hepatocytes. By analogy to other tumor-associated viruses, any strategies related to cell cycle progression have the potential to go awry and contribute to multistep transformation. This chapter reviews the evidence supporting a role for HBV replication in the multistep process leading to HCC.

Introduction

The link between chronic HBV infection and hepatocellular carcinoma (HCC), first suggested in 1981,[1] is now firmly established. Worldwide, HCC ranks as the fifth most common cause of cancer deaths in men. In HBV-endemic areas, the lifetime risk for HCC among chronic HBV carriers is 15 to 100 times that of non-HBV patients. The International Agency for Research on Cancer (IARC) lists HBV as an environmental carcinogen, capable of causing liver cancer. However, the mechanism by which chronic HBV contributes to cancer is not well understood. Unlike other environmental carcinogens that are chemical in nature, HBV is an infectious agent, and so its role in tumor formation must take into account the complicated issue of virus-host interactions. The complex events that occur over the decades of chronic infection eventually lead to cancer, and glimpses of how this occurs have been gathered using a variety of experimental systems.

Lessons from Other Tumor Viruses

It is estimated that approximately 20% of human cancers have an associated viral factor. When considering the possible mechanism(s) by which HBV contributes to HCC formation, it is helpful to review the general principles learned from studies of other tumor-associated viruses. These include that tumor-associated viruses (1) establish chronic infection in their natural host, (2) seldom act as complete carcinogens, (3) cause tumors in a subset of infected people, (4) form tumors after a long latency period, (5) form tumors either directly or indirectly, and (6) modulate growth regulatory pathways in the cell (reviewed in Ref. 2). There is now strong evidence to show that these criteria fit well with the characteristics of HBV-associated HCC.

The importance of the role of virus-host interaction in the outcome of a chronic viral infection cannot be over-emphasized. As part of their strategy for robust and sustained replication, tumor viruses have evolved the ability to usurp critical cell growth regulatory pathways, including the novel inactivation of tumor suppressor genes. It is generally believed that cellular transformation by a virus is an accidental outcome of the viral strategy to enhance its replication. Viruses that replicate well in their hosts will infect larger numbers of cells, increasing the opportunity for "accidental" events that lead to an increased risk for cancer. In support of this idea, several studies show that increased HBV replication (measured by polymerase chain reaction [PCR] quantitation of viral copy number) is associated with an increased risk of HCC (reviewed in Ref. 3).

HBV and HCC

The mechanism by which chronic HBV infection causes HCC is likely complex. HBV establishes chronic infection in 90% of infected infants and 10% of infected adults, and this significant difference in the outcome of infection is believed to be due to an inability of the immature immune system of infants to eliminate the virus. However, only a subset of chronic carriers develops HCC, and this risk is greatly increased in parts of the world in which infected patients are additionally exposed to environmental carcinogens. Tumors typically arise after decades of infection, and this is thought to reflect the need for a cell to accumulate sufficient mutations to result in a growth advantage. Although HBV replication is not directly cytopathic, the cycles of immune-mediated cell killing of virus-infected hepatocytes and the subsequent compensatory regeneration are important in providing the selection process for the emergence of cells containing mutations that provide some growth advantage over cells that lack such mutations.

A Model for HBV-associated HCC

Since HCCs usually develop over decades, it is difficult to establish the precise role of HBV in carcinogenesis. A model for the mechanism(s) of HBV-associated HCC was proposed in 1996[4] and is now supported by experimental evidence from many laboratories. Central to the model is ongoing chronic virus replication (Fig. 1), which provides a continuous source of viral antigen in the liver. The host immune response to virus-infected hepatocytes contributes to the cycles of cell death and regeneration, thus driving the selection process for cells containing mutations that provide a growth advantage. The host immune response is therefore predicted to play a critical role in the development of HCC. In addition, some viral gene products have properties that are potentially oncogenic. Integration of viral DNA is also ongoing, and while this occurs at random locations in the host genome and only portions of the viral genome become integrated, the integration process itself is highly mutagenic and can lead to the loss of tumor suppressor genes. There is also evidence that HBV alters the ability of the cell to repair DNA damage. All of these factors contribute to a general genetic instability that, in the presence of environmental carcinogens and other HCC cofactors, eventually leads to the emergence of liver cancer. The data that support this general model of HBV-associated HCC are reviewed in this chapter.

Immune-mediated Pathogenesis

During chronic infection, ongoing HBV replication provides a target for immune-mediated cell killing. The ability of terminally differentiated hepatocytes to undergo compensatory regeneration both restores liver mass and provides a selection process for the outgrowth of cells that may contain mutations offering a growth advantage. This link between chronic liver inflammation and cancer has been shown for other infectious agents such as hepatitis C virus (HCV), and is thought to be a central driving force for the mechanism by which HBV causes cancer.

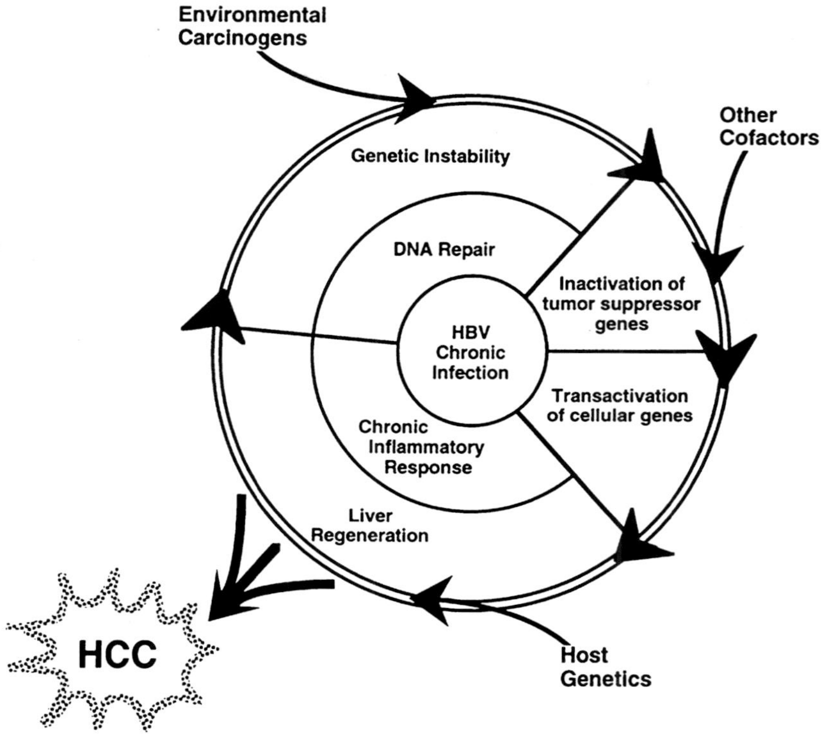

Fig. 1. Model for HBV and HCC. Multiple factors influence the development of liver cancer, and chronic HBV replication is central to this process. The continued production of viral antigens drives the immune response, leading to cell death and compensatory liver regeneration. The latter provides an opportunity for errors during DNA replication, and may select for cells with a growth advantage. HBV may also affect the ability of the cell to repair damaged DNA, allowing mutations to accumulate over time and increasing the likelihood that tumor suppressor genes, such as p53, become inactivated. The ability of HBx to transactivate cellular targets may contribute to HCC, but the precise targets important in the process are unknown. Environmental carcinogens are also important HCC cofactors and may become more damaging if viral infection has disabled the cellular DNA repair system. Host genetics is likely important, both in determining the host immune response and/or cancer susceptibility. The accumulation of genetic changes over years of chronic HBV infection leads eventually to the development of HCC. Reproduced with permission from Butel *et al.*[4]

Early Immune Response to HBV Infection

The early host response to virus infection has been studied in the only animal model of HBV infection, the chimpanzee. Whereas inoculation of chimpanzees with HCV results in the activation of many cellular innate immune response genes, inoculation with HBV had no measurable effect on the host genes during the entry and expansion phases of viral infection of the liver.[5] This unexpected result indicates that, early in infection, HBV has the ability to avoid detection by the host's innate immune response. The mechanism by which HBV acts as a "stealth" virus is unknown, and it is similarly unclear if this viral mechanism is related to the mechanism by which HBV is able to maintain chronic replication.

Immune-mediated Clearance of HBV

During chronic HBV infection, virus replication occurs in the majority of hepatocytes in the liver. This was convincingly demonstrated by immuno-staining of livers of woodchucks infected with the highly related woodchuck hepatitis virus (WHV). Essentially, 100% of hepatocytes in chronically infected woodchucks contain detectable viral antigen. Clearance of HBV infection is mediated by cytokines produced by the innate and adaptive immune responses (reviewed in Ref. 6). In particular, tumor necrosis factor (TNF) α, interferon (IFN) γ, and IFN α/β are important in viral clearance, but do not directly kill virus-infected cells.

The main effector cells responsible for elimination of HBV-infected cells are the CD8+ cytotoxic T lymphocytes (CTLs), which can be specific for any of the viral proteins. Although virus is cleared from some infected hepatocytes without concomitant cell death, other cells die and release alanine transaminase (ALT) into the serum. In general, there is a weaker CTL response during chronic HBV infection, and this may contribute to the ability of HBV to persist. It is not clear why or how many HBV-infected cells appear to escape immune elimination. There is speculation that the virus is able to modulate the immune response, but the lack of a convenient small-animal model makes it difficult to test a specific hypothesis. Of relevance to HCC is the ability of the virus to continue replicating, and the ability of the liver to restore lost hepatocytes through compensatory regeneration.

Immune Pathogenesis During Chronic Infection

Because HBV is able to infect only humans and chimpanzees, it is challenging to design experiments that provide a mechanistic understanding of how HBV causes HCC. Transgenic mice that express all or parts of the HBV genome have provided important insights. Mice that encode the entire HBV genome produce virus particles that are infectious in chimpanzees. The livers of these mice are similar to the livers of chronically infected humans in that the majority of hepatocytes express significant (detectable) amounts of viral antigen. However, studies of viral pathogenesis in these mice are complicated by the fact that the transgenic mouse recognizes viral antigens as self-antigens, and does not mount an immune response.

There is now direct evidence for the importance of the immune response in viral pathogenesis. Adoptive transfer of HBV-immunized splenocytes from syngeneic non-transgenic mice to immunologically tolerant HBV transgenic mice led to clearance of HBV surface antigen, while control-immunized splenocytes did not. Clearance of the virus by the splenocytes was accompanied by a spike in ALT levels, indicating liver damage by the immune system. Because the immune cells continued to attack the transgenic hepatocytes producing HBsAg, these mice developed chronic inflammatory disease that lead to the development of HCC.[7] This provides direct confirmation that the immune response to HBV-infected cells can lead to cell death, regeneration, and HCC.

Although normal adult hepatocytes rarely divide, the loss of hepatocytes by immune-mediated killing leads to a carefully regulated process in which neighboring hepatocytes re-enter the cell cycle and divide. These cycles of cell death and regeneration, triggered by a continual source of viral antigens, may provide a selection process for cells containing mutations that provide a growth advantage over neighboring cells. In addition, any effect HBV replication may have on the cell cycle, perhaps as part of the viral strategy to benefit replication in quiescent hepatocytes, may be anticipated to upset the normal events of liver regeneration and thus contribute to pathogenesis.

Effect of HBx on Liver Regeneration

HBV replication occurs in non-dividing hepatocytes thought to reside in the G0 stage of the cell cycle. By analogy with other tumor viruses, HBV may need to stimulate hepatocytes to create a cellular environment more favorable for virus replication. Given the limited genetic information encoded by HBV, it is a particularly attractive idea that the sole viral regulatory protein, HBx, might function to stimulate the hepatocyte. For example, it is unclear whether the quiescent hepatocyte would have sufficient deoxynucleoside triphosphates required for virus genome replication. Alternatively, virus replication in hepatocytes may activate a cellular response that triggers cell cycle progression. If this activation is detrimental to virus replication, it would lead to a viral strategy to enforce a checkpoint halting cell cycle progression. There is experimental evidence for both scenarios, and since deregulation of the cell cycle is a hallmark of cancer, this might represent a viral replication strategy that accidentally triggers changes leading to cancer.

Several studies have established that HBx has an effect on the early stages of the cell cycle in transfected cells grown in culture. However, the effect is complex and HBx has been shown to both promote cell cycle progression and induce a G1/S block that inhibits the growth of cells (reviewed in Ref. 8). The experimental variables responsible for such differing results remain undefined. In the livers of transgenic mice, HBx expression leads to an increase in S phase cells in the livers of 10- to 14-day old mice, but has no obvious effect on the hepatocytes of adult mice. This indicates that the effect of HBx on cell cycle progression may depend on factors present in dividing hepatocytes.[9,10] Since the passage of hepatocytes from G1 to S phase is dependent on growth factors,[11] one interpretation for the effect of HBx on hepatocytes of young livers is that growth factors present in the neonate liver overcome a G1 restriction point, thereby allowing

HBx to push cells into S phase. This would also be observed for HBx in cultured cells in the presence of serum.

To determine the effect of HBx on cell cycle progression *in vivo*, several groups have used the partial hepatectomy (PH) model in which two-thirds of the liver is removed from HBx transgenic mice and their non-transgenic littermates. Following PH, the liver is regenerated through a carefully organized sequence of events that culminates in a peak in DNA synthesis 40 hours post surgery. One study revealed evidence that HBx could promote premature entry into the cell cycle in the regenerating liver. Despite this early deregulation, by eight days post PH, the regenerated liver mass was indistinguishable between HBx and control mice.[12] Other studies using different lineages of HBx mice concluded that HBx expression instead induced G1 arrest, resulting in the inability of the livers to regenerate.[13,14] It is difficult to reconcile these very different conclusions for HBx function during liver regeneration. Differences among these studies include the age at which PH was performed, the genetic background of the mice, and the time points post PH at which animals were sacrificed and livers removed for analysis. Only one study examined time points within the first 48 hours post PH.[12] The mice in another study had a background of steatosis,[14] which is reported to be an independent risk factor for the failure of liver regeneration. While the question of whether HBx alters compensatory regeneration seems straightforward, the results of these studies reveal the complexity and limitations of current experimental models. In summary, studies have demonstrated that HBx does indeed alter early liver regeneration, but the precise effect of HBx remains unclear.

Liver Regeneration in HBV Transgenic Mice

Additional information on HBV and liver regeneration can be gleaned by PH studies on mice that express the entire HBV genome and produce virus particles. Following PH in these mice, there was a transient inhibition of HBV replication detected at the post-translational level, perhaps due to PH-induced inflammatory cytokines. HBV RNA levels were unaffected by the PH, leading to the conclusion that expression of the HBV transgene was independent of the cell cycle.[15] A similar study examined liver regeneration in mice transgenic for a partial HBV genome that encoded the *preS*, *S*, and *X* genes.[16] Although the livers of these mice eventually regenerated as well as those of non-transgenic controls, early liver regeneration was delayed, as evidenced by decreased liver/body weight ratio and fewer BrdU- and PCNA-positive cells at 24, 48, and 72 hours after PH. However, no difference in apoptosis was observed compared to controls.

In summary, studies in HBV transgenic mouse models provide evidence that the immune response to chronically infected cells is a critical factor for the development of HCC. In addition, the subsequent cell death and compensatory regeneration are important for liver homeostasis, but can be altered by the expression of HBV gene products. Over time, and in the presence of other cofactors such as environmental carcinogens or coinfection with other hepatitis viruses, these events are predicted to create an environment in which there is opportunity for DNA errors that become fixed mutations as the cells divide.

Role of Viral Integration in HCC

The observation that HBV DNA is integrated in most HCCs in HBV-endemic areas provided the rationale for investigating the role of HBV integration in HCC. By analogy to studies with other tumor-associated viruses, it was considered that HBV integration events result in at least three possibilities: (1) expression of a viral-encoded oncogene, (2) insertional inactivation of a tumor suppressor gene, or (3) insertional activation of a proto-oncogene. There is limited experimental support for two of these scenarios in the woodchuck and human systems of hepadnavirus-associated HCC.

General Features of HBV Integration

Integration of HBV DNA occurs early in infection. Viral integration into the host chromosome can be detected in the livers of patients with acute HBV infection using sensitive PCR-based methods. It is thought that viral integration continues to occur during the decades-long chronic HBV replication. While retroviruses encode an integrase protein to ensure preservation of the entire viral genome during integration, HBV lacks such a protein so the region of HBV serving as the virus junction varies among different inserts. The entire HBV genome is rarely found intact within a viral insert. In addition, HBV integration occurs at random locations throughout the genome. It is hypothesized that only cells containing viral inserts that confer some growth advantage would expand to eventually form a tumor.

It is recognized that the majority of HCCs from HBV-endemic areas contain integrated HBV DNA. The integrated DNA is usually detected by Southern blot hybridization of tumor DNA, a method that provides sensitivity to identify single-copy genes. The presence of viral inserts in HCCs provides evidence for the clonal expansion of hepatocytes containing the original viral integration. Viral inserts are present in one to six copies per tumor. Determining whether the viral insert contributed to the development of that HCC, or merely marks a clonally expanded cell, has proved a considerable challenge.

Integrated HBV DNA is from all regions of the viral genome, although individual viral inserts usually only retain a portion of the genome, often in rearranged form. There are several reports that the *X* gene serves as a preferred junction within the integrated viral genome. Given the cancer cofactor role for HBx (discussed below), the idea that all or part of HBx is retained in viral inserts suggests a functional significance to the HBx sequences. An alternative explanation centers on the mechanism by which HBV DNA integrates into the chromosome. During virus replication, portions of the viral genome near the direct repeat regions are present as single-stranded (ss) DNA ends. These ssDNA regions are positioned within the *X* gene and serve as favored templates for non-homologous recombination with the chromosomal DNA. In support of this idea, there is an increased frequency of hepadnavirus DNA integration in cells treated with inhibitors of DNA repair.[17] It is difficult to determine whether the *X* gene is found at the virus-cell junction because it provides some growth advantage to the cell, or because of the nearby location of viral ssDNA that is a preferred template for integration.

Lessons Learned from the WHV System

Much of our understanding of the role of HBV in HCC is learned from studies of the related woodchuck virus (WHV). Infection of woodchucks during the neonatal period results in a high incidence of chronic infection that leads to HCC. This animal model has proven extremely valuable for our current understanding of HBV pathology. Similar to HBV integration in human cells, WHV integration occurs randomly throughout the host genome. However, the tumors that arise often contain viral integrations that activate the *myc* family of oncogenes (c-*myc*, N-*myc*, and, predominantly, N-*myc*2, a retroposon not found in the human genome). The integration of a subgenomic WHV fragment containing a viral enhancer element in the proximity of these oncogenes and proto-oncogenes is sufficient to activate their expression (Fig. 2a, see box labeled Enh). In one study, more than 40% of the woodchuck HCCs contained a viral insertion site either within 3-kb of the N-*myc*2 gene or in the 3′-noncoding region of the gene.[18] This positioning favors the idea that N-*myc*2 expression was activated by the viral enhancer, which is able to activate heterologous promoters in a position- and orientation-independent manner. Importantly, WHV-activated N-*myc*2 is tumorigenic in transgenic mice. In summary, there is abundant evidence that insertional activation by WHV is important in HCC formation in woodchucks. This raises the hope for a similar unifying molecular pathway of insertional activation by HBV in human HCCs.

HBV Integration in Human HCCs

To determine a possible etiologic role for viral integration in human HCCs, several laboratories have cloned viral inserts from end-stage tumors and examined both the viral sequences and the flanking cellular DNAs. Despite extensive analysis of many HCCs, there is no evidence for insertional activation of the *myc* family of oncogenes in human HCCs, indicating there are fundamental differences between HCC formation in woodchucks versus humans. Indeed, there does not appear to be any unifying molecular pathway altered by HBV

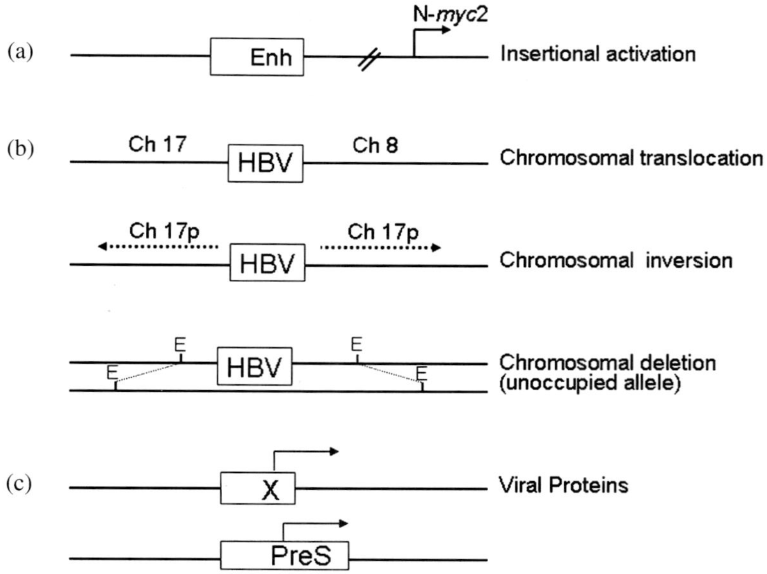

Fig. 2. Integrated hepadnaviral DNA cloned from HCCs. Analysis of HCC DNA reveals subgenomic viral fragments (shown above as boxes labeled Enh [for enhancer], HBV, X, and PreS) integrated into chromosomal DNA (shown as single black lines). *Panel A,* Woodchuck tumors often reveal viral inserts containing a viral enhancer sequence that activates the N-myc2 pseudo-proto-oncogene. *Panel B,* The integration of HBV subgenomic fragments is highly mutagenic to the host chromosome, and leads to chromosomal translocations, inversions, and deletions. *Panel C,* In some inserts, viral ORFs are preserved and may serve as templates for expression of individual viral genes or novel gene products produced during rearrangements of the subgenomic fragments. Evidence suggests that individual viral inserts may contribute to multistep HCC formation, often in ways unique to that tumor.

integration events. Instead, a very complex picture of HBV integration and HCC has emerged.

There is evidence that some viral inserts can directly disrupt important cellular growth regulatory genes. Indeed, HCCs from two unrelated patients were shown to contain viral inserts directly within the retinoic acid receptor β (RARβ) gene and the cyclin A gene. Given the normal function of these genes in cell growth regulation, it is likely that their deregulation contributed to multistep hepatocarcinogenesis in these tumors. However, the RARβ and cyclin A genes are not commonly inactivated in other HCCs, and so these particular examples of insertional inactivation appear unique to specific HBV-positive tumors.

A new, more sensitive approach has been used to clone greater numbers of viral inserts. PCR analysis of HCC DNA using one viral primer and one primer specific for human *Alu* repeat sequences (*Alu*-PCR; present in flanking cellular DNA) was used to examine virus-cell junctions in 22 HCCs. Subsequent sequencing of the virus-cell junction identified 16 different cellular growth regulatory genes at or near the site of integration.[19] This could be interpreted to suggest that integration-mediated inactivation of genes occurs commonly in HCCs. Alternatively, it is not known whether these viral insertions led to a growth advantage for the cell containing that insert, as evidenced by clonal expansion. Important characterization of the cells containing these virus-cell junctions was not performed due to insufficient amounts of tissue.

While insertional activation of cellular genes is not a common outcome of HBV integration in HCCs, analysis of viral inserts from many tumors has shown that the site of viral integration often contains gross chromosomal alterations that include chromosomal translocations, deletions, and inverted repeats (Fig. 2B). The importance of such chromosomal rearrangements to multistep tumorigenesis depends on whether cellular regulatory genes are located nearby. For example, a nearby tumor suppressor gene may be deleted during a chromosome translocation or inverted duplication caused by viral insertion, and this could provide a growth advantage to the cell containing the insert. This has been shown for an HBV-mediated chromosomal inversion that led to the loss of chromosome 17p, which includes the p53 tumor suppressor gene (reviewed in Ref. 20). Interestingly, loss of p53 occurs commonly in human HCCs, although a viral etiology is not always evident.

There are additional challenges to determining the contributions of viral inserts to hepatocarcinogenesis. As HBV-positive HCCs frequently contain multiple inserts, a definitive role for viral integration in the formation of a tumor can only be reached by either studying tumors with a single viral insert or by carefully examining each HBV insert from a tumor containing more than one insert. Since the integration-mediated chromosomal changes may affect cellular genes distantly located from the integration site, definitive results are often lacking in these studies.

In summary, during chronic HBV replication, portions of the viral genome become integrated into the host chromosome. While the majority of these inserts may occur in regions of the chromosome where there is no impact on the cell, other inserts can activate (or inactivate) nearby genes, leading to a growth advantage for the cell containing that insert. The pattern of viral inserts is complex, and no unifying molecular pathways seem activated (or inactivated) in human HCCs. Although viral integration likely contributes overall to HCC formation, its specific role likely differs among individual tumors.

Potential Oncogenic Properties of HBV Gene Products

Although there is no compelling evidence that HBV encodes an acutely acting oncogene responsible for hepatocarcinogenesis, three of the seven proteins produced by the four HBV open reading frames (ORFs) display properties that suggest their potential involvement in HCC. These three proteins include the largest surface antigen (preS1), a truncated version of preS2 (preS2^t), and the viral regulatory protein, HBx.

PreS1

The HBV surface (S) ORF utilizes three in-frame ATG initiation codons to generate the large (preS1), medium (preS2), and small (S) surface antigens that comprise the outer surface of the mature viral particle. The ratio of these three forms of viral surface antigen within the particle is regulated during natural infection, and in HBV transgenic mice that replicate the virus. However, the expression of preS1 may become deregulated, leading to a dramatically overexpressed preS1 protein. High levels of preS1 can inhibit secretion of S protein, and accumulation of this protein ultimately can damage hepatocytes (reviewed in Ref. 21). In transgenic mice, the overexpression of preS1 leads to chronic inflammation and subsequent HCC by 12 months of age. The mechanism of transformation is thought to be related to the amount of preS1, the extent of hepatocellular damage, and the subsequent inflammation and regeneration. It has also been reported that a small portion of total preS1 may possess transactivator function, and this is predicted to accelerate pathogenesis in mice that overexpress preS1. In summary, while preS1 expression is part of normal HBV replication, when its expression is deregulated there is potential for hepatocyte alterations that may contribute to the formation of HCC.

PreS2 Truncations

The random viral integration that occurs during the decades of chronic HBV infection (discussed above) can lead to truncated viral proteins that possess new functions not present in the original full-length protein. Specifically, viral integrations that truncate the preS2 ORF between nucleotide 221 and 573 generate a novel protein with transactivator function (reviewed in Ref. 22). There is experimental data that indicate this is biologically important. Transgenic mice that express truncated preS2 in their livers have an increased hepatocyte proliferation rate, and tumors develop in older animals.[23] Thirty percent of human HCCs examined in one study contained truncated preS2 ORFS capable of transactivator function,[24] suggesting that these novel proteins may represent an important viral factor in a subset of HCCs.

Regulatory HBx Protein

The 17-kDA HBx protein is encoded by all mammalian hepadnaviruses, and is required for WHV replication in the woodchuck model[25] and for HBV replication in HepG2 cells and in hydrodynamically injected mice.[26] The precise role provided by HBx in virus replication has not been defined, but it is well known that HBx is a broadly acting transcriptional activator capable of activating expression from both viral and cellular promoters (reviewed in Ref. 8). Of the many identified targets of HBx, it remains unknown which are critical for virus replication and, therefore, suspect for a role in HCC. HBx localizes to both the cytoplasm and the nucleus (reviewed in Ref. 8), and is likely to have differing functions depending on its subcellular localization.

Many studies have focused on whether HBV encodes an oncogene. Initially, this idea seems not to have much merit. The viral genome contains only four ORFs, none of which share sequence homology with known oncogenes. In addition, the long latency between HBV infection and the emergence of HCC suggests the lack of a viral transforming gene. While there remains strong interest in the regulatory HBx protein as a potential viral oncogene in cell culture, it is generally accepted that HBx does not behave as an oncogene *in vivo* (discussed below).

HBx and the Cell Cycle

Several laboratories have reported that HBx is able to affect cell cycle progression (see above section on the effect of HBx and liver regeneration). This has been reported for HBx both expressed in transfected cells in culture and in livers of mice under certain conditions. It is thought that HBx may have evolved an ability to modulate the normal quiescent (G0) hepatocyte to better support virus replication. A major difficulty with this idea is that, depending on the study, HBx has seemingly opposite effects on the cell. In some studies HBx promotes G0/G1 progression while in other studies HBx enforces a G1/S cell cycle checkpoint that can be fatal for the cells. The factors contributing to such disparate results are not well understood, but could include differences in cell lines used, growth state of the cells, and expression levels of HBx. In general, HBx is not well tolerated in rapidly dividing HeLa cells. Results from several *in vivo* studies in transgenic mice are equally perplexing, with some studies showing that HBx induces premature cell cycle entry and others concluding that HBx induces a G1/S checkpoint that prevents liver regeneration. Possible variables in these *in vivo* studies include the genetic background of the mice, and the presence or absence of steatosis. The expression level of HBx may also contribute to variable results in these studies. There are few if any HBx antibodies that react equally well with all subtypes of HBx, thus preventing a direct comparison of HBx levels in cells or in the livers of transgenic mice. If HBx does function to induce cell cycle progression to facilitate virus replication, then one can envision that, if deregulated, this viral strategy has the potential to contribute to transformation. For example, premature cell cycle entry in the presence of environmental carcinogens may lead to S phase progression prior to the complete repair of DNA adducts.

Activation of Signal Transduction Pathways

There is abundant data showing that HBx is able to stimulate cytoplasmic signaling pathways (reviewed in Ref. 8). This function of HBx represents a mechanism by which HBV could enhance cellular factors needed for virus replication without the need to induce cell cycle progression. In plasmid-transfected cells grown in culture, HBx activates a variety of cytoplasmic signaling pathways. No direct interaction between HBx and any of the protein kinases in these pathways has yet been confirmed, and so it is thought that upstream activators such as Src kinases may be responsible for the effect of HBx. Inhibitors of cytoplasmic calcium release also inhibit HBV replication, so a role of calcium activation by HBx has been proposed as a potential activator of Src.[27] The effect of calcium may be either direct or indirect, and further studies are required to solidify this HBx function as it relates to virus replication. Many reports have shown that HBx localizes to both the nucleus and the cytoplasm, with a subset of cytoplasmic-localized HBx found in the mitochondrial membrane.[28] In summary, HBx has the ability to stimulate cytoplasmic signal transduction pathways, possibly as a strategy to "activate" hepatocytes to better support virus replication. However, the underlying mechanisms remain to be identified, and it is still not clear how these activated pathways would contribute to viral replication and pathogenesis.

HBx and Apoptosis

Many viruses encode proteins that either promote or inhibit programmed cell death (apoptosis) (reviewed in Ref. 29). Since chronic HBV infection leads to cycles of immune-mediated cell death and regeneration, it is intriguing to hypothesize that HBx might also act to modulate apoptosis. The ability of HBx to induce apoptosis would allow for virus spread while avoiding the immune response. In addition, increased apoptotic cell death might contribute, over time, to pathogenesis and HCC by increasing the amount of compensatory regeneration that occurs. Alternatively, HBx-induced inhibition of apoptosis might be desirable for virus maturation, and could contribute to HCC by permitting the survival of cells bearing mutations that would normally be candidates for apoptosis. Studies examining the effect of HBx on apoptosis have been performed in a variety of cell types using many different apoptotic stimuli (microinjection of p53, treatment with TNFα, anti-Fas, DNA damage agents, etc.). However, the findings have been ambiguous. Under some experimental conditions HBx inhibits apoptosis, while in cell culture HBx appears to sensitize cells to death by DNA damaging agents (etoposide, doxycycline) (reviewed in Ref. 8).

The effect of HBx on hepatic apoptosis *in vivo* has been studied in the livers of transgenic mice. In the livers of 15-day-old mice, HBx expression is associated with a twofold increase in apoptosis.[30,31] However, this pro-apoptotic effect of HBx is no longer apparent in the mice by 30 days of age,[32] unless there is accompanying preneoplasia.[33] The biological relevance of these observations is uncertain, as the studies were not performed in the context of virus replication. Thus, it remains possible that HBx may influence apoptosis as part of its strategy to enhance virus replication. Designing experiments that provide definitive results in the context of virus replication *in vivo* remains a challenge.

HBx and DNA Repair

Another approach to understanding HBx function in virus replication includes identifying cellular proteins that bind to HBx. One such binding partner, damaged DNA binding protein 1 (DDB1),[34] is critical for several pathways in the cell, including DNA replication and repair. The role of DDB1 in DNA repair raises the possibility that HBx expression might contribute to HCC formation by inadvertently interfering with the repair of damaged DNA. Indeed, HBx can interfere with the ability of HepG2 cells to repair damaged DNA.[35] The effect of HBx on the repair of DNA lesions *in vivo* was examined in HBx double-transgenic mice containing the lambda phage transgene, which permits the determination of relative DNA mutation frequency in the high molecular weight DNA of the liver. In these studies, HBx expression did not alter the accumulation of spontaneous DNA mutations,[32] but did lead to a slight elevation in DNA mutation frequency in the livers of mice treated with liver-specific carcinogens.[10] The features of these mutations suggest that HBx acts as a tumor promoter, rather than as a complete carcinogen. This idea is consistent with observations that HBx can promote cell cycle progression under certain conditions.

HBx and Genome Instability

Genetic instability is a hallmark of cancer, including HCC. There is evidence that HBV can both directly and indirectly contribute to genetic instability in hepatocytes. Evidence for a direct mechanism involves the regulatory HBx protein. One of the HBx-interacting proteins identified is called HBx inhibitory protein (HBXIP), and was discovered by Melegari *et al.*[36] One normal function of HBXIP is as a regulator of centrosome dynamics and cytokinesis. Expression of HBx from a strong cytomegalovirus early promoter in cervical carcinoma (HeLa) cells is associated with excessive centrosome replication and associated nuclear abnormalities.[37] It is difficult to reconcile how this function of HBx might be important for virus replication, although such a function might clearly be involved in viral pathogenesis if HBx expression were to become deregulated.

A second mechanism by which HBx may contribute to genome instability is suggested by reports that HBx can induce both regional hypermethylation and global hypomethylation in Chang liver (HeLa) cells.[38] Hypermethylation would be predicted to silence genes, and is consistent with the observation that HBx may induce a G1/S cell cycle restriction point. Alternatively, hypomethylation is consistent with elevated gene expression and may be related to the mechanism by which HBx activates viral and cellular gene expression. In either instance, inappropriate cellular gene expression over time might contribute to genetic instability, depending on the normal function of the genes affected. Importantly, these observations need to be extended to normal hepatocytes. It is important to note that none of these features of genome instability observed in cell culture have been observed in the livers of HBx transgenic mice, and so their role in viral pathogenesis in humans awaits confirmatory data.

Does HBx Cause HCC in Transgenic Mice?

The above properties of HBx suggest this viral protein has many functions that could be related to pathogenesis and HCC. Several independent laboratories have created HBx transgenic mice in order to study the function(s) of HBx *in vivo*. HBx expression in the mice is driven by viral or cellular (liver-specific) promoters, and the consequences of HBx expression studied on different genetic backgrounds. The majority of HBx transgenic mice have normal liver histology (Table 1), a phenotype consistent with the fact that HBV replication is non-cytopathic to hepatocytes. However, subsets of HBx mice develop HCCs, and this has been the subject of much discussion. The lineages of HBx mice that develop HCC are reported to have confounding factors of spontaneous HCC in non-transgenic littermates[39] and hepatic steatosis.[14] Other factors, such as HBx expression level and duration of expression, have been proposed to explain the presence of HCC in some but not all lineages of mice. However, these explanations are difficult to address experimentally due to the lack of antibody reagents.

Despite the controversy regarding whether HBx alone causes HCC, it is generally agreed that HBx expression leads to an elevated HCC incidence in transgenic mice that are additionally treated with other cancer cofactors (Table 1). This provides powerful evidence that HBx

Table 1. HBx Transgenic Mice[a]

Regulatory Region[b]	Genetic Background[c]	Pathology Observed (% of Total Mice)	Role of HBx in HCC	Reference
Untreated mice (spontaneous HCC)				
ATX	ICR (Outbred)	None	None	Lee *et al.* (1990)
MUP	C57BL6/SJL	None	None	Guidotti *et al.* (1990)
MUP	C57BL6	None	None	Dandri *et al.* (1996)
HBV	CD-1 (Outbred)	HCC (100%)	Cofactor[d]	Kim *et al.* (1991)
HBV	C57BL6/6J	None	None	Reifenberg *et al.* (1997)
INS	C57BL6/6J	None`	None	Reifenberg *et al.* (1997)
PEX7	C57BL6/6	None	None	Billet *et al.* (1995)
SVX	C57BL6/6	None	None	Billet *et al.* (1995)
ATIII	C57BL6/6	None	None	Billet *et al.* (1995)
WAP	NMRI (inbred)	None	None	Klein *et al.* (2003)
ALB	C57BL6	HCC (86%)	Cofactor[e]	Wu *et al.* (2006)
Mice treated with cancer cofactor				
ATX	ICR (Outbred)	HCC with DEN	Cofactor	Slagle *et al.* (1996)
MUP	C57BL6	HCC with DEN	Cofactor	Dandri *et al.* (1996)
HBV	C57BL6/DBA2	HCC with *myc	Cofactor	Terradillos *et al.* (1997)
ATIII	C57BL6/DBA2	HCC with *myc	Cofactor	Terradillos *et al.* (1997)

(*Continued*)

Table 1. (*Continued*)

Regulatory Region[b]	Genetic Background[c]	Pathology observed (% of Total Mice)	Role of HBx in HCC	Reference
ATX	ICR (outbred)	HCC with HCV	Cofactor	Keasler *et al.* (2006)
WAP	NMRI (albino)	Mammary carcinoma in p53 +/– mice	Cofactor	Klein *et al.* (2003)
HBV	C57BL6	HCC with DEN	Cofactor	Zhu *et al.* (2004)

[a]Expression of HBx may be in the context of a portion of the HBV genome, or as the X gene.

[b]Regulatory regions used to drive HBx expression included ATX (human alpha-1-antitrypsin), MUP (major urinary protein; MUP-X mice by Dandri *et al.* expressed the woodchuck X protein), HBV (native HBx promoter/enhancer), INS (rat insulin), PEX7 (HBV X native promoter plus HBV Enhancer I), SVX (SV40 early region promoter/enhancer), ATIII (antithrombin III), WAP (whey acidic protein), ALB (albumin).

[c]Genetic background of the mice used to generate transgenics.

[d]Spontaneous HCCs were reported in non-transgenic controls.

[e]Accumulation of hepatocyte fat (steatosis) may have contributed to HCCs.

Note: A study by Yu *et al.* (1999) reporting HCC in HBx mice was not reproduced by Dirsch *et al.* (2004) and so is not included.

expression can serve as a cancer cofactor in mice. Does HBx function similarly in humans? It is important to recall that HBx is required for virus replication, and so is being expressed continually during the decades of chronic virus replication that precede the development of HCC. Therefore, it is quite possible that HBx plays a similar cancer cofactor role in chronically infected humans, particularly those living in areas of the world with chronic exposure to environmental carcinogens.

HBV-associated HCC: Other Mechanisms

HCC develops in multiple steps over several decades. There is evidence that chronic HBV replication may contribute either directly or indirectly to that process. Additional genetic changes are found in HCCs, and it remains unclear how these may be linked to HBV.

MicroRNAs and HCC

MicroRNAs (miRNAs) are small, noncoding RNAs approximately 22 nucleotides in length that specifically target protein-coding mRNAs. These regulatory miRNAs are complementary to a short sequence in the mRNAs they regulate, leading to binding, processing, and degradation of the target mRNAs. It is estimated that each miRNA recognizes approximately 100–200 different mRNA targets. This suggests that a single miRNA can control the expression of a significant number of cellular genes, and provides a novel mechanism for post-transcriptional regulation of gene expression. It is known that miRNAs are important in several critical cellular processes including proliferation, differentiation, cell death, and carcinogenesis.

Cellular miRNAs may be usurped by viruses in their strategy to enhance viral replication. For example, cellular miR-122, a regulator of genes that control fatty acid and cholesterol biosynthesis in the liver, also positively regulates hepatitis C virus (HCV) replication by binding to the 5′ untranslated region of the HCV genomic RNA (reviewed in Ref. 40). However, it is unclear how this viral strategy would trigger events that contribute to an altered miRNA profile in the cell. It is similarly unknown whether HCV and HBV will share in their targeting of specific miRNA pathways in the liver.

Both the expression levels of different miRNAs and the targets of those miRNAs may be cell-type dependent. Not surprisingly, examination of miRNA expression levels in specific cancer types has revealed tumor-specific miRNA "signatures," which may additionally predict particular clinical outcomes (reviewed in Ref. 41). It is not clear what could cause the deregulation of normal miRNA expression that would lead to cancer. It has been suggested that transcriptional deregulation, epigenetic alterations, mutations, DNA copy number abnormalities, and defective miRNA biogenesis may all lead to altered miRNA expression. Of note, many of these mechanisms of genetic deregulation are also present in HBV-infected cells, which raises the possibility that HCCs may similarly undergo alterations in miRNA regulation.

There is accumulating evidence that HCCs have a specific miRNA signature distinct from that in non-tumor liver tissue. Relative to expression in non-tumor tissue, HCCs revealed several elevated miRNAs: miR-21[42–44] (reviewed in Ref. 41) and the miR-17–92 polycistron.[43] Other miRNAs showed decreased expression in HCCs, relative to their expression in non-tumor liver tissue. These include miR-122a[42,44] (reviewed in Ref. 41), and miR-223.[45] Interestingly, miR-21 is now known to be overexpressed in breast and hepatocellular cancer, cholangiocarcinoma, and gliomas (reviewed in Ref. 41). Consistent with a role in HCC formation, overexpression of miR-21 is associated with increased cellular proliferation in several HCC-derived cell lines, including HepG2, PLC/PRF-5, SK-HEP-1, and SNU-182. In contrast, the suppression of miR-21 by miRNA inhibitors caused growth suppression in the same cell lines.[44] Although not all targets of a given miRNA are known, one target of the miR-21 is the phosphatase and tensin homolog (PTEN) tumor suppressor gene. Overexpression of miR-21 is predicted to decrease PTEN expression, effectively inactivating this tumor suppressor gene and contributing to hepatocarcinogenesis.

Finally, there are several examples of viruses that encode their own miRNAs, including herpes simplex virus type 1, human cytomegalovirus, Epstein-Barr virus, Kaposi's Sarcoma associated herpes virus, and simian virus 40 (reviewed in Ref. 40). It is significant that each of these viruses has been associated with human cancer, although the link between viral miRNAs and cancer is not yet established. A computational approach was used to identify a potential miRNA within the HBV genome.[46] This miRNA was not predicted to have cellular targets, but was hypothesized to target the viral polymerase mRNA, perhaps to reduce virus replication in order to evade the host immune system. It is intriguing to consider that viral miRNAs or altered cellular miRNA expression profiles play a role in HCC formation. Further examination of different miRNA targets and expression levels could lead to a better understanding of how miRNAs contribute to HCC formation and may yield diagnostic markers for predicting disease severity and progression.

Tumor-specific DNA Loss in HCC

There is evidence that human HCCs frequently undergo loss of chromosome arms as the tumor develops. Tumor-specific chromosomal losses in HCC samples have been reported on 1p, 4q, 6q, 8p, 13q, 16p, 16q, and 17p (reviewed in Ref. 47). The loss of specific chromosomal regions in a significant proportion of tumors suggests that a tumor suppressor gene may have resided in the deleted region. Given the ability of HBV to integrate and cause chromosomal rearrangements (Fig. 2B), it is not surprising that HBV-positive HCCs have a higher rate of chromosomal abnormalities relative to HCCs associated with other risk factors. In some instances, the loss of a chromosome region containing a tumor suppressor gene can be linked to a specific viral integration event. However, not all such allele losses in HCCs can be linked to HBV inserts. There may be other mechanisms of HBV-induced chromosomal instability in addition to integration-mediated allele losses. A recent study suggested HBx expression leads to defects in chromosome segregation by affecting the S phase progression of the cell cycle.[48] This may be an indirect effect of HBV on the host chromosome stability.

Epigenetic Silencing by Promoter Hypermethylation

Promoter regions of genes contain CpG islands that are often the target for gene silencing by hypermethylation. This mechanism has been documented in HCCs for several tumor suppressor genes, including $p16^{INK4a}$, E-cadherin, and $p14^{ARF}$ (reviewed in Ref. 49). In the instance of $p16^{INK4a}$, silencing is due mainly to promoter hypermethylation, with aberrant methylation also being observed in both cancerous and non-cancerous tissues of chronic HBV infected livers (reviewed in Ref. 49). HBV DNA can also be methylated in liver cells, and this could represent a mechanism by which viral expression is down-regulated. Evidence that the virus itself is responsible for this altered methylation is slowly accumulating. There is one report that HBx expression is associated with both DNA methylation and decreased DNA methyltransferase activity.[38] Global hypomethylation, predicted to enhance gene expression, has been reported for cells expressing HBx. While the link between HBV DNA methylation and HCC is unclear, methylation status of host genes is being investigated as a predictor of HCC prognosis.

Cooperation With Other Risk factors

Chronic HBV infection is but one risk factor associated with the development of HCC. Others include chronic infection with HCV, ethanol consumption, and genetic diseases that lead to an accumulation of iron in hepatocytes. There is evidence that chronic HBV can worsen liver disease associated with chronic HCV in patients coinfected with both viruses. A meta-analysis of 32 studies that involved 4,560 cases of HCCs and 6,988 control patients revealed an increased risk of HCC for patients who were coinfected with both HBV and HCV, relative to those patients chronically infected with either virus alone.[50] This is further supported by evidence in transgenic mice where there is an increased HCC development in HCV/HBx transgenic mice compared to HCV/WT mice, suggesting that HBx may act as a cancer cofactor with HCV.[51]

Another HCC risk factor present in HBV-positive patients is exposure to environmental carcinogens. Aflatoxins are potent hepatocarcinogens produced by the mold *Aspergillus flavus,* a common contaminant of stored foodstuffs in many parts of the world where chronic HBV is also endemic. Aflatoxins and HBV have a synergistic effect on the development of HCC. A possible role for HBx in this synergy is suggested by a study in which HBx transgenic mice exposed to aflatoxin B1 had increased mutation frequency relative to that measured in treated non-transgenic control littermates.[52]

Prevention of HBV-related HCC

Vaccination has proven an effective way to prevent HBV infection and, therefore, chronic HBV infection. The incidence of HCC in Taiwanese children has decreased by nearly half in the years following mandatory immunization with recombinant HBV surface (S) antigen.[53] This is the first example of a "cancer vaccine" and provides definitive evidence that HBV is a cancer virus. However, the vaccine is not effective in those patients already chronically infected with HBV and who face a high risk of dying from liver disease, including HCC. In addition, an estimated 10% of immunized adults do not make antibodies that protect them from subsequent infection (vaccine non-responders).

It is known that the level of chronic virus replication in patients correlates with risk of HCC (reviewed in Ref. 3). Therefore, therapeutic interventions that lower the levels of replicating virus are anticipated to lower the risk of complications. Current antiviral therapies include IFN-α and nucleoside analogs, such as Lamivudine and Adefovir. IFN-α is beneficial for treating chronic HBV infection, but has shown no significant effect on preventing HCC. There is evidence, however, that the nucleoside analog Lamivudine may reduce the incidence of HCC. In the woodchuck model, the use of therapeutic vaccines in combination with antiviral treatments induced specific immune responses and transiently reduced viral loads in woodchucks with chronic WHV infection. The development of other antiviral treatments such as immunotherapies designed to activate the host HBV-specific T cell response and adoptive T cell transfer may uncover more effective ways to treat chronic HBV infection (reviewed in Ref. 54).

A new HBV treatment of potential utilizes RNA interference technology. While miRNAs (discussed above) are encoded in the genome, small interfering RNAs (siRNAs) are exogenous RNAs. Both target mRNAs for degradation by cellular machinery, resulting in the silencing of the target gene expression. Synthetic siRNAs targeting the HBV genome are predicted to lead to the degradation of viral mRNAs and pregenomic RNA, resulting in a decreased viral copy number. Indeed, principal studies in cell culture and small-animal models have had success in decreasing HBV levels (reviewed in Ref. 55). Targets often include the HBx sequence, which is common to each of the HBV mRNAs. This allows a single siRNA to promote the degradation of all viral mRNAs. Other studies have also used siRNAs to target the viral polymerase and surface antigen sequences (reviewed in Ref. 56).

An alternative therapeutic approach involves using siRNAs specific to the cell instead of to the viral mRNAs. In these instances, siRNAs are targeted to mRNAs encoding cellular oncogenes or other genes involved in carcinogenesis that are up-regulated in HCCs (reviewed in Ref. 55). With the promise of siRNA treatments comes several challenges such as the innate immune response to the synthetic siRNA, non-specific interactions with cellular genes, difficult delivery of the siRNAs, dosage regulation, and variable cellular gene expression within HCCs, making it hard to target a specific sequence (reviewed in Ref. 55).

Summary

The process by which chronic HBV infection leads to the development of HCC is very complex and occurs in multiple steps. A model for this process was proposed more than 10 years ago (Fig. 1) and is now supported by data from laboratories around the world. Central to this model is the high level of virus replication that occurs in chronically infected livers. While virus replication is itself not harmful to hepatocytes, the host immune response to virus-infected cells leads to liver damage, cell death, and compensatory regeneration. Liver regeneration is an evolutionarily conserved process that serves to replenish lost cells, but the presence of virus replication (and, specifically, HBx) may deregulate steps in the process, increasing the likelihood of mutations in liver DNA. During the decades of virus replication, subgenomic portions of the viral genome integrate at random locations in the genome, causing gross chromosomal changes (Fig. 2B). These viral inserts can mediate the loss of tumor suppressor genes and, in the woodchuck model of HCC, activate the expression of the *myc* family of oncogenes. A great deal of attention has focused on the regulatory HBx protein, which may contribute to HCC as a cancer cofactor (Table 1). Other mechanisms likely contribute to HCC, including inactivation of tumor suppressor genes by miRNAs, and silencing of growth regulatory genes by promoter methylation. The latter mechanisms are relatively new observations, and hold promise for a better understanding of multistep HCC formation and the specific role of HBV. It is likely that the role of HBV may differ among different HCCs, and this reflects the complex nature of the liver and its ability to regenerate, as well as the variable risk factors that can combine to ultimately form HCC. Future research must target ways to decrease virus replication in patients with chronic HBV infection. In this regard, newly discovered siRNAs offer hope as a therapy.

References

1. Beasley RP, Hwang LY, Lin CC, Chien CS. (1981) Hepatocellular carcinoma and hepatitis B virus. A prospective study of 22 707 men in Taiwan. *Lancet* **2**: 1129–1133.
2. Butel JS. (2000) Viral carcinogenesis: revelation of molecular mechanisms and etiology of human disease. *Carcin* **21**: 405–426.
3. Fattovich G, Bortolotti F, Donato F. (2008) Natural history of chronic hepatitis B: Special emphasis on disease progression and prognostic factors. *J Hepatol* **48**: 335–352.
4. Butel JS, Lee T-H, Slagle BL. (1996) Is the DNA repair system involved in hepatitis-B-virus-mediated hepatocellular carcinogenesis? *Trends Microbiol* **4**: 119–124.
5. Wieland S, Thimme R, Purcell RH, Chisari FV. (2004) Genomic analysis of the host response to hepatitis B virus infection. *Proc Natl Acad Sci USA* **101**: 6669–6674.
6. Chang JJ, Lewin SR. (2007) Immunopathogenesis of hepatitis B virus infection. *Immunol Cell Biol* **85**: 16–23.
7. Nakamoto Y, Guidotti LG, Kuhlen CV, *et al.* (1998) Immune pathogenesis of hepatocellular carcinoma. *J Exp Med* **188**: 341–350.
8. Bouchard MJ, Schneider RJ. (2004) The enigmatic X gene of hepatitis B virus. *J Virol* **78**: 12725–12734.
9. Terradillos O, Billet O, Renard CA, *et al.* (1997) The hepatitis B virus X gene potentiates c-myc-induced liver oncogenesis in transgenic mice. *Oncogene* **14**: 395–404.
10. Madden CR, Finegold MJ, Slagle BL. (2001) Hepatitis B virus X protein acts as a tumor promoter in the development of diethylnitrosamine-induced preneoplastic lesions. *J Virol* **75**: 3851–3858.
11. Loyer P, Cariou S, Glaise D, *et al.* (1996) Growth factor dependence of progression through G1 and S phases of adult rat hepatocytes *in vitro*. *J Biol Chem* **271**: 11484–11492.
12. Hodgson AJ, Keasler VV, Slagle BL. (2008) Premature cell cycle entry induced by hepatitis B virus regulatory HBx protein during compensatory liver regeneration. *Canc Res* **68**: 10341–10348.

13. Tralhao JG, Roudier J, Morosan S, *et al.* (2002) Paracrine *in vivo* inhibitory effects of hepatitis B virus X protein (HBx) on liver cell proliferation: An alternative mechanism of HBx-related pathogenesis. *Proc Natl Acad Sci USA* **99**: 6991–6996.
14. Wu BK, Li CC, Chen HJ, *et al.* (2006) Blocking of G1/S transition and cell death in the regenerating liver of Hepatitis B virus X protein transgenic mice. *Biochem Biophys Res Commun* **340**: 916–928.
15. Guidotti LG, Matzke B, Chisari FV. (1997) Hepatitis B virus replication is cell cycle independent during liver regeneration in transgenic mice. *J Virol* **71**: 4804–4808.
16. Dong Z, Zhang J, Sun R, *et al.* (2007) Impairment of liver regeneration correlates with activated hepatic NKT cells in HBV transgenic mice. *Hepatology* **45**: 1400–1412.
17. Petersen J, Dandri M, Bürkle A, *et al.* (1997) Increase in the frequency of hepadnavirus DNA integrations by oxidative DNA damage and inhibition of DNA repair. *J Virol* **71**: 5455–5463.
18. Wei Y, Fourel G, Ponzetto A, *et al.* (1992) Hepadnavirus integration: mechanisms of activation of the N-myc2 retrotransposon in woodchuck liver tumors. *J Virol* **66**: 5265–5276.
19. Paterlini-Brechot P, Saigo K, Murakami Y, *et al.* (2003) Hepatitis B virus-related insertional mutagenesis occurs frequently in human liver cancers and recurrently targets human telomerase gene. *Oncogene* **22**: 3911–3916.
20. Slagle BL, Lee TH, Butel JS. (1992) Hepatitis B virus and hepatocellular carcinoma. [Review]. *Prog Med Virol* **39**: 167–203.
21. Chisari FV, Ferrari C. (1995) Hepatitis B virus immunopathogenesis. *Annual Review of Immunology* **13**: 29–60.
22. Hildt E, Hofschneider PH. (1998) The PreS2 activators of the hepatitis B virus: Activators of tumour promoter pathways. *Recent Results Cancer Res* **154**: 315–329.
23. Hildt E, Munz B, Saher G, *et al.* (2002) The PresS2 activator MHBs^t of hepatitis B virus activates c-raf-1/Erk2 signaling in transgenic mice. *EMBO* **21**: 525–535.
24. Schluter V, Meyer M, Hofschneider PH, *et al.* (1994) Integrated hepatitis B virus X and 3′ truncated preS/S sequences derived from human hepatomas encode functionally active transactivators. *Oncogene* **9**: 3335–3344.
25. Zoulim F, Saputelli J, Seeger C. (1994) Woodchuck hepatitis virus X protein is required for viral infection *in vivo*. *J Virol* **68**: 2026–2030.
26. Keasler VV, Hodgson AJ, Madden CR, Slagle BL. (2007) Enhancement of hepatitis B virus replication by the regulatory X protein *in vitro* and *in vivo*. *J Virol* **81**: 2656–2662.
27. Bouchard MJ, Wang L-H, Schneider RJ. (2002) Calcium signaling by HBx protein in hepatitis B virus DNA replication. *Science* **294**: 2376–2378.

28. Takada S, Shirakata Y, Kaneniwa N, Koike K. (1999) Association of hepatitis B virus X protein with mitochondria causes mitochondrial aggregation at the nuclear periphery, leading to cell death. *Oncogene* **18**: 6965–6973.
29. Teodoro JG, Branton PE. (1997) Minireview: regulation of apoptosis by viral gene products. *J Virol* **71**: 1739–1746.
30. Pollicino T, Terradillos O, Lecoeur H, *et al.* (1998) Pro-apoptotic effect of the hepatitis B virus X gene. *Biomed Pharmacother* **52**: 363–368.
31. Terradillos O, Pollicino T, Lecoeur H, *et al.* (1998) p53-independent apoptotic effects of the hepatitis B virus HBx protein *in vivo* and *in vitro*. *Oncogene* **17**: 2115–2123.
32. Madden CR, Finegold MJ, Slagle BL. (2000) Expression of hepatitis B virus X protein does not alter the accumulation of spontaneous mutations in transgenic mice. *J Virol* **74**: 5266–5272.
33. Koike K, Moriya K, Iino S, *et al.* (1994) High-level expression of hepatitis B virus HBx gene and hepatocarcinogenesis in transgenic mice. *Hepatology* **19**: 810–819.
34. Lee T-H, Elledge SJ, Butel JS. (1995) Hepatitis B virus X protein interacts with a probable cellular DNA repair protein. *J Virol* **69**: 1107–1114.
35. Becker SA, Lee TH, Butel JS, Slagle BL. (1998) Hepatitis B virus X protein interferes with cellular DNA repair. *J Virol* **72**: 266–272.
36. Melegari M, Scaglioni PP, Wands JR. (1998) Cloning and characterization of a novel hepatitis B virus X binding protein that inhibits viral replication. *J Virol* **72**: 1737–1743.
37. Fujii R, Zhu CJ, Wen YF, *et al.* (2006) HBXIP, cellular target of hepatitis B virus oncoprotein, is a regulator of centrosome dynamics and cytokinesis. *Cancer Res* **66**: 9099–9107.
38. Park IY, Sohn BH, Yu E, *et al.* (2007) Aberrant epigenetic modifications in hepatocarcinogenesis induced by hepatitis B virus X protein. *Gastroenterology* **132**: 1476–1494.
39. Kim C-M, Koike K, Saito I, *et al.* (1991) HBx gene of hepatitis B virus induces liver cancer in transgenic mice. *Nature* **351**: 317–320.
40. Gottwein E, Cullen BR. (2008) Viral and cellular microRNAs as determinants of viral pathogenesis and immunity. *Cell Host Microbe* **3**: 375–387.
41. Shi XB, Tepper CG, deVere White RW. (2008) Cancerous miRNAs and their regulation. *Cell Cycle* **7**: 1529–1538.
42. Ladeiro Y, Couchy G, Balabaud C, *et al.* (2008) MicroRNA profiling in hepatocellular tumors is associated with clinical features and oncogene/tumor suppressor gene mutations. *Hepatology* **47**: 1955–1963.
43. Connolly E, Melegari M, Landgraf P, *et al.* (2008) Elevated expression of the miR-17-92 polycistron and miR-21 in hepadnavirus-associated hepatocellular carcinoma contributes to the malignant phenotype. *Am J Pathol* **173**: 856–864.

44. Meng F, Henson R, Wehbe-Janek H, *et al.* (2007) MicroRNA-21 regulates expression of the PTEN tumor suppressor gene in human hepatocellular cancer. *Gastroenterology* **133**: 647–658.
45. Wong QW, Lung RW, Law PT, *et al.* (2008) MicroRNA-223 is commonly repressed in hepatocellular carcinoma and potentiates expression of Stathmin1. *Gastroenterology* **135**: 257–269.
46. Jin WB, Wu FL, Kong D, Guo AG. (2007) HBV-encoded microRNA candidate and its target. *Comput Biol Chem* **31**: 124–126.
47. Buendia MA. (2000) Genetics of hepatocellular carcinoma. *Seminars in Cancer Biology* **10**: 185–200.
48. Martin-Lluesma S, Schaeffer C, Robert EI, *et al.* (2008) Hepatitis B virus X protein affects S phase progression leading to chromosome segregation defects by binding to damaged DNA binding protein 1. *Hepatology* **48**: 1467–1476.
49. Tischoff I, Tannapfe A. (2008) DNA methylation in hepatocellular carcinoma. *World J Gastroenterol* **14**: 1741–1748.
50. Donato F, Boffetta P, Puoti M. (1998) A meta-analysis of epidemiological studies on the combined effect of hepatitis B and C virus infections in causing hepatocellular carcinoma. *Int J Cancer* **75**: 347–354.
51. Keasler VV, Lerat H, Madden CR, *et al.* (2006) Increased liver pathology in hepatitis C virus transgenic mice expressing the hepatitis B virus X protein. *Virology* **347**: 466–475.
52. Madden CR, Finegold MJ, Slagle BL. (2002) Altered DNA Mutation Spectrum in Aflatoxin B1-Treated Transgenic Mice that Express the Hepatitis B Virus X Protein. *J Virol* **76**: 11770–11774.
53. Chang MH, Chen CJ, Lai MS, *et al.* (1997) Universal hepatitis B vaccination in Taiwan and the incidence of hepatocellular carcinoma in children. Taiwan Childhood Hepatoma Study Group. *N Engl J Med* **336**: 1855–1859.
54. Bohne F, Protzer U. (2007) Adoptive T-cell therapy as a therapeutic option for chronic hepatitis B. *J Viral Hepat* **14**(Suppl 1): 45–50.
55. Arbuthnot P, Thompson LJ. (2008) Harnessing the RNA interference pathway to advance treatment and prevention of hepatocellular carcinoma. *World J Gastroenterol* **14**: 1670–1681.
56. Chen Y, Cheng G, Mahato RI. (2008) RNAi for treating hepatitis B viral infection. *Pharm Res* **25**: 72–86.

Chapter 5

Treatment of Chronic Viral Hepatitis B and Drug Resistant Variants

Tin Nguyen[*,†], Paul Desmond[*] and Stephen Locarnini[†]

Abstract

Chronic hepatitis B (HBV) infection represents a global health problem, with an estimated 400 million people affected worldwide. Potential long-term sequelae include cirrhosis, hepatic decompensation, and hepatocellular carcinoma. The paradigm for treatment of chronic HBV is continually evolving with the advent of newer medications, an increased understanding of the natural history of chronic infection, and improved laboratory assay sensitivity. The decision to initiate therapy is dependent on multiple factors including serum ALT, HBV viral load, liver histology, and other patient variables. Antiviral drug resistance is becoming a major clinical challenge in patients with chronic hepatitis B, and can result in ongoing progression of liver disease. Strategies to help prevent or minimize resistance include the use of highly potent oral antiviral agents with a high genetic barrier to resistance, ongoing patient education, regular on-therapy monitoring of HBV viral load, and appropriate timing of testing for genotypic resistance. The common pathways of antiviral

[*] Gastroenterology Department, St Vincent's Hospital, Melbourne

[†] Victorian Infectious Diseases Reference Laboratory, Melbourne, Australia

Email: stephen.locarnini@mh.org.au

drug resistance are predictable, and its application is advantageous to the treating physician when determining salvage treatment options.

Introduction

Chronic hepatitis B virus (HBV) infection is a major worldwide health problem. Despite the availability of an effective vaccine since 1982, the global disease burden and economic costs attributable to HBV infection is staggering. The World Health Organization estimates that of the two billion people exposed to HBV, 400 million people have chronic disease.[1] More than 75% of infected patients are located in Asia, where the disease is endemic in many countries.[2] Chronic HBV infection predisposes patients to complications of cirrhosis (20–30%), hepatic decompensation, and hepatocellular carcinoma (5–10%),[3] and accounts for over a million deaths annually.[4]

The ultimate goal of HBV treatment is the prevention or delay of complications arising from long-standing infection. This can be achieved by rapid and durable suppression of viral replication. Although antiviral therapy for chronic HBV is associated with improved outcomes,[5] treatment is often long term and possibly even lifelong in the setting of hepatitis e antigen (HBeAg) negative disease and cirrhosis. Clinical experience with oral antiviral nucleos(t)ide analogues (NA) such as lamivudine and adefovir highlight the problem of cumulative resistance rates over time.

Antiviral drug resistance poses a major clinical challenge in patients with chronic hepatitis B. The emergence of antiviral resistant HBV strains not only results in virologic and biochemical breakthrough, but can lead to loss of histologic improvements gained during treatment, hepatic decompensation, and even death. With a limited spectrum of currently available drugs, the development of antiviral drug resistance not only limits alternative salvage treatment options, but can also lead to the emergence of multi-drug resistant organisms with a potential significant public health impact.

This chapter will address the mechanisms of antiviral resistance, diagnosis of emerging resistance, clinical consequences of resistance,

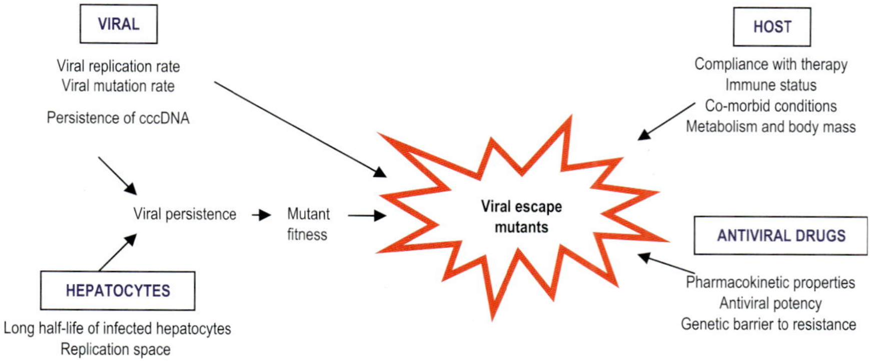

Fig. 1. Selection of drug resistant mutants.

and strategies to both prevent and manage this increasingly difficult problem.

Selection of Drug Resistant Mutants

The natural history of HBV infection is both variable and complex, and is still not completely defined and understood. The interplay of factors — including the particular host, virus strains, hepatocyte heterogeneity, as well as antiviral drug selection — can contribute to antiviral persistence, development of viral escape mutants, and eventual clinical antiviral resistance (Fig. 1).

Viral Factors

HBV is an enveloped virus containing a partially double-stranded, relaxed circular DNA genome of approximately 3,200 base pairs.[6] Following viral entry into a susceptible hepatocyte and transport of the viral genome to the nucleus, there are two important virologic events in the HBV life cycle: formation of covalently closed circular DNA (cccDNA), and reverse transcription of viral pregenomic RNA (pgRNA) into the HBV viral genome.[7]

Chronic HBV infection and viral persistence is maintained by transcriptional templates of cccDNA. HBV cccDNA resides in the hepatocyte nucleus as a stable episome, and serves as the transcriptional template for viral pgRNA, as well as other viral transcripts. It is estimated that 5–50 copies of cccDNA are present within an infected cell.[8] cccDNA does not integrate into the host genome, and is found as a viral minichromosome that persists during hepatocyte division.[9] It has a long half-life of at least 30–60 days.[10,11] A stable intranuclear pool of cccDNA is maintained by recycling of newly formed replicating core-associated HBV DNA in the cytosol back into the nucleus.[12]

Studies in the duck HBV model have demonstrated that initial cccDNA formation from incoming virions cannot be prevented by current oral antiviral agents.[13,14] Unfortunately, this suggests that ongoing hepatocyte HBV infection can still occur during "effective" antiviral therapy. Furthermore, although cccDNA levels decline during therapy, even the most potent antiviral drug does not completely eradicate cccDNA.[14,15] This is because the decline in cccDNA is due to a non-cytolytic pathway, involving initially suppression of viral replication, and then reduced intranuclear recycling of HBVDNA. cccDNA itself is not eradicated. Thus, in conjunction with its inherent stability and long life, it is likely that only hepatocyte death can completely eradicate cccDNA. cccDNA persistence explains the high relapse rate with treatment cessation, as well as the potential for disease reactivation in immunosuppressed patients with apparently resolved infection.

The second key event in the HBV life cycle is reverse transcription of viral pgRNA into the viral DNA genome. Unlike DNA polymerases, which have a 3′5′exonuclease that acts as a proofreading mechanism, the HBV reverse transcriptase lacks this and is thus error prone. This consequently results in a mutational rate of approximately 10^{-4} base substitutions/site/year.[16] Coupled with a high HBV replication rate of approximately 10^{11} virions per day,[17] single base pair mutations could theoretically occur at each nucleotide within the HBV genome each day. However, the viability of many of these spontaneous mutants is limited, primarily because the HBV genome is arranged into four overlapping but frame shifted open reading frames with highly conserved structural regions.

Despite these constraints, a diverse number of viral quasispecies exist prior to therapy, including those potentially associated with antiviral resistance. The dominant viral quasispecies prior to therapy is usually wild-type. By definition, this wild-type species is that which is best adapted to the host environment. However, the viral quasispecies population evolves over time and is dependent on the viral mutation rate, viral load, selective pressure of the immune response, and the replication fitness of virus in the presence of antiviral therapy.

Host Factors

The immune response plays a crucial role in the immunopathogenesis of HBV infection. Immune status, age of infection, and route of transmission all determine the eventual outcome of HBV infection. This is evidenced by high rates of chronic infection with perinatal transmission, childhood acquired infection, or in immunocompromised hosts (e.g. coinfection with HIV). In contrast, 95% of immunocompetent adults infected with hepatitis B have a self-limiting illness, and eventually clear infection, while 95% of neonates born to HBeAg-positive mothers become chronically infected without perinatal vaccination.

Recent studies have revealed new insights into the role of the innate immune system in chronic hepatitis B.[18] There is emerging evidence suggesting that viral proteins may target pathogen recognition receptors such as the Toll-Like receptor (TLR) family to escape immune surveillance. Patients with HBeAg positive disease demonstrate down-regulation of TLR2 on hepatocytes, Kupffer cells, and peripheral blood mononuclear cells, in comparison to HBeAg negative disease. In contrast, TLR2 and tumor necrosis factor α (TNFα) can be up-regulated in patients with HBeAg negative disease.[19,20] Further studies are required in patients on antiviral treatment to determine the changes within the innate immune system during development of antiviral resistance.

The natural history of chronic hepatitis B infection consists of four phases: immune-tolerant, immune-clearance, immune-control, and immune-escape.

The immune-tolerant phase is characterized by positivity for hepatitis B e antigen (HBeAg), high levels of viral replication, normal serum ALT (typically lower limit of normal), and normal/near-normal histological profile. Patients in this phase are usually young, and this period can last for longer than 20 years until immune tolerance is lost. Usually patients progressing to chronic infection in adulthood do not undergo an immune-tolerant phase, and instead enter directly into an immune-clearance phase.

In patients with chronic HBV infection acquired early in life, the immune-clearance phase usually occurs at age 20–40. It is characterized by increasing levels of serum ALT, decreased HBV DNA levels, and histologic damage. During this phase, HBV is under intense immune pressure, which can result in the emergence of dominant precore (PC) and basal core promoter (BCP) variant quasispecies, and eventual HBeAg seroconversion (HBeAg to anti-HBe).

The *in vitro* effect of PC and BCP mutations in both the wild-type and lamivudine (LMV) resistant mutation background has been examined in a number of studies including a cloned replication-competent HBV (genotype A, adw2) vector system and HBV baculovirus system (genotype D, ayw).[21,22] These studies demonstrate that the PC stop codon mutation in a wild-type background did not affect viral replication. However, the PC mutation appeared to restore replication efficacy in LMV resistant mutants, without affecting *in vitro* sensitivity to LMV. The presence of BCP mutations was associated with higher levels of viral replication in comparison to both wild-type virus alone, as well as in combination with LMV resistant mutants.[22] Other clinical studies have found conflicting associations between PC variant mutations with the development of LMV resistance.[22–24] This has been attributed to differences in genotype among the study population, although the mechanisms underlying this have not been fully elucidated.

Evidence for the importance of the immune system is also demonstrated in patients treated for HBV. During the immune-clearance phase, patients can spontaneously develop HBeAg seroconversion at a rate of 5–10% per year.[25] Initiation of immunomodulator therapy with interferon (IFN) improves HBeAg seroconversion rates up to 30%.[26,27] High pre-treatment ALT levels (above twice normal) and low HBV DNA titers are an important predictor of response to IFN therapy, highlighting the requirement for an already stimulated or "primed" immune response.[28] Oral nucleos(t)ide analogues (NA) have also been demonstrated to improve HBeAg seroconversion rates. Studies have shown that viral suppression from these drugs is associated with the restoration of CD4 and CD8 T cells against HBV epitopes.[29] This may then lead to HBeAg seroconversion and sustained immunological control of viral replication.

Finally, other host factors that contribute to the development of resistance include adherence to therapy, co-morbid conditions such as HIV infection and genetic factors such as differences in drug metabolism.

Antiviral Drug

Mechanism of Action of Oral Antivirals Agents in Chronic HBV

In HIV medicine, the strategic development of highly active antiretroviral therapy (HAART) was designed to target multiple and different antiviral viral targets. The treatment landscape in chronic hepatitis B is different, however, as all current nucleos(t)ide analogues essentially target one site: the HBV polymerase/reverse transcriptase. These drugs can be classified into three groups depending on structure and mechanism of action (Table 1).

Pharmacokinetics of Oral Antiviral Agents in Chronic HBV

The intracellular activity of oral antiviral agents used in chronic HBV can be affected by multiple factors, including absorption into systemic circulation, intracellular uptake, presence of activating enzymes, and competition from endogenous nucleotides.[32]

The bioavailability and absorption of oral antivirals may be affected by food. For example, the maximal serum concentration (Cmax) and area under the curve (AUC) of entecavir following a meal is reduced by 44–46% and 18–20%, respectively. Thus, entecavir should be taken on an empty stomach for a maximal antiviral effect

Table 1. Mechanism of Action of Nucleos(t)ide Analogues in Chronic Hepatitis B[30,31]

	Drug	Structure	Mechanism of Action
L-nucleoside analogues	Lamivudine	Enantiomer of 2′-3′ dideoxy-3′ thiacytidine.	Acts on reverse transcriptase and DNA polymerase. Inhibit synthesis of both the viral minus strand and plus strand.
	Telbivudine	L-thymidine analogue	
	Emtricitabine	Fluorinated analogue of cytidine	
	Clevudine	Thymidine analogue	Acts on DNA polymerase. Mainly inhibits plus strand synthesis.
Acyclic Phosphonates (nucleotide analogues)	Adefovir	Nucleotide analog of adenosine monophosphate	Acts on primase, reverse transcriptase and DNA polymerase.
	Tenofovir	Pro drug of adenosine monophosphate	
Cyclopentane ring group	Entecavir	Carboxyl analogue of 2′deoxyguanosine	Acts on primase and reverse transcriptase. DNA synthesis inhibited at three steps: HBV DNA polymerase priming, reverse transcription of the minus HBVDNA strand from pg RNA, and the synthesis of the HBVDNA plus strand.

LMV: Lamivudine; LdT: Telbivudine; FTC: emtricitabine; CLD: clevudine; ADF: adefovir; TDF: tenofovir; ETV: entecavir.

(at least two hours after a meal and two hours before the next meal). Conversely, tenofovir administration is recommended with food, although absorption may be increased on either an empty stomach or with a high fat meal. Lamivudine, adefovir, and telbivudine may be taken with or without food.

The bioavailability of oral nucleos(t)ide analogues is not affected significantly by hepatic impairment. However, oral nucleos(t)ide analogues undergo renal metabolism and require dose reduction with severe renal impairment.

Following absorption and distribution to the liver, these drugs enter hepatocytes via a competitive process through cellular transporters on the hepatocyte cell surface membrane. Upon entry into the intracellular compartment, cellular kinases are required to phosphorylate the drugs into an active form. It is often forgotten that oral antiviral agents used in the treatment of chronic HBV infection are pro-drugs, and that the phosphorylation step is competitive between all nucleos(t)ide analogues (NA). This property may have implications for the potential future use of combination therapy in HBV and requires further study.

Hepatocytes

HBV replication is not directly cytopathic toward hepatocytes. It is the host immune system's HLA class I antigen restricted CD8 cytotoxic T cells response mediated against viral antigens on hepatocytes that appears to be responsible for hepatic injury. The degree of immune system activation is manifested by serum ALT levels; high baseline serum ALT levels have been associated with the development of LMV resistance.[23,33] It has been suggested that higher ALT levels results in increased regeneration of non-infected hepatocytes, which function as an expanded "replication space" through which resistant viral mutants can emerge and take a stronger "footing."[34]

Molecular Basis of Antiviral Drug Resistance

Primary and Secondary Mutations

The HBV polymerase open reading frame consists of four domains: terminal protein, spacer, polymerase/reverse transcriptase (POL/RT), and RNase H.[35] The POL/RT can further be divided into seven conserved subdomains (A–G). Mutations associated with antiviral drug resistance typically occur in subdomains A–E of the POL/RT domain. The YMDD primary catalytic sequence is in the C domain.

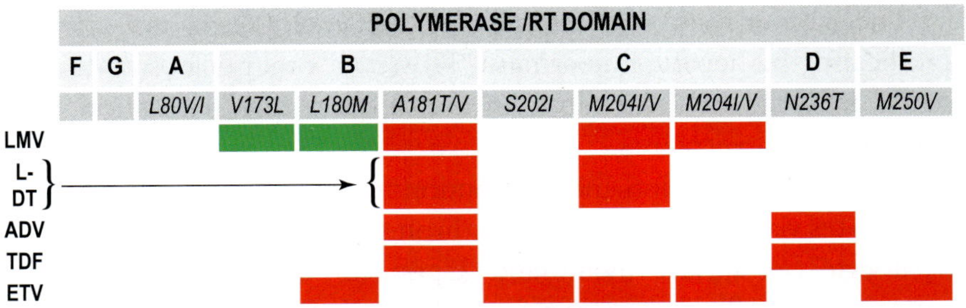

Fig. 2. Mutations associated with antiviral drug resistance. Primary mutations are highlighted in red, and secondary in green. LMV (lamivudine), L-dT (telbivudine), ADV (adefovir), TDF (tenofovir), ETV (entecavir).

Mutations selected during antiviral therapy can be classified as either primary or secondary. Primary resistant mutations are those that directly reduce the sensitivity of the viral polymerase to a particular antiviral agent. Also, the replication fitness of the mutant virus is reduced. Secondary mutations are compensatory and restore the replication efficacy of the mutant virus (see Fig. 2). In addition, other mutations with as yet unclear or no effect on viral replication may develop.

The primary mutation associated with resistance to lamivudine and other L-nucleosides is rtM204I/V, located in the YMDD locus of the C domain.[30,36,37] Secondary mutations that may be co-selected include rtL180M and rtV173L in the B domain.[38,39] It is important to note that rtM204I/V is also implicated in antiviral resistance to entecavir. rtA181V (domain B) and rtN236T (domain D) are primary mutations conferring resistance to adefovir.[40]

Lamivudine and adefovir have different chemical structures and mechanisms of action, and thus have different proposed mechanisms of resistance (Table 2).

Pathways of Antiviral Drug Resistance

The chemical classification of NA is helpful because it determines the common pathways of antiviral drug resistance. These pathways are generally predictable, and to date, five have been characterized (see Table 3).

Table 2. Mechanisms of Resistance to Lamivudine and Adefovir

Drug	Mechanism of Resistance
Lamivudine	• Primary mutation causes steric hindrance, altering ability of HBV POL/RT to bind lamivudine relative to natural substrate[41] • Altered catalytic efficiency of mutant POL/RT[42] • Increased excision of lamivudine as a chain terminator by pyrophosphorolysis or an ATP dependent process[43]
Adefovir	• Altered HBV POL/RT triphosphate binding site for adefovir[44]

Table 3. Pathways of Antiviral Resistance in Chronic Hepatitis B

Pathway	Mutation	Associated Resistance
L-nucleoside	rtM204V/I	Lamivudine Emtricitabine Telbivudine Clevudine
Acyclic phosphonate	rtN236T	Adefovir Tenofovir
"Shared"	rtA181T/V	L-nucleosides (see above) Acyclic phosphonates (see above)
Naïve entecavir resistance	rtL180M + rtM204V with one of rtT184, S202 or M250	Entecavir
Multi-drug resistance	Complex patterns e.g. rtA181T + rtI233V + rtN236T + rtM250L	Multi-drug

L-nucleoside Pathway

The L-nucleoside drugs include lamivudine, emtricitabine, telbivudine, and clevudine. Treatment with L-nucleosides can lead to the primary resistant mutation rtM204V/I. rtL180M and rtV173L are secondary mutations restoring replication efficacy. The rates of cumulative genotypic resistance rates to these drugs are listed in Table 4. Lamivudine is the most well studied L-nucleoside as it is associated

Table 4. Annual Resistance Rates for Oral Antiviral Agents in Chronic Hepatitis B

	Cumulative Resistance Rate (% of patients)				
Drug	**1 yr**	**2 yrs**	**3 yrs**	**4 yrs**	**5 yrs**
Lamivudine[45,46]	23	46	55	71	80
Adefovir[47] (HBeAg neg)	0	3	11	18	29
Entecavir[48]	0.2	0.5	1.2	1.2	1.2
Emtricitabine[49,50]	13	18	—	—	—
Telbivudine[37] (HBeAg pos) (HBeAg neg)	4.4 2.7	21.6 8.6	—	—	—
Tenofovir[51]	0	—	—	—	—
Adefovir (LMV resistant)[52]	0–18	38.3	—	—	—
Adefovir/LMV[53] (LMV resistant)	1	2	4	4	—
Entecavir (LMV resistant)[54]	6	15	35	39.5	—

with 80% genotypic resistance rates following up to five years of continuous use. Once rtM204V/I and virologic breakthrough occurs, drugs with cross-resistance within this group should *not* be introduced as salvage therapy.

Acyclic Phosphonate Pathway

Treatment with adefovir can select out rtN236T. Tenofovir is a nucleotide analogue that is structurally similar to adefovir, and has been licensed in the treatment of HIV as monotherapy or in a dual tablet Truvada (tenofovir 300 mg and emtricitabine 200 mg). Tenofovir has much higher potency than adefovir, simply because it is a pro-drug of a similarly active agent that can be given at a much higher dosage due to less nephrotoxicity.[30] Week 48 data has not yet shown HBV mutations that confer resistance to tenofovir monotherapy. Longer-term studies with tenofovir are required to determine whether rtN236T or other primary mutations such as rtA181T/V will be selected.

Shared Pathway

The HBV Pol isolated from most patients with antiviral resistance to lamivudine is rtM204V/I, and thus a salvage option is to add on adefovir therapy. However, the "shared pathway" that selects out rtA181T/V can confer resistance to both L-nucleosides (e.g., lamivudine) and acyclic phosphonates (e.g., adefovir). rtA181T/V is seen in 40% of adefovir and 5% of lamivudine failure, respectively.[55]

Naive Entecavir Resistance Pathway

Entecavir is a potent oral antiviral agent in treatment-naive patients. *In vitro* studies have shown it to be 100 to 300 times more potent that lamivudine.[31] In both HBeAg positive and negative disease, it has been shown to have superior rates of biochemical, virologic, and biochemical improvement in comparison to lamivudine. The five-year cumulative genotypic resistance rate with entecavir in previously treatment-naive patients is only 1.2%. This is because of the high genetic barrier to resistance of entecavir. Virologic breakthrough to entecavir requires at least three substitutions, including rtM204V and rtL180M associated with lamivudine resistance, and an additional substitution at either rtS202I, rtT184G and/or rtM250V.[56] In the setting of previous LMV resistance, sequential salvage monotherapy with entecavir is associated with 39.5% genotypic resistance rates after four years of entecavir therapy.[54]

Therapy for Chronic HBV in Treatment-naive Patients

Goals of Treatment for Chronic HBV Infection

The major long-term goal of antiviral treatment in the management of chronic hepatitis B is to prevent or delay the onset of complications associated with infection. It is now recognized that active viral replication drives the development of complications. Thus the cornerstone of HBV therapy is to achieve rapid and durable suppression of viral replication.[5]

It is likely that the target viral load reduction required to achieve histologic improvement is different from the reduction necessary to prevent the emergence of antiviral resistance. Previous studies with all current antiviral therapies have demonstrated histologic improvement following 48 weeks of therapy.[57] A viral load reduction to <4 $\log_{10}$ copies/mL may be a surrogate marker of histologic improvement. Observational data from a Taiwanese cohort of 3,582 patients indirectly supports this, showing a significant increase in the risk of cirrhosis in untreated patients with a baseline HBV viral load of >4 $\log_{10}$ copies/mL.[58]

A critical issue is whether a target viral load threshold exists below which the emergence of antiviral resistance does not occur. Ideally, complete suppression to undetectable levels by PCR is preferable, although this may not be always achievable in clinical practice. A recent phase III trial comparing telbivudine versus lamivudine stratified patients into different week-24 viral load levels and examined genotypic resistance rates at week 48. A direct relationship was demonstrated between HBV DNA levels at week-24 and subsequent emergence of drug resistance. Among HBeAg-positive patients treated with telbivudine, genotypic resistance of 1%, 2%, 8%, and 11% respectively were observed with HBV viral loads of undetectable, <3 $\log_{10,}$ 3–4 $\log_{10}$ and ≥4 $\log_{10}$ copies/mL.[59] However, following two years of therapy, cumulative genotypic resistance rates of 4%, 25%, 29%, and 30% respectively were observed based on the 24-week viral load level.[60] What is important to note here is that resistance can still develop with telbivudine despite undetectable viral loads. In clinical practice, a viral load target of 3 $\log_{10}$ copies/mL (equivalent to approximately 2.2 $\log_{10}$ IU/mL) may be a reasonable viral threshold target for minimizing emergence of resistance, with the caveat that a potent drug with a high genetic barrier to resistance is used.

Indications for Treatment Initiation

Viral persistence, antiviral resistance, and clinical manifestations of chronic HBV infection are due to the interplay of hepatocytes, viral factors, and the host immune response. It is perhaps not surprising then that the decision to treat HBV infection is determined by the degree of viral replication, serum ALT, and degree of liver disease on histology.

A number of regional consensus guidelines (Asia-Pacific, European and two American) have been published.[30,61–63] Subtle differences exist, and may be in part due to cost, availability, and possibly genotypic differences. The Asia-Pacific consensus statements for the management of chronic hepatitis B were recently developed and published by a panel of experts in the region, and have incorporated new data on the natural history and treatment options in chronic HBV infection (Tables 5 and 6).

Combination Therapy

Combination therapy is the mainstay of treatment in HIV, where drugs with different mechanisms of action and targets are used to minimize the development of resistance. In chronic hepatitis C, combination therapy with pegylated IFN and ribavirin has a synergistic effect, both improving sustained virologic response rates, and reducing long term relapse rates. However, although combination therapy in chronic HBV remains a focus of intense research, there is still insufficient data to support its use as first line therapy.

A number of studies have evaluated the combination of IFNα and LMV versus LMV or IFNα monotherapy.[64–67] While the on-treatment viral suppression was greater in the combination group, there was no significant difference in the sustained off-treatment responses when compared to the IFNα monotherapy arm. Furthermore, combination IFNα and LMV has not been proven to be more efficacious in previous IFNa non-responders.[68]

A potential benefit of combination therapy over monotherapy is the reduction of drug resistance. The combination of LMV and telbuvidine (LdT), both L-nucleosides and thus competing for the same cell transporters and phosphorylation systems, has been proven to be inferior to LdT monotherapy in all parameters of response.[36] Another trial comparing LMV versus lamivudine/adefovir combination revealed no improvement in HBeAg seroconversion rates at week 104.[69] However, the combination of lamivudine/adefovir was associated with greater suppression of viral replication and lower rates of resistance to LMV.

Table 5. HBV Treatment Algorithm for Patients With Compensated Disease

	HBV DNA (IU/mL)	ALT	Treatment Strategy
HBeAg Positive	<20,000	Normal	No treatment Monitor every 3–6 months — HBV DNA, HBeAg ALT
		Normal	No treatment Monitor every 3–6 months — HBV DNA, HBeAg, ALT Consider biopsy if >40 years — treat if moderate inflammation or fibrosis
	<20,000	1–2 × ULN	No treatment Monitor every 3–6 months — HBV DNA, HBeAg, ALT Consider biopsy if >40 years — treat if moderate inflammation or fibrosis
	<20,000	2–5 × ULN	Treatment if persistent (3–6 months), or risk of hepatic decompensation First line options: IFN based therapy, ETV, LMV, LdT, ADV
	<20,000	>5 × ULN	Treatment indicated — consider observation for 3 months for HBeAg seroconversion if no concerns for decompensation First line options: IFN based therapy, ETV, LMV, LdT Adefovir not first-line due to decreased potency
HBeAg Negative	<2,000	Normal	No treatment Monitor every 3–6 months — HBV DNA, HBeAg, ALT
	>2,000	Normal	No treatment Monitor every 3–6 months — HBV DNA, HBeAg, ALT

(*Continued*)

Table 5. (*Continued*)

	HBV DNA (IU/mL)	ALT	Treatment Strategy
HBeAg Negative	>2,000	1–2 × ULN	No treatment Monitor every 3–6 months — HBV DNA, HBeAg, ALT Consider biopsy if >40 years — treat if moderate inflammation or fibrosis
	>2,000	>2 × ULN	Treatment if persistent (3–6 months), or risk of hepatic decompensation First line options: IFN based therapy, ETV, LMV, LdT, ADF Long-term oral antiviral treatment usually required

Patients on treatment should have three-monthly HBV DNA and ALT monitoring to assess response. Patients at risk of hepatocellular carcinoma should have six-monthly screening with AFP and ultrasound. Adapted from Liaw *et al.*[63]

Table 6. HBV Treatment Algorithm for Patients With Cirrhosis

	HBV DNA (IU/mL)	Recommendation
Compensated	<2,000	No treatment Monitor every 3–6 months — HBV DNA, HBeAg, ALT
	>2,000	Treatment indicated irrespective of ALT level First line options include ETV, LMV, LdT Adefovir and IFN based therapies can also be considered if no previous hepatic flare
Decompensated	—	Consider liver transplantation Treatment indicated First line options include ETV, LMV, LdT, ADV IFN contraindicated

Adapted from Liaw *et al.*,[63] with kind permission from Springer Science and Business Media.

The era of highly potent antiviral agents such as entecavir and tenofovir arrived. Entecavir monotherapy as primary therapy is associated with only 1.2% genotypic resistance following five years of continuous monotherapy. With the availability of highly potent drugs with a high genetic barrier to resistance, there is currently insufficient need to warrant combination therapy as initial first-line therapy. The results of an ongoing study comparing entecavir versus entecavir/tenofovir combination are eagerly awaited.

Finally, it is important to consider that the mechanism of action and intracellular metabolism of current antiviral drugs in chronic HBV may limit their use in a combination therapy cocktail. Drugs within the same class may compete for entry in the cell. Upon cellular entry these drugs must all be phosphorylated into an active form, a competitive process with common enzymatic pathways between drugs. In addition, once active, all current oral antiviral drugs must then compete for the target site on the HBV POL/RT. Thus, it is possible that the intracellular concentrations of active drug and antiviral efficacy of current antiviral agents are different when used as combination therapy compared to monotherapy. These factors are often not considered by treating physicians, and require further studies to determine appropriate dosing.

Monitoring and Diagnosis of Antiviral Drug Resistance

Antiviral drug resistance is becoming a major clinical challenge for patients with chronic hepatitis B. The emergence of antiviral resistance can be indicated by a number of factors: an increasing HBV viral load despite compliance ($\geq$1.0 log IU/mL increase in viral load from nadir), identification of genotypic markers associated with antiviral resistance, increasing serum ALT, and clinical deterioration such as hepatic decompensation.

Patients should have regular three-monthly assessments to monitor clinical status and compliance with therapy, as well as determination of HBV viral load and serum ALT. Renal function should be

Table 7. Definitions of Response During Treatment of Chronic HBV[30,70]

Antiviral treatment effect: sustained ≥ 1 $\log_{10}$ IU/mL reduction of HBV DNA from baseline levels during therapy and within 3 months of treatment initiation.

Virologic response: decrease in serum HBV DNA to undetectable levels by sensitive PCR assays, with HBeAg seroconversion in patients who were initially HBeAg positive.

Biochemical response: normalization of serum ALT.

Histologic response: decrease in histology activity index by at least 2 points, and no worsening of fibrosis score when compared to pre-treatment biopsy.

Table 8. Definitions Related to Antiviral Resistance During Chronic HBV Treatment[30,70]

Primary antiviral treatment failure: failure of drug to initially reduce HBV DNA by ≥ 1 $\log_{10}$ IU/mL within 3 months of treatment initiation.

Virologic breakthrough: an increase in serum HBV DNA by ≥ 1 $\log_{10}$ IU/mL above nadir after achieving a virologic response during continued treatment.

Secondary antiviral treatment failure: rebound of HBV replication of ≥ 1 $\log_{10}$ IU/mL in patients with an initial antiviral treatment effect, as confirmed by two consecutive laboratory assessments one month apart.

Genotypic resistance: *In vitro* detection of mutations shown to confer resistance to NA in question.

Phenotypic resistance: *In vitro* confirmation that the detected mutation decreases susceptibility to the NA in question (indicated by increased inhibitory concentration).

assessed in patients on adefovir or tenofovir therapy, and HBeAg positive patients should be monitored for HBeAg seroconversion.

In clinical practice, serum HBV DNA testing with sensitive assays is the best investigation for monitoring antiviral efficacy. A reduction in serum HBV DNA always precedes biochemical and histologic response, and conversely virologic breakthrough always precedes loss of these responses (see Tables 7 and 8 for definitions). Regular HBV DNA testing allows determination of an initial antiviral treatment effect and subsequent durability, and monitoring for the development of virologic breakthrough.

Following commencement of antiviral therapy, a week-12 HBV viral load measurement is important to assess for either an initial antiviral treatment effect, or primary antiviral treatment failure. Switching of drug therapy should be considered with primary antiviral treatment failure.

Following this, week 24 is a key time-point for measurement of viral load to reassess therapy and characterize virologic responses. It should be noted that week 48 data may be more useful with drugs of lower potency, such as adefovir, if they have a high genetic barrier to resistance. Nonetheless, patients with suboptimal responses at week 24 may be at increased risk of developing antiviral resistance, and consideration should be given to add-on therapy.[61]

A virologic breakthrough is defined by an increase in serum HBV DNA by >1 $\log_{10}$ IU/mL above nadir after achieving a virologic response during continued treatment.[30] Given that virologic breakthrough may occur months before a biochemical breakthrough,[30] vigilant viral-load monitoring provides the opportunity for early initiation of salvage therapy. This is particularly important in high-risk groups, such as patients with cirrhosis, in whom a hepatitis flare can be complicated by hepatic decompensation and even death.

Patients compliant with therapy who develop virologic breakthrough should ideally be tested for genotypic resistance in order to identify any possible drug resistant viruses (see Fig. 3). Genotypic resistance is defined as the detection of mutations that have been demonstrated *in vitro* to confer resistance to the NA in question. The detection of these mutations facilitates appropriate decision-making with regard to salvage "add-on" therapy. It is important to note, however, that in clinical trials up to 30–50% of patients with virologic breakthrough did not harbour resistant mutations.[71] This is usually attributable to poor compliance with therapy, and highlights the importance of patient education and counseling both prior and during antiviral therapy.

HBV Viral Load Assays

Multiple commercial HBV DNA assays are currently available, each with their own inherent advantages and disadvantages (see Table 9).

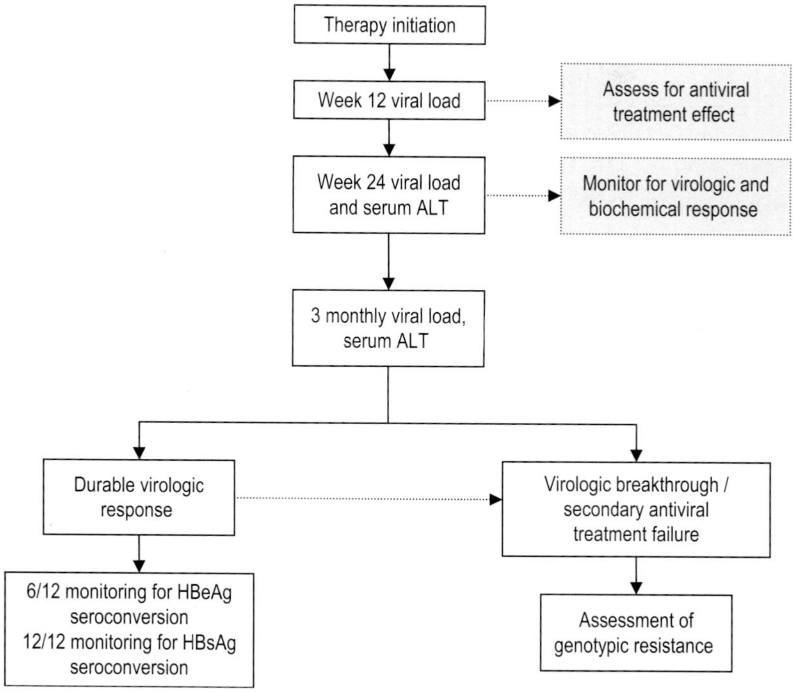

Fig. 3. Optimal monitoring of patients on oral antiviral therapy for chronic HBV.

Previous deficiencies in HBV DNA testing have included the lack of standardization between assays and limitations in dynamic range. The World Health Organization has now defined an international standard for HBV DNA testing calibrated to IU/mL,[72] and this should now be used to ensure comparability between trials as well as patients, if different assays are used.

An ideal assay should have a broad dynamic range. This not only facilitates early detection of virologic breakthrough but also adequate assessment of an antiviral treatment effect. Real-time PCR assays have recently become available. While still expensive, these assays have improved reproducibility compared with other technologies, and exhibit a far broader dynamic range (from <10 up to 10^9 IU/mL). It is therefore likely that they will soon become the standard for HBV DNA testing, not only in clinical trials but in standard clinical practice.

Table 9. Available HBV DNA Assays

Assay	Brand	Range (IU/mL)	PROs and CONs
Signal amplification	Digene Capture I (DIGENE CORP)	No conversion	Generally widely available, and only small sample volume required
	Digene Capture 2	$1.0 \times 10^5 - 3.4 \times 10^8$	Poor lower range of quantification
	Ultrasensitive Digene Capture 2	$1.6 \times 10^3 - 1.0 \times 10^8$	
	Versant (BAYER)	$3.5 \times 10^2 - 1.8 \times 10^7$	
Target amplification (PCR)	Cobas Amplicor (ROCHE)	$5 \times 10^1 - 4 \times 10^4$	Rapid turnaround Limited upper range of quantification, requires dilution Requires larger sample volume
Real-time PCR	Cobas Taqman	$3 \times 10^1 - 1.1 \times 10^8$	Still expensive and not yet widely available Broad dynamic range
	Abbot (ABBOT)	$1 \times 10^1 - 1.0 \times 10^9$	
	Real Art (ARTUS BIOTECH)	$2 \times 10^1 - 1.0 \times 10^8$	

Genotypic Resistance Testing

The most common method used to detect mutations associated with antiviral resistance is direct (or population) sequencing of HBV DNA. A limitation of this method is that the mutant virus must be present in at least 20% of the viral quasispecies pool to be detected. Other methods for detection of genotypic resistance include restriction fragment length polymorphism (RFLP) analyses, reverse hybridization assays (INNO-LiPA), cloning and sequencing, and mutation-specific real-time PCR. Like HBV viral load assays, each of these has advantages and disadvantages. For example, while some assays may be more sensitive than population sequencing, they may only allow detection of previously characterized mutations.[71]

The rtA181T / sW172 Mutant*

Due to the overlapping reading frames of the HBV genome, the development of rtA181T/V also results in a stop codon in the overlapping S region (sW172*). This results in the truncation of the last 55 amino acids of the surface protein C-terminal hydrophobic region. *In vitro* analysis demonstrates that this mutant is defective in surface protein secretion. However, it also appears to have a dominant inhibitory effect on wild-type virion secretion.[73]

The clinical significance of this is that the development of rtA181T/V *may not* be accompanied by a corresponding rise in HBV viral load. Thus, the traditional concept of virologic breakthrough as a marker for the development of genotypic resistance may not be applicable to rtA181T/V. Given regular viral load assessments are the cornerstone of monitoring patients on therapy, the emergence of rtA181T/V will clearly have important implications in future definitions of virological drug resistance.

Management of Antiviral Resistance

There are a number of published consensus guidelines outlining the management of antiviral resistance (see Fig. 4).[30,61] These have focused on recommendations for changing or switching therapy in patients who have developed resistance to monotherapy. While these recommendations are useful, some exceptions to the general rules exist (see below), and consequently clinicians need to understand the patterns of drug resistance with HBV infection so as to be able to determine the most appropriate salvage add-on treatment.

As a general principle, salvage "add-on" or "switch to" drugs with similar cross-resistance profiles should not be used (see Table 1). The advantage of "add-on" therapy is that continuation of the original drug will have an ongoing effect on a wild-type virus, and that the combination of two drugs may reduce the replication fitness of the virus. The potential disadvantage is that secondary compensatory mutants may develop over time.

Prevention of resistance

- Avoid unnecessary treatment
- First-line therapy should have high potency with low rate of drug resistance
- Adherence to therapy important

↓

Monitoring

- 3-monthly monitoring of HBV DNA with sensitive assay, ALT, HBeAg status, clinical status
- Alternative therapy in patients with primary antiviral treatment failure
- In patients with virologic breakthrough, check compliance, and confirm resistance with genotypic testing

↓

Timing of salvage therapy

- Earlier initiation of salvage therapy better

↓

	AASLD Guidelines[30]	**US Guidelines**[61]
LMV	Add ADV or TDF Switch to Truvada* Switch to EVT (pre-existing LMV-R resistance predisposes to ENT)	Add ADV (preferable to switch) Switch to ETV (increased risk of subsequent ETV) In future, potential add-on TDF or switch to Truvada
ADF	Add LMV Switch to Truvada Switch to or add ETV	Add LMV (preferable to switch) Switch to ETV (if no prior LMV resistance) In future, potential switch to Truvada
ETV	Switch to or add ADF or TDF	Add or switch to ADV or TDF
LdT	Add ADF or TDF Switch to Truvada Switch to ETV (pre-existing LdT-R resistance predisposes to ETV)	

LMV, lamivudine; ADF, adefovir; ETV, entecavir; TDF, tenofovir, LdT, telbivudine

* Truvada (combination emtricitabine 200 mg and tenofovir 300 mg)

Fig. 4. Management of antiviral resistant HBV.

Resistance to L-nucleoside

The bulk of existing data on salvage options for resistance to L-nucleosides is for lamivudine. In this situation add-on therapy is the most preferable. Lampertico *et al.* recently showed that add-on adefovir therapy in lamivudine resistance resulted in durable four-year virologic and biochemical responses with no virologic or clinical breakthroughs.[53] The cumulative rate of *de novo* rtA181T mutation at four years was 4%,[53] compared with up to 38% adefovir resistance (either rtN236T or rtA181T) at two years with adefovir switch.[52,74] Furthermore, for optimal suppression of viral replication, adefovir should be added to lamivudine as soon as genotypic resistance is detected.[75]

It is important to note that while adefovir has no cross-resistance with the L-nucleoside resistance pathway of rtM204V/I, the shared resistance pathway of rtA181T/V is associated with resistance to both adefovir and lamivudine (40% and 5% of adefovir and lamivudine failure respectively).[55] In this setting one option is to add on tenofovir to lamivudine. Tenofovir has much higher potency than adefovir, simply because it is a pro-drug of the similarly active agent that can be given at a much higher dosage due to less nephrotoxicity.[30] Early studies have demonstrated that tenofovir is effective rescue therapy in lamivudine refractory patients with a suboptimal response to adefovir switch monotherapy.[76]

Consensus guidelines have also included switching to entecavir as a potential salvage treatment option. Unfortunately, the typical L-nucleoside resistance pathway (rtM204VI) also encompasses entecavir in lamivudine-refractory patients.[55] *In vitro* studies demonstrate that entecavir is less effective against lamivudine-resistant HBV strains in comparison to wild-type: 20- to 30-fold higher concentrations are required for polymerase inhibition.[31] This is because lamivudine resistance already preselects two of the three mutations required for entecavir resistance.[77] Virologic breakthrough to entecavir requires at least three substitutions, but these include two lamivudine-resistant mutations (M204V and L180M), and an additional substitution at either S202I, T184G, and/or M250V[56]. In clinical practice, switching to entecavir in this setting results in cumulative virologic breakthrough rates at two, three, and four years of 11%, 27%, and 39% respectively.[54] Consequently, accumulating evidence suggests that entecavir switch is not an optimal treatment for patients with lamivudine resistance.

Acyclic Monophosphate Resistance

Week-48 data has not yet shown HBV mutations that confer resistance to tenofovir monotherapy. Clearly, longer-term studies are required to determine this. However, the cumulative probability of resistance to adefovir over one, two, three, four, and five years is almost exponential, at 0%, 3%, 11%, 18%, and 29% respectively.[47,78] These data indicate that while the rate of antiviral resistance may initially be low, once established, a progressive rise in rates of resistance is inevitable. These patients should be managed with the addition of either an L-nucleoside or entecavir. The caveat to L-nucleoside add-on therapy would be that the primary mutation to adefovir is rtN236T, and not the "shared" pathway mutation of rtA181T/V.

Entecavir Resistance

Entecavir is a potent oral antiviral agent in treatment-naive patients. *In vitro* studies have shown it to be 100 to 300 times more potent than lamivudine.[31] The five-year genotypic resistance rate with entecavir monotherapy in the treatment-naive patients is only 1.2%. In this setting, add-on acyclic phosphonate therapy (adefovir or tenofovir) should be initiated as there is no cross-resistance.

Clinical Consequences of Antiviral Drug Resistance

The potential consequences of antiviral drug resistance are outlined in Table 10.

There are a number of potential public health concerns with the emergence of antiviral drug resistance. Firstly, the transmission of HBV infection with drug resistant mutants to previously seronegative patients has been reported.[83] This is of concern as the efficacy of first-line treatment options may be affected in these patients. Secondly, due to the overlapping reading frame between the polymerase and envelope genes (HBsAg), mutations selected in the polymerase during oral antiviral therapy can also result in mutant HBsAg. It has been demonstrated that

Table 10. Potential Consequences of Antiviral Drug Resistance in Chronic HBV[45,46,79–81]

Virologic	Virologic breakthrough and rebound Reduced HBeAg seroconversion rates
Biochemical	Biochemical breakthrough
Histologic	Histologic progression of disease
Clinical	Hepatic flare and decompensation Increased recurrence post liver transplantation (viral load is strongest predictor of HBV recurrence post liver transplant)[82]
Public health	Alteration in HBsAg antigenicity Transmission of drug resistant HBV Development of multi-drug resistant HBV population

rtM204V + rtL180M + rtV173L selected during lamivudine therapy results in HBsAg (sE164D + sI195M) with altered antigenicity, and that this HBsAg has reduced anti-HBs binding *in vitro*.[84] The implication is that these mutants could represent vaccine escape mutants.

Prevention of Antiviral Resistance

The prevention of antiviral drug resistance can be considered in terms of appropriate initiation of therapy, selection of initial antiviral agent, and close monitoring. Adherence to current treatment consensus guidelines avoids unnecessary treatment. For example, treatment should be deferred in patients in the immune-tolerant phase of infection. On initiation of therapy, drugs with high potency and a high genetic barrier are preferable. Furthermore, to increase the pharmacological barrier to resistance, patients should be counseled on the importance of adherence to therapy as well as on appropriate dosing in relation to food (to maximize drug bioavailability).

Once on therapy, patients should be monitored with regular three-monthly viral loads. Ideally, sequencing for genotypic resistance should be performed in patients who are adherent to therapy and develop a virologic breakthrough. Alternative therapy should be considered in the setting of primary antiviral treatment failure, and salvage therapy with virologic breakthrough associated with genotypic resistance. Sequential monotherapy in the setting of antiviral resistance should be avoided.

Conclusion

Antiviral drug resistance in chronic hepatitis B poses a major clinical challenge and public health issue. Clinicians need to understand the emerging patterns of drug resistance so as to be able to determine the most appropriate salvage add-on treatment. These patterns are in general predictable and should be considered advantageous, rather than a hindrance to treating physicians. Ongoing updates of algorithms incorporating indications for treatment, new antiviral drugs, viral-load monitoring, and use of genotypic sequencing are required. Only in this way can the impact of potential future multi-drug resistance be minimized.

References

1. Lee WM. (1997) Hepatitis B virus infection. *N Engl J Med* **337**:1733–1745.
2. Gust ID. (1996) Epidemiology of hepatitis B infection in the Western Pacific and South East Asia. *Gut* **38** (Suppl 2): S18–23.
3. Okuda H. (2007) Hepatocellular carcinoma development in cirrhosis. *Best Pract Res Clin Gastroenterol* **21**: 161–173.
4. World Health Organization. Fact sheet no. 204. Hepatitis B. (2000). Available at: http://www.who.int/mediacentre/factsheets/fs204/en/.
5. Liaw YF, Sung JJ, Chow WC, *et al.* (2004) Lamivudine for patients with chronic hepatitis B and advanced liver disease. *N Engl J Med* **351**: 1521–1531.
6. Locarnini S. (2004) Molecular virology of hepatitis B virus. *Semin Liver Dis* **24** (Suppl 1): 3–10.
7. Tuttleman JS, Pourcel C, Summers J. (1986) Formation of the pool of covalently closed circular viral DNA in hepadnavirus-infected cells. *Cell* **47**: 451–460.
8. Zhang YY, Zhang BH, Theele D, *et al.* (2003) Single-cell analysis of covalently closed circular DNA copy numbers in a hepadnavirus-infected liver. *Proc Natl Acad Sci USA* **100**: 12372–12377.
9. Newbold JE, Xin H, Tencza M, *et al.* (1995) The covalently closed duplex form of the hepadnavirus genome exists in situ as a heterogeneous population of viral minichromosomes. *J Virol* **69**: 3350–3357.

10. Zhu Y, Yamamoto T, Cullen J, *et al.* (2001) Kinetics of hepadnavirus loss from the liver during inhibition of viral DNA synthesis. *J Virol* **75**: 311–322.
11. Addison WR, Walters KA, Wong WW, *et al.* (2002) Half-life of the duck hepatitis B virus covalently closed circular DNA pool *in vivo* following inhibition of viral replication. *J Virol* **76**: 6356–6363.
12. Ganem D, Prince AM. (2004) Hepatitis B virus infection-natural history and clinical consequences. *N Engl J Med* **350**: 1118–1129.
13. Delmas J, Schorr O, Jamard C, *et al.* (2002) Inhibitory effect of adefovir on viral DNA synthesis and covalently closed circular DNA formation in duck hepatitis B virus-infected hepatocytes *in vivo* and *in vitro*. *Antimicrob Agents Chemother* **46**: 425–433.
14. Kock J, Baumert TF, Delaney WEt, *et al.* (2003) Inhibitory effect of adefovir and lamivudine on the initiation of hepatitis B virus infection in primary tupaia hepatocytes. Hepatology **38**: 1410–1418.
15. Sung JJ, Wong ML, Bowden S, *et al.* (2005) Intrahepatic hepatitis B virus covalently closed circular DNA can be a predictor of sustained response to therapy. *Gastroenterology* **128**: 1890–1897.
16. Girones R, Miller RH. (1989) Mutation rate of the hepadnavirus genome. *Virology* **170**: 595–597.
17. Nowak MA, Bonhoeffer S, Hill AM, *et al.* (1996) Viral dynamics in hepatitis B virus infection. *Proc Natl Acad Sci USA* **93**: 4398–4402.
18. Ratnam D, Visvanathan K. (2008) New concepts in the immunopathogenesis of chronic hepatitis B: the importance of the innate immune system. *Hepatology International* **2** (Suppl 1): 12–18.
19. Visvanathan K, Skinner NA, Thompson AJ, *et al.* (2007) Regulation of Toll-like receptor-2 expression in chronic hepatitis B by the precore protein. *Hepatology* **45**: 102–110.
20. Riordan SM, Skinner NA, Kurtovic J, *et al.* (2006) Toll-like receptor expression in chronic hepatitis C: correlation with pro-inflammatory cytokine levels and liver injury. *Inflamm Res* **55**: 279–285.
21. Chen RY, Edwards R, Shaw T, *et al.* (2003) Effect of the G1896A precore mutation on drug sensitivity and replication yield of lamivudine-resistant HBV *in vitro*. *Hepatology* **37**: 27–35.
22. Tacke F, Gehrke C, Luedde T, *et al.* (2004) Basal core promoter and precore mutations in the hepatitis B virus genome enhance replication efficacy of Lamivudine-resistant mutants. *J Virol* **78**: 8524–8535.
23. Thompson AJ, Ayres A, Yuen L, *et al.* (2007) Lamivudine resistance in patients with chronic hepatitis B: role of clinical and virological factors. *J Gastroenterol Hepatol* **22**: 1078–1085.
24. Shin JW, Chung YH, Choi MH, *et al.* (2005) Precore stop codon mutation of hepatitis B virus is associated with low breakthrough rate following long-term lamivudine therapy. *J Gastroenterol Hepatol* **20**: 844–849.

25. Niederau C, Heintges T, Lange S, *et al.* (1996) Long-term follow-up of HBeAg-positive patients treated with interferon alfa for chronic hepatitis B. *N Engl J Med* **334**: 1422–1427.
26. Hoofnagle JH, Peters M, Mullen KD, *et al.* (1988) Randomized, controlled trial of recombinant human alpha-interferon in patients with chronic hepatitis B. *Gastroenterology* **95**: 1318–1325.
27. Tine F, Liberati A, Craxi A, *et al.* (1993) Interferon treatment in patients with chronic hepatitis B: a meta-analysis of the published literature. *J Hepatol* **18**: 154–162.
28. Wong DK, Cheung AM, O'Rourke K, *et al.* (1993) Effect of alpha-interferon treatment in patients with hepatitis B e antigen-positive chronic hepatitis B. A meta-analysis. *Ann Intern Med* **119**: 312–323.
29. Boni C, Bertoletti A, Penna A, *et al.* (1998) Lamivudine treatment can restore T cell responsiveness in chronic hepatitis B. *J Clin Invest* **102**: 968–975.
30. Lok AS, McMahon BJ. (2007) Chronic hepatitis B. *Hepatology* **45**: 507–539.
31. Levine S, Hernandez D, Yamanaka G, *et al.* (2002) Efficacies of entecavir against lamivudine-resistant hepatitis B virus replication and recombinant polymerases *in vitro*. *Antimicrob Agents Chemother* **46**: 2525–2532.
32. Shaw T, Locarnini SA. (1995) Hepatic purine and pyrimidine metabolism: implications for antiviral chemotherapy of viral hepatitis. *Liver* **15**: 169–184.
33. Yuen MF, Sablon E, Hui CK, *et al.* (2001) Factors associated with hepatitis B virus DNA breakthrough in patients receiving prolonged lamivudine therapy. *Hepatology* **34**: 785–791.
34. Zhou T, Saputelli J, Aldrich CE, *et al.* (1999) Emergence of drug-resistant populations of woodchuck hepatitis virus in woodchucks treated with the antiviral nucleoside lamivudine. *Antimicrob Agents Chemother* **43**: 1947–1954.
35. Poch O, Sauvaget I, Delarue M, Tordo N. (1989) Identification of four conserved motifs among the RNA-dependent polymerase encoding elements. *EMBO J* **8**: 3867–3874.
36. Lai CL, Leung N, Teo EK, *et al.* (2005) A 1-year trial of telbivudine, lamivudine, and the combination in patients with hepatitis B e antigen-positive chronic hepatitis B. *Gastroenterology* **129**: 528–536.
37. Leung N. (2008) Recent data on treatment of chronic hepatitis B with nucleos(t)ide analogues. *Hepatology International* **2**: 163–178.
38. Delaney WEt, Yang H, Westland CE, *et al.* (2003) The hepatitis B virus polymerase mutation rtV173L is selected during lamivudine therapy and enhances viral replication *in vitro*. *J Virol* **77**: 11833–11841.
39. Ono SK, Kato N, Shiratori Y, *et al.* (2001) The polymerase L528M mutation cooperates with nucleotide binding-site mutations, increasing hepatitis B virus replication and drug resistance. *J Clin Invest* **107**: 449–455.
40. Angus P, Vaughan R, Xiong S, Yang H, Delaney W, Gibbs C, Brosgart C, Colledge D, Edwards R, Ayres A, Bartholomeusz A, Locarnini S. Resistance to

adefovir dipivoxil therapy associated with the selection of a novel mutation in the HBV polymerase. Gastroenterology 2003; 125: 292–297.

41. Das K, Xiong X, Yang H, *et al.* (2001) Molecular modeling and biochemical characterization reveal the mechanism of hepatitis B virus polymerase resistance to lamivudine (3TC) and emtricitabine (FTC). *J Virol* **75**: 4771–4779.
42. Doo E, Liang TJ. (2001) Molecular anatomy and pathophysiologic implications of drug resistance in hepatitis B virus infection. *Gastroenterology* **120**: 1000–1008.
43. Naeger LK, Miller MD. (2001) Mechanisms of HIV-1 nucleoside reverse transcriptase inhibitor resistance: is it all figured out? *Curr Opin Investig Drugs* **2**: 335–339.
44. Bartholomeusz A, Tehan BG, Chalmers DK. (2004) Comparisons of the HBV and HIV polymerase, and antiviral resistance mutations. *Antivir Ther* **9**: 149–160.
45. Lai CL, Dienstag J, Schiff E, *et al.* (2003) Prevalence and clinical correlates of YMDD variants during lamivudine therapy for patients with chronic hepatitis B. *Clin Infect Dis* **36**: 687–696.
46. Leung NW, Lai CL, Chang TT, *et al.* (2001) Extended lamivudine treatment in patients with chronic hepatitis B enhances hepatitis B e antigen seroconversion rates: results after 3 years of therapy. *Hepatology* **33**: 1527–1532.
47. Hadziyannis SJ, Tassopoulos NC, Heathcote EJ, *et al.* (2005) Long-term therapy with adefovir dipivoxil for HBeAg-negative chronic hepatitis B. *N Engl J Med* **352**: 2673–2681.
48. Gish RG, Lok AS, Chang TT, *et al.* (2007) Entecavir therapy for up to 96 weeks in patients with HBeAg-positive chronic hepatitis B. *Gastroenterology* **133**: 1437–1444.
49. Lim SG, Ng TM, Kung N, *et al.* (2006) A double-blind placebo-controlled study of emtricitabine in chronic hepatitis B. *Arch Intern Med* **166**: 49–56.
50. Gish RG, Trinh H, Leung N, *et al.* (2005) Safety and antiviral activity of emtricitabine (FTC) for the treatment of chronic hepatitis B infection: a two-year study. *J Hepatol* **43**: 60–66.
51. Marcellin P, Buti M, Krastev Z, *et al.* (2007) A randomized, double-blind, comparison of tenofovir DF (TDF) for the treatment of HBeAg-negative chronic hepatitis B (CHB): study GS-US-174-0102. *Hepatology* **46** (Supp 1) 80A.
52. Chen CH, Wang JH, Lee CM, *et al.* (2006) Virological response and incidence of adefovir resistance in lamivudine-resistant patients treated with adefovir dipivoxil. *Antivir Ther* **11**: 771–778.
53. Lampertico P, Vigano M, Manenti E, *et al.* (2007) Low resistance to adefovir combined with lamivudine: a 3-year study of 145 lamivudine-resistant hepatitis B patients. *Gastroenterology* **133**: 1445–1451.
54. Colonno R, Rose R, Pokornowski K, *et al.* (2007) Four year assessment of entecavir resistance in nucleoside naive and lamivudine refractory patients. *Journal of Hepatology* **46** (Suppl 1) S294.

55. Locarnini S. (2008) Primary resistance, multidrug resistance, and cross-resistance pathways in HBV as a consequence of treatment failure. *Hepatology International* **2**: 147–151.
56. Tenney DJ, Rose RE, Baldick CJ, *et al.* (2007) Two-year assessment of entecavir resistance in Lamivudine-refractory hepatitis B virus patients reveals different clinical outcomes depending on the resistance substitutions present. *Antimicrob Agents Chemother* **51**: 902–911.
57. Hoofnagle JH, Doo E, Liang TJ, *et al.* (2007) Management of hepatitis B: summary of a clinical research workshop. *Hepatology* **45**: 1056–1075.
58. Iloeje UH, Yang HI, Su J, *et al.* (2006) Predicting cirrhosis risk based on the level of circulating hepatitis B viral load. *Gastroenterology* **130**: 678–686.
59. Lai CL, Gane E, Liaw YF, *et al.* (2007) Telbivudine versus lamivudine in patients with chronic hepatitis B. *N Engl J Med* **357**: 2576–2588.
60. Han SH, Lai CL, Gane E, *et al.* (2007) Telbivudine GLOBE Trial at Year Two: efficacy, safety, and predictors of outcome in patients with chronic hepatitis B digestive diseases week. Washington DC.
61. Keeffe EB, Dieterich DT, Han SH, *et al.* (2006) A treatment algorithm for the management of chronic hepatitis B virus infection in the United States: an update. *Clin Gastroenterol Hepatol* **4**: 936–962.
62. Proceedings of the European Association for the Study of the Liver (EASL) International Consensus Conference on Hepatitis B. (2003) September 14–16, 2002. Geneva, Switzerland. *J Hepatol* 39 (Suppl 1): S1–235.
63. Liaw YF, Leung N, Kao JH, *et al.* (2008) Asian-Pacific consensus statement of the management of chronic hepatitis B: a 2008 update. *Hepatology International* **2:** 263–283.
64. Lau GK, Piratvisuth T, Luo KX, *et al.* (2005) Peginterferon Alfa-2a, lamivudine, and the combination for HBeAg-positive chronic hepatitis B. *N Engl J Med* **352**: 2682–2695.
65. Janssen HL, van Zonneveld M, Senturk H, *et al.* (2005) Pegylated interferon alfa-2b alone or in combination with lamivudine for HBeAg-positive chronic hepatitis B: a randomised trial. *Lancet* **365**: 123–129.
66. Chan HL, Leung NW, Hui AY, *et al.* (2005) A randomized, controlled trial of combination therapy for chronic hepatitis B: comparing pegylated interferon-alpha2b and lamivudine with lamivudine alone. *Ann Intern Med* **142**: 240–250.
67. Marcellin P, Lau GK, Bonino F, *et al.* (2004) Peginterferon alfa-2a alone, lamivudine alone, and the two in combination in patients with HBeAg-negative chronic hepatitis B. *N Engl J Med* **351**: 1206–1217.
68. Schiff ER, Dienstag JL, Karayalcin S, *et al.* (2003) Lamivudine and 24 weeks of lamivudine/interferon combination therapy for hepatitis B e antigen-positive chronic hepatitis B in interferon nonresponders. *J Hepatol* **38**: 818–826.
69. Sung JJ, Lai JY, Zeuzem S, *et al.* (2008) Lamivudine compared with lamivudine and adefovir dipivoxil for the treatment of HBeAg-positive chronic hepatitis B. *J Hepatol* **48**: 728–735.

70. Locarnini S, Hatzakis A, Heathcote J, *et al.* (2004) Management of antiviral resistance in patients with chronic hepatitis B. *Antivir Ther* **9**: 679–693.
71. Degertekin B, Lok AS. (2008) Monitoring antiviral resistance in patients receiving nucleos(t)ide analog therapies for hepatitis B: which method should be used? *J Hepatol* **48**: 892–894.
72. Saldanha J, Gerlich W, Lelie N, *et al.* (2001) An international collaborative study to establish a World Health Organization international standard for hepatitis B virus DNA nucleic acid amplification techniques. *Vox Sang* **80**: 63–71.
73. Warner N, Locarnini S. (2008) The antiviral drug selected hepatitis B virus rtA181T/sW172* mutant has a dominant negative secretion defect and alters the typical profile of viral rebound. *Hepatology* **48**: 88–98.
74. Lee YS, Suh DJ, Lim YS, *et al.* (2006) Increased risk of adefovir resistance in patients with lamivudine-resistant chronic hepatitis B after 48 weeks of adefovir dipivoxil monotherapy. *Hepatology*; **43**: 1385–1391.
75. Lampertico P, Vigano M, Manenti E, *et al.* (2005) Adefovir rapidly suppresses hepatitis B in HBeAg-negative patients developing genotypic resistance to lamivudine. *Hepatology* **42**: 1414–1419.
76. Van Bommel F, Zollner B, Sarrazin C, *et al.* (2006) Tenofovir for patients with lamivudine-resistant hepatitis B virus (HBV) infection and high HBV DNA level during adefovir therapy. *Hepatology* **44**: 318–325.
77. Tenney DJ, Levine SM, Rose RE, *et al.* (2004) Clinical emergence of entecavir-resistant hepatitis B virus requires additional substitutions in virus already resistant to lamivudine. *Antimicrob Agents Chemother* **48**: 3498–3507.
78. Marcellin P, Chang TT, Lim SG, *et al.* (2003) Adefovir dipivoxil for the treatment of hepatitis B e antigen-positive chronic hepatitis B. *N Engl J Med* **348**: 808–816.
79. Liaw YF, Chien RN, Yeh CT, *et al.* (1999) Acute exacerbation and hepatitis B virus clearance after emergence of YMDD motif mutation during lamivudine therapy. *Hepatology* **30**: 567–572.
80. Dienstag JL, Goldin RD, Heathcote EJ, *et al.* (2003) Histological outcome during long-term lamivudine therapy. *Gastroenterology* **124**: 105–117.
81. Mutimer D, Pillay D, Shields P, *et al.* (2000) Outcome of lamivudine resistant hepatitis B virus infection in the liver transplant recipient. *Gut* **46**: 107–113.
82. Mutimer D, Pillay D, Dragon E, *et al.* (1999) High pre-treatment serum hepatitis B virus titre predicts failure of lamivudine prophylaxis and graft re-infection after liver transplantation. *J Hepatol* **30**: 715–721.
83. Thibault V, Aubron-Olivier C, Agut H, Katlama C. (2002) Primary infection with a lamivudine-resistant hepatitis B virus. *AIDS* **16**: 131–133.
84. Torresi J. (2002) The virological and clinical significance of mutations in the overlapping envelope and polymerase genes of hepatitis B virus. *J Clin Virol* **25**: 97–106.

Chapter 6

Virion Secretion of Naturally Occurring Core Antigen Variants of Human Hepatitis B Virus

*Chiaho Shih**

Introduction

Human hepatitis B virus (HBV) encodes a core antigen (HBc or HBcAg) with a molecular weight near 22 kD. This HBc protein plays multiple roles in HBV life cycle. It can self-assemble into an icosahedral particle, which provides a shell to protect the encapsidated RNA pregenome (pgRNA) and the reverse transcribed DNA genome. The HBc protein is also known to be able to bind DNA and RNA through the arginine rich domain at its C-terminus. A very important role of HBc capsids is to serve as a vehicle for intracellular trafficking of HBV genome, including the import of relaxed circular DNA (RC) to the nucleus, and targeting to the endoplasmic reticulum for envelope formation and virion secretion (reviewed in Ref. 1). HBc is also known to be a major target for cytotoxic T lymphocytes (CTL). Therefore, HBc is highly versatile and plays an important role in immune-mediated clearance and pathogenesis (reviewed in Ref. 2; also see the chapter by Milich in this book). For the readers interested in HBsAg and HBeAg variants, the chapter by David Milich is recommended.

* Institute of Biomedical Sciences, Academia Sinica, Taipei, Taiwan. E-mail: cshih@ibms.sinica.edu.tw

Highly Frequent Naturally Occurring HBV Core Variants

HBV Core Mutations Associated with Liver Damage

In chronic hepatitis B patients, HBc has been found to accumulate a lot of mutations, including internal deletion and missense mutations.[3–6] A constant frustration for researchers in the field is to elucidate the functional significance of these naturally occurring HBc mutations.[7] Often, no strong phenotype can be associated with these HBc variants. This is in part due to the highly versatile roles of HBc. It is difficult to predict what kind of available assays should be used for a particular kind of variant in order to obtain the most informative results.

The most frequent mutation of HBc occurred at amino acid 97, changing from an isoleucine to a leucine (I97L) in HBV genotypes B and C (e.g., references cited in Ref. 8). This HBc mutation was first found in the serum samples of Japanese hepatitis B patients.[3] HBV core DNA sequences were compared between 10 asymptomatic healthy carriers and 20 patients with chronic liver disease. No significant changes in the nucleotide or amino acid sequences were noted in the 10 healthy carriers. However, clustering changes in amino acids 84–101 were found in 15 of the 20 chronic liver disease patients. These 15 patients with significant HBc mutations had advanced liver disease, such as chronic active hepatitis and cirrhosis, whereas the other five patients without significant HBc mutations exhibited only mild liver disease, such as chronic persistent hepatitis. HBc 84–101, which includes amino acid 97, is likely to be a target site for cytotoxic T lymphocytes.[3] A similar study was performed by using different clinical cohorts, including five fulminant, 10 severe fatal exacerbation, and nine self-limited acute hepatitis patients.[4] No significant mutation was found in the nine self-limited acute hepatitis patients. In contrast, significant mutations clustering HBc 84–99 or HBc 48–60 were found in 14 out of 15 patients with fulminant hepatitis or severe exacerbation. These data suggest that these core variants may be the cause of acute exacerbation and severe liver disease. Alternatively, they might be the consequence of immune attack and represent survivors of immune selection.

To address the cause-effect or temporal relationship between the emergence of HBV variants and the onset of liver disease, Dr. Chun-Jen Liu from Professor Ding-Shinn Chen's group at National Taiwan University Hospital performed prospective analysis of full-length HBV sequences from 14 patients before, during, and after a single episode of acute exacerbation (AE). It was concluded that most exacerbations were preceded by an upsurge of serum HBV identical to the pre-existing HBV strain at baseline. After exacerbation, about half of the patients were repopulated by a different dominant strain of HBV variants, suggesting a role of immune selection for the emergence of HBV variants.[9] By transfection analysis, they detected no difference in the replication potential between HBV strains at baseline versus those emerged after exacerbation. Liu also compared full-length sequences of HBV genomes in the longitudinal samples of patients with consecutive episodes of acute exacerbation.[10] In two patients with repeated AE, viral variants that emerged after each AE were different, and it was proposed that the AE could be caused by the dysfunction of host immune surveillance leading to reactivation of pre-existing virus.[10] The exact mechanism of AE with regard to host immunity and HBV variants, particularly core variants, deserves further investigation.

HBV Core Mutations Associated with Hepatoma Patients

Before 1994, most if not all the studies of HBV variants had always been based on serum samples from patients. Seiyu Hosono, a graduate student from Japan in my laboratory at the University of Pennsylvania, Philadelphia, was interested in comparing HBV variants between the liver and serum compartments. In collaboration with Professors Chue-Shue Lee and Czau-Siung Yang at the National Taiwan University Hospital, Hosono extracted total DNA from paired samples of hepatocellular carcinoma (HCC) and adjacent non-tumor liver tissues from Taiwanese HCC patients, and pre-screened for HBV-replicating liver samples by Southern blot analysis. It was a general misconception that HCC contains no active HBV DNA replication, when in fact approximately 25% of HCC samples (7/27) exhibited free replicative forms of

HBV DNA.[11] In part because the liver is a highly vasculated organ, replicating HBV DNA present in HCC samples was often interpreted as a cross-contamination from the adjacent non-tumor tissues. However, in some rare occasions, we observed a much stronger HBV DNA replication signal in the HCC tumors, and much weaker or no HBV DNA signal in the adjacent non-tumors (C. Shih and S. Hosono, unpublished results). The HBc gene from liver samples positive for viral replication were further examined by Hosono using PCR cloning and sequencing. Approximately 70% of HBV-related hepatoma samples could detect a few copies of integrated HBV DNA.[12] However, relative to the great majority of the replicating HBV DNA, the integrated HBV DNA represents an insignificant minority, and thus could not be scored after PCR amplification, cloning, and sequencing. In our comparative studies, integrated HBV DNAs usually contain an average of fourfold less frequent mutation than the replicating HBV DNA in hepatomas, suggesting that integration in general probably occurs at earlier clinical stages.[6]

During his initial studies of HBV pre-core hotspot mutations at pre-core codon 28, Hosono accidentally noted on his sequencing gel another frequent core mutation P5T at HBc codon 5, changing from proline to threonine. This P5T mutation is only five amino acids (about 15 nucleotides) away from the well-known pre-core mutational hotspot, and therefore could not go unnoticed by Hosono. This finding was rather unexpected for us at that time, since frequent core mutations were supposed to cluster only within the three so-called mutation clustering regions (MCR): HBc 48–60, HBc 84–101, and HBc 147–155 (Fig. 1).[3,4,13,14] In his subsequent studies of the entire core gene, Hosono identified a few additional frequent mutations outside the MCR, including mutations V13L, V13A, and P130T. These mutations were often present in both hepatoma and adjacent non-tumor liver tissues. We reported these novel and frequent mutations of HBc at the 1994 International Meeting of the Molecular Biology of Hepatititis B Virus at Pasteur Institute (Fig. 1).[5,6,15] Here, an important question to ask is, is it

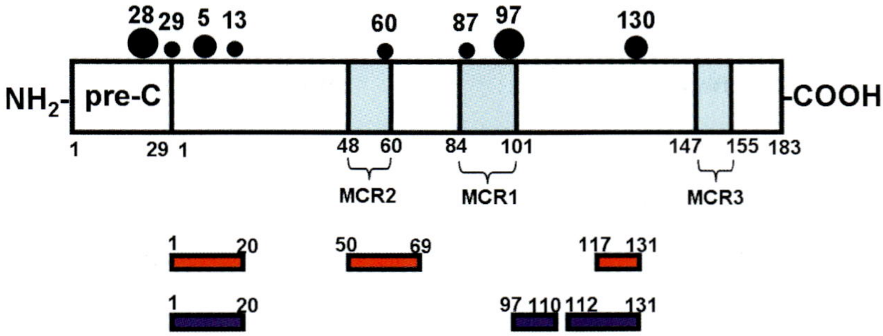

Fig. 1. Mutational hotspots of HBV core antigen (HBc) in chronic carriers. The missense mutation from phenylalanine or isoleucine to leucine at amino acid 97 of HBc is the most frequent naturally occurring mutation. The sizes of the black dots reflect their respective estimated average mutation frequencies. Large dots, more than 45%; medium dots, 20–35% dots, and small dots, fewer than 20% of patients are found to have virus with a predominant mutation at this position (adapted from Ref. 6). For clarity, this cartoon is not drawn to scale. Green: MCR (mutation clustering regions based on Japanese patients) (Ref. 13); Red: major T cell epitopes of HBc restricted by human class II HLA (based on Italian patients) (Ref. 22); Purple: major T cell epitopes of woodchuck hepatitis B virus core antigen (Ref. 26).

possible that these novel mutations are in fact directly oncogenic or hepatoma patient–specific, and thus could not have been identified previously in the serum samples from non-hepatoma patients with chronic liver diseases?[3,4,13,14]

In the following years after the HBV meeting at the Pasteur Institute, these frequent HBc mutations, including those novel mutations outside the Japanese reported MCR regions, were found in the serum samples of Chinese hepatitis B patients without hepatomas,[16,17] as well as in African and French HCC developed on noncirrhotic livers.[18] Therefore, these frequent HBc mutations have no direct association with either hepatoma or cirrhosis. Clinically, there is no detectable difference in the mutation frequency or profiles of HBc between patients with and without HBeAg seroconversion, interferon treatment, and interferon responders versus non-responders.[17] These naturally occurring hotspot mutations of HBc are summarized in the cartoon of Fig. 1.

Immunoevasion of Naturally Occurring HBV Variants

For HBc variants to emerge in patients, there must be some selective advantages associated with such mutations. In the example of mutation I97L, since both isoleucine and leucine are hydrophobic and structural isomers, it is hard to imagine that such a subtle mutation could result in any significant effect on HBV. Because these hotspot HBc mutations coincide with T cell epitopes (Fig. 1), they could represent immune escape mutations from $CD4^+$ Th cells or $CD8^+$ CTL.[19,20,21] If so, it is reasonable to speculate that these variant epitopes could result in reduced binding with MHC molecules or reduced recognition by T cell receptors (TCR). For example, as shown in Fig. 1, mutations P5T and V13A/L coincide with a class II HLA restricted T cell epitope HBc 1–20. Similarly, mutations I59T and L60V coincide with another T cell epitope HBc 50–69, and mutation P130T coincides with a T cell epitope HBc 117–131.[6,22] While these three T cell epitopes are all class II HLA restricted, we noted that P130T coincides with HBc 130–140, which is a CTL epitope restricted by class I HLA-Cw9,[3] as well as a Th1 epitope restricted by class II HLA-DRw52.[6,23] DRw52 is a highly frequent allele found in 55% of Taiwanese[24] and 78% of Italians.[25] The core antigens of HBV and woodchuck hepatitis B virus (WHV) are highly homologous to each other. In the woodchuck animal model, T cell epitope mapping with overlapping peptides covering the entire WHV core antigen (WHcAg) revealed T cell responses of acutely WHV-infected woodchucks to WHcAg 1–20, WHcAg 100–119, and WHcAg 112–131.[26] Again, mutation P5T coincides with WHcAg 1–20, and mutation P130T coincides with WHcAg 112–131. More detailed mapping revealed a very potent T cell epitope WHcAg 97–110. All woodchucks immunized with WHc 97–110 were protected from WHV infection.[26] The most frequent HBc mutation, I97L, coincides with the most potent T cell epitope, WHc 97–110. Furthermore, naturally occurring core internal deletion (CID) variants, spanning across HBc amino acid 97, were also observed in chronic hepatitis B patients,[6,8,16,27–33] as well as in woodchucks.[34,35] It is also noteworthy that an HLA-A11 restricted $CD8^+$ CTL epitope was mapped to HBc 88–96, which is adjacent to amino acid 97.[23]

Can we demonstrate further the immune escape nature of these naturally occurring HBc mutations using human T cell clones? Acute hepatitis patients can mount a strong and polyclonal immune response to HBV. However, they are very rare these days. In contrast, while chronic hepatitis B patients are prevalent, they are immunologically tolerant to HBV. Therefore, the hypothesis of immune escape mutation of naturally occurring HBc cannot be easily tested using T cell clones from chronic hepatitis B patients with highly polymorphic HLA backgrounds. Furthermore, it was reported that CTL escape variants were not common in chronic hepatitis B.[36] Considering the multi-specific and polyclonal nature of immune response to HBV in acute patients, one would expect a low probability for HBV to acquire simultaneously multiple immune escape mutations. In a recent report, intranasal infection of Balb/c mice was conducted by using three different recombinant adenoviruses: AdHBV-WT, AdHBV-L60V, and AdHBV-I97L. Infected mice produced anti-HBc efficiently to comparable levels. In contrast, AdHBV-WT induced a stronger T cell proliferation response than AdHBV-L60V and AdHBV-I97L. It was concluded that the variant HBcAg had no influence on humoral immune response, but could dampen the T cell response to HBcAg, suggesting that L60V and I97L were immune escape mutants.[37]

Similar to the core gene, frequent mutations of HBsAg accumulate in T cell epitopes restricted by HLA class I in natural infection.[38] Previously, Pei-Ching Tai, a PhD student from Taiwan in my laboratory at the University of Texas Medical Branch at Galveston, demonstrated that the internally deleted M envelope protein, encoded by a frequent naturally occurring HBV preS2 deletion variant, cannot be recognized by a neutralizing antibody in Western blot and immunofluorescence analyses.[39] This polyclonal neutralizing antibody was raised against the polypeptide of preS2 120–145 (*ayw*), and shown to be virus-neutralizing in a chimpanzee experiment.[40] Therefore, this preS2 deletion variant does appear to be a mutant escaping from immune surveillance.

In addition to immune escape from adaptive immunity (neutralizing antibody and T cell epitopes), are there any HBV variants capable of immune escape from innate immunity? This issue is particularly relevant considering the importance of cytokines in HBV clearance via a noncytolytic antiviral mechanism.[41–43] Different HBV genotypes can be considered as variants of each other. The natural history of chronic HBV infection varies among different genotypes in different geographical regions.[44] Patients infected with HBV genotypes A and B have a better response to interferon-based therapy than those with genotypes C and D.[45] Similarly, HCV genotype 1 is known to be more resistant to interferon therapy than other genotypes (see chapters in this book by Liu and Chen, Yu and Chuang, and Tanaka and Mizokami, respectively). Recently, the HBcAg of mutant I97L, compared to the WT HBcAg, was reported to suppress the interferon-induced MxA gene transcription.[46] The relationship between HBV variants and innate immunity remains to be further explored in the future.

Immature Virion Secretion

Deficiency in Viral DNA Replication of ayw *Mutant F97L*

To address the issue of the functional significance of naturally occurring HBc variants, Thomas Ta-Tung Yuan, another PhD student from Taiwan in my laboratory at the University of Pennsylvania and the University of Texas Medical Branch, set out to study an Italian HBV mutant F97L, which changes from phenylalanine to leucine at amino acid 97 of HBc.[47,48] The HBc mutant F97L of subtype *ayw* (genotype A) prevalent in Europe is the counterpart of HBc mutant I97L of subtypes *adr* and *adw* (genotypes B and C) prevalent in Southeast Asia.

Yuan created the HBc mutant F97L by site-directed mutagenesis of wild-type (WT) HBV *ayw* (a generous French gift from Dr. Pierre Tiollais at Pasteur Institute)[49] and compared the viral replication between WT HBV and its congenic derivative F97L. After transfection of these wild-type and F97L HBc mutant DNA into Huh7 hepatoma cells, he extracted core particle-associated viral DNA and performed Southern blot analysis. Because both phenylalanine and leucine are hydrophobic, it was anticipated that no major phenotypic difference between WT HBV and mutant F97L could be detected. To our surprise, mutant F97L exhibited highly significant reduction in viral DNA replication, particularly plus-strand DNA synthesis. The most surprising discovery came from his studies of the virion secretion behavior of WT HBV and mutant F97L.

The Dogma of Genome Maturation Signal for Virion Secretion

In 1982, Summers and Mason proposed a model of genome maturation signal in their studies of duck hepatitis B virus.[50] This model described the phenomenon that mature capsids containing mature genome of RC DNA can be preferentially released into the medium as mature virions (density around 1.24 g/cm^3). In contrast, immature capsids containing RNA or single strand (SS) DNA are preferentially retained inside the cells. A genome maturation signal can be generated upon capsid maturation, which can facilitate virion secretion. The selective export of virions containing mature genome has received support by several independent studies.[51–54] In addition to the enveloped infectious virions, HBV also secretes several different kinds of subviral particles, such as HBsAg particles (without genome and core protein), naked core particles (without envelope), and empty virions (without genome). To date, the rules governing the secretion of these subviral particles are less clear.

Immature Virion Secretion of ayw *Mutant F97L*

In the case of human HBc mutant F97L, Yuan observed an excessive amount of immature virions containing SS DNA in the medium, in addition to mature virions containing RC DNA (Fig. 2). Owing to its lower molecular weight, HBV SS DNA migrates faster than the partially double-strand RC DNA. A legitimate concern is whether the excessive amount of SS DNA of immature virions of mutant F97L could actually reflect an adventitious degradation product of RC DNA occurring during the experimental procedures. We consider this possibility highly unlikely for several reasons: (1) This immature secretion phenomenon was highly reproducible only in mutant F97L, and has never been observed in WT HBV; and (2) When the plus-strand specific riboprobe was used in the Southern blot analysis, only

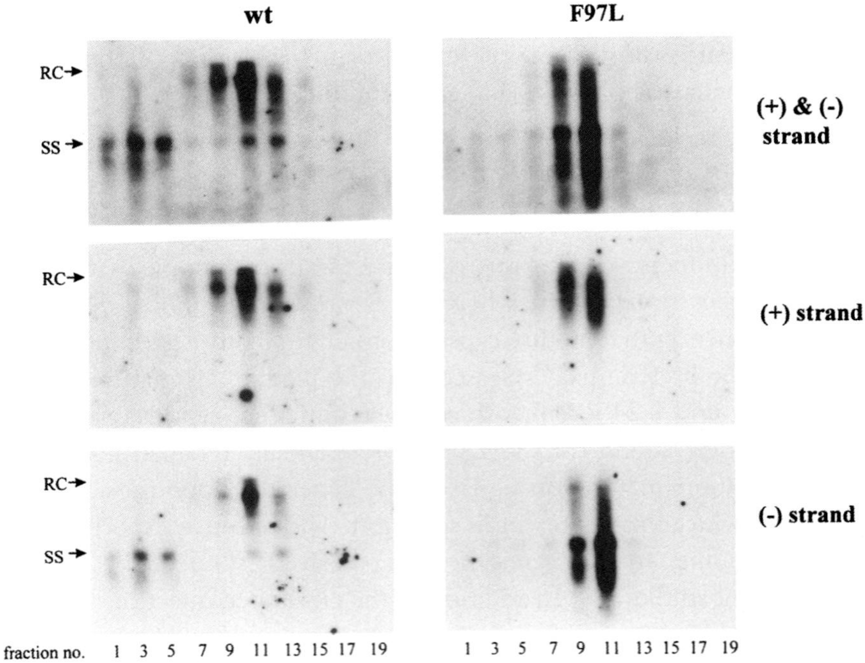

Fig. 2. Gradient centrifugation analysis of extracellular viral particles reveals the secretion of immature minus-strand, ss DNA forms in the Dane particle fractions (1.24 g/cm^3) for mutant F97L. Conditioned media were collected on days 3 and 5 post transfection. Viral particles were first purified from the media through a 20% sucrose cushion; this was followed by isopycnic centrifugation in a gradient of 20 to 50% (wt/vol) cesium chloride. The extracellular HBV DNA in each fraction was analyzed by Southern blot analysis with a ds-specific 3.1-kb HBV DNA probe (upper panels), a plus-strand-specific riboprobe (middle panels), and a minus-strand-specific riboprobe (lower panels) (adapted from Yuan *et al.*, 1999, *J Virol* **73**: 5731–5740; with permission from the American Society for Microbiology).

the higher molecular weight RC DNA of mutant F97L could be detected, and the lower molecular weight regions of SS DNA of mutant F97L contained no plus-strand specific signals on the Southern blot. This result indicates that the RC DNA of mutant F97L has remained structurally intact throughout the experimental procedures. We coined this surprising discovery of mutant F97L an "immature secretion phenotype." By a *cis-trans* genetic test, Yuan demonstrated that the change of nucleotide sequences *per se* at amino acid 97 does not contribute to immature secretion. By complementation and gradient centrifugation analyses, Yuan demonstrated that the core protein of mutant F97L alone is necessary and sufficient for immature virion secretion. Like WT HBV, the secretion of immature virions (density also around 1.24 g/cm^3) also depends on the presence of an envelope protein.[55] In brief, the pleiotropic effects of mutant F97L, including the intracellular phenotype of plus-strand DNA deficiency and the extracellular phenotype of immature secretion, can be observed reproducibly in several different experimental settings. To our knowledge, this was the first demonstration of a phenotypic defect in virion secretion of a highly frequent naturally occurring HBV variant.

The finding of immature secretion raised a number of questions: 1) Is this a general phenomenon that can be extended to other HBV genotypes? 2) Is there any cause-effect relationship between the intracellular plus-strand DNA deficiency and the extracellular immature virion secretion, as both phenotypes share a common feature of excessive amount of immature SS DNA? 3) What could be the mechanism behind the immature secretion phenomenon? 4) Is this a general phenomenon that can be extended to HBc mutations at a position other than amino acid 97?

Dissociation of Extracellular Immature Secretion from Intracellular DNA Replication in adr *Mutant I97L*

To address the first question, we repeated the experiment we had done with subtype *ayw* mutant F97L in Huh7 cells, by using subtype *adr* mutant I97L in HepG2 cells (a generous Japanese gift of *adr* plasmid was from Dr. K. Koike at the Cancer Institute, Tokyo).[56] Indeed, to our satisfaction, we could reproduce the immature secretion phenomenon with the subtype *adr* mutant I97L in HepG2 cells. However, unexpectedly, unlike with mutant F97L, Yuan observed no significant difference in viral DNA replication between WT HBV and mutant I97L in HepG2 cells. This result served to dissociate the extracellular phenotype of immature virion secretion from the intracellular phenotype of viral DNA replication. In other words, immature secretion of virions containing SS DNA of mutant I97L is not consequential to any defect in plus-strand DNA synthesis and RC DNA formation.[57]

The immature secretion phenomenon of mutant I97L described here was later confirmed by Schormann *et al.*,[58] using a genotype A HBV plasmid. Since this immature secretion phenotype is based on the tissue culture system, it is natural to ask whether a similar phenomenon may be found *in vivo*? Indeed, it was observed in woodchuck hepatitis B virus in one woodchuck treated with acyclovir.[59] Such a drug-induced immature secretion is worth further investigation in the future. Furthermore, virion-like particles containing SS DNA were also found in sera containing snow geese hepatitis B virus (SGHBV).[60] It is therefore possible that the immature secretion phenomenon can be demonstrated in patients in the near future.

Kinetic Studies of Immature Virion Secretion

Sophie Le Pogam, a postdoctoral fellow from France, joined the laboratory at a time when Yuan was finishing up his postdoctoral fellowship. She asked what could be the mechanism for immature virion secretion, if this phenotype was unrelated to intracellular viral DNA synthesis. She tested the possibility of whether immature virion secretion could actually reflect "premature virion secretion." In her hypothesis, she wondered whether the secretion of mature virions of mutant F97L could occur any earlier or faster than that of WT HBV? In addition, she wondered whether in mutant F97L the secretion of immature virions containing SS DNA could occur earlier or faster than that of mature virions containing RC DNA? The answer is neither! She did very careful comparison of virion secretion kinetics between WT HBV and mutant F97L via gradient centrifugation and Southern blot analyses.[61] She found no difference in the secretion kinetics of mature virions between WT HBV and mutant F97L. Nor did she find any difference in the secretion kinetics between mature and immature virions of mutant F97L.

A Low-Level Virion Secretion Phenotype of Core Mutants P5T and L60V

Can the immature secretion phenomenon be extended to any other naturally occurring HBc mutants, such as P5T and L60V (Fig. 1)? To address this issue, Le Pogam *et al.* characterized HBc mutants P5T and L60V by gradient centrifugation analysis. She found no immature secretion associated with mutants P5T or L60V. Instead, a completely novel phenotype dubbed "low-level secretion" or "low secretion" was found. Relative to WT HBV, a significantly reduced amount of mature virions containing RC DNA was secreted into the medium of Huh7 cells transfected with mutants P5T or L60V. Importantly, similar intracellular steady state levels of HBc protein and viral DNA synthesis were observed between WT HBV, mutants L60V and P5T. Therefore, the extracellular low-secretion phenotype is independent from the intracellular viral DNA replication and HBc protein stability. This low-secretion phenotype can be rescued by cotransfection of mutant L60V or P5T and a WT core protein expression vector. Furthermore, by complementation analysis, this low-secretion phenotype appeared to be solely dependent on the mutant core protein L60V or P5T.[62]

Since the mutant core protein is responsible for the low-secretion phenotype, is it due to the loss of the parental WT amino acid, or the acquisition of a new mutant amino acid? To address this issue, Le Pogam *et al.* characterized mutants P5T, P5A, and P5S, changing from proline to threonine, alanine, and serine, respectively. All three mutants exhibited the low-secretion phenotype. This result indicates that it is the loss of proline, rather than the acquisition of a threonine, that is directly responsible for the low-level secretion of mature virions in naturally occurring mutant P5T.[62] Based on this study, both helical (L60V and I97L) and non-helical regions (P5T) on the HBc fold are important for envelopment and virion secretion.

The low-secretion phenotype is not limited to naturally occurring variants in patients. It was also observed by engineering artificial insertions or deletions of the HBc gene. For example, an extension at the N-terminus of HBc by 10 or 23 amino acids can still allow viral DNA replication; yet, it effectively inhibited virion secretion.[63] Similarly,

Volker Bruss' laboratory from Germany reported that deletions at amino acid positions 12 or 134, or insertion at amino acid 141 of HBc, can allow formation of nucleocapsids and viral DNA synthesis, but failed to support virion secretion.[64] By single amino acid substitution, a total of 11 HBc mutants were created and shown to allow nucleocapsid formation, but blocked envelopment and virion secretion. These mutations map to a ring-like groove around the base of the spike. One of these 11 mutants from the Bruss laboratory, L60A, lent support for our previous naturally occurring mutant L60V.[62,65] Further detailed mutagenesis on 6 of these 11 mutants by the same research group confirmed that these residues are indeed important for envelopment and virion secretion.[66] Taken together, a very precise conformation of HBc particles is required for productive interaction with the envelopment machinery leading to virion secretion.

Naturally Occurring and Engineered Compensatory Mutations

Compensatory Mutation of HBc P130T for Immature Virion Secretion

HBV variants are known to contain multiple naturally occurring mutations throughout the entire genome. As shown in Fig. 1, while I97L is the most frequent mutation of HBc, mutations P5T and P130T are also highly frequent in natural infection. We asked if there is any synergistic or compensatory effect on HBV when more than one single mutations are present on the same HBc. Since I97L and P130T are frequently associated with each other in natural infection, Yuan characterized the double mutant I97L/P130T and single mutant P130T. Interestingly, the immature secretion defect of mutant I97L can be offset by another frequent HBc mutation P130T.[67] In contrast, the single mutant P130T by itself exhibited only somewhat enhanced accumulation of the full-length RC DNA, and no apparent extracellular phenotype in virion secretion at all. In summary, while HBc I97L is the first mutant known to exhibit immature virion secretion, HBc P130T is the first mutant known to be compensatory for the immature virion secretion.

Compensatory Mutation of HBc P5T for Immature Virion Secretion

As mentioned above, naturally occurring HBV variants often contain multiple mutations spreading throughout the entire genome, including core, envelope, polymerase, and X genes. Our approach so far has relied heavily on the construction of site-directed mutants in the context of a WT HBV genetic background. However, naturally occurring mutants are often in the context of an HBV variant with multiple mutations. In order to study the functional significance of an HBc mutation in a context more closely mimicking naturally occurring HBV variants, rather than mimicking the WT HBV, we acquired an HBV variant clone from Dr. Yu-Mei Wen in Shanghai, China.[68] In addition to mutation I97L, this *adr* variant clone contains multiple additional mutations, such as an HBc mutation P5T, enhancer II mutations at nucleotide 1762 and 1764, mutations truncating the X protein and abrogating the production of the M envelope protein.

Pong Kian Chua, a postdoctoral fellow from Singapore, joined the lab at the time when Le Pogam was about to complete her postdoctoral fellowship. Chua characterized this variant clone (dubbed "Shanghai *adr*" in my laboratory) by gradient centrifugation analysis. Chua was surprised when he first discovered that this Shanghai *adr* exhibited mature virion secretion like a WT HBV, despite its HBc I97L mutation. There must be a suppressor mutation somewhere on this Shanghai *adr* that can offset the immature secretion of mutation I97L. We found no P130T in Shanghai *adr*. So, where could we locate such a compensatory (suppressor) mutation? Chua's first guess was to bet on the mutation P5T, which had earlier been shown by Le Pogam to result in a low-secretion phenotype.[62]

Under this same HBV variant context of Shanghai *adr*, Chua systematically compared the virion secretions among four different genotypes: P5T/I97L, T5P/I97L, P5T/L97I, and T5P/L97I in HepG2 cells. Mutant T5P/L97I, which contains wild-type amino acids at position 5 and 97, exhibited mature virion secretion as expected from a WT HBV. Consistent with the previous work by

Le Pogam *et al.*[62] mutant P5T/L97I, which contains a single mutation at amino acid 5, exhibited a low-secretion phenotype. Again, as predicted from the previous work by Yuan *et al.*[57] mutant T5P/I97L, which contains a single mutation at amino acid 97, exhibited immature secretion. Most interestingly, the double mutant P5T/I97L, which is the parental genotype of Shanghai *adr*, exhibited neither immature nor low secretion. Instead, it behaved like a WT HBV with mature virion secretion. This experiment clearly established mutation P5T as another compensatory mutation for immature virion secretion.[69] Conversely, I97L can be considered as a compensatory mutation for the low-secretion phenotype of mutation P5T. In other words, I97L and P5T appeared to be mutually compensatory for each other, at least in the variant context of Shanghai *adr*. Can this result be extended to the WT HBV context? Indeed, Chua demonstrated that the mutually compensatory effect between I97L and P5T can be reproduced in a WT *adr* context (dubbed Tokyo *adr*).[56]

Engineered Compensatory Mutation pre-S1-A119F for Immature Virion Secretion.

Only approximately 76% of HBV 97L variants also contain compensatory mutations P5T or P130T.[69] Since core-envelope interaction is believed to be important for envelopment and virion secretion, we speculated that some naturally compensatory mutations could actually reside in the envelope genes (HBs). To further test this possibility, Le Pogam introduced mutations into the cytoplasmic loops of the preS1 domain of the large envelope protein. She focused on pre-S1 hydrophobic amino acids 108, 112, and 119, and identified a compensatory mutant preS1-A119F, which can successfully rescue the immature secretion of mutant I97L. Other mutants, such as A119I, A119L, I108F, L112I, and L112F, failed to rescue the immature secretion of mutant I97L.[61] Therefore, the putative intermolecular interaction between core and the preS1 domain of the large envelope protein plays an important role in HBV virion secretion.

Compensatory Envelope Mutation W74L for the Low Virion Secretion of Envelope Mutant L77R

Encouraged by Le Pogam's successful demonstration of the compensatory effect of the engineered preS1-A119F mutation on the immature virion secretion of I97L, Pong Kian Chua sought to identify naturally occurring compensatory mutations in the envelope protein gene. Although we did not find any mutations similar to preS1-A119F in the envelope genes, we studied instead two novel naturally occurring mutations (W74L and L77R) in the cytosolic loop of the small S envelope protein. Like the low-secretion phenotype of HBc mutation P5T, HBs mutation L77R alone resulted in more than 10-fold reduced secretion of virions. Further characterizations of mutant L77R revealed multiple phenotypes: (1) significantly reduced levels of secreted virions; (2) reduced levels of intracellular and extracellular HBsAg by enzyme immunoassay; (3) increased accumulation of intracellular small S envelope protein by Western blot; (4) enhanced intracellular accumulation of full-length RC form DNA; (5) highly restricted distribution of HBsAg in the ER and Golgi. In contrast to HBsAg, the secretion of HBeAg was normal in L77R-transfected cells. All of these aforementioned pleiotropic phenotypes of mutant L77R can be rescued by a second naturally occurring S envelope mutation nearby, W74L.[70] Similar to our mutation L77R, mutation G145R was reported to result in a few-fold reduction of virion secretion.[71,72] Unlike L77R on the cytosolic loop, mutation G145R is located topologically within an immuno-dominant loop (HBs 101–163) on the lumen side of the ER. Like G145R, additional naturally occurring mutations in the small envelope gene, such as I110M and G119E, also resulted in defective virion secretion.[73] It was also noted that both I110M and G119E exhibited greatly reduced levels of HBsAg, which is known to be required for virion assembly and secretion.[74–76] Another mutation M133T, changing from methionine to threonine, was reported to secrete virions with a higher efficiency, and is compensatory for the virion secretion defect of mutants I110M and G119E.[73] Mutation M133T is highly frequent in HBsAg-negative patients.[77,78] In summary, virion secretion defect is observed not only in naturally occurring core variants, but also in envelope variants.

Furthermore, we and others demonstrated the existence of naturally occurring compensatory mutations of HBV core and envelope genes for naturally occurring virion secretion defect.

Host Factors and ESCRT Involved in Virion Secretion

As discussed so far, HBV virion secretion depends on productive core-envelope interaction. In addition to viral factors, are there any host factors that could influence virion secretion? Chua would like to pursue this issue. A few years earlier, it was demonstrated that a host factor Vps4 is important for HIV virion secretion.[79] Vps4 is a cellular ATPase known to be involved in ESCRT (endosomal sorting complex required for transport). However, HIV and HBV are very different viruses, despite the fact that both contain a reverse transcriptase. Virion secretion of HIV is mediated by budding through plasma membrane, while HBV is mediated by budding through the intracellular membrane. By cotransfection of a dominant negative Vps4 mutant K173Q (or E228Q) with an HBV genomic dimer, Chua demonstrated that both intracellular HBV viral DNA replication and extracellular virion secretion were greatly reduced when cotransfected at a ratio as high as 20 to 1 (e.g., 10 μg HBV: 0.5 μg Vps4-K173Q). This inhibitory effect was observed in both *adr* and *ayw* subtypes in both Huh7 and HepG2 cells.[80] The potent reduction effect on HBV by mutant Vps4 is not due to cytotoxicity, as shown by the MTT assay. Our results of Vps4 on HBV virion secretion were confirmed by two independent reports in the following year.[81,82] In addition to Vps4, Reinhild Prange's group from Germany demonstrated that ESCRT-III complex-forming CHMP proteins were essential to virion secretion. Depletion or overexpression of γ2-adaptin, a ubiquitin receptor, also resulted in the inhibition of HBV virion secretion.[81,83] It is hypothesized that Nedd4, a ubiquitin ligase, could recognize the PPAY motif on HBV core particles, and target HBV core particles to the membrane of the endosomal multivesicular body (MVB) for budding via the bridging effect of γ2-adaptin.

This dose-dependent effect of Vps4 on intracellular HBV DNA replication was somewhat surprising to us initially,[80] as Vps4 has no known effect on HIV-1 intracellular replication.[79] By endogenous polymerase assay (EPR), HBV DNA synthesis was shown to be significantly diminished by dominant negative mutant Vps4.[81] However, using a dot blot analysis, Prange's group found no inhibitory effect of mutant Vps4 on HBV DNA replication.[81] It remains to be determined whether the discrepancy could be related to differences in the methods being used (e.g., dot blot by Lambert versus Southern blot by Chua; or dot blot by Lambert versus EPR by Lambert). It is also worth mentioning that a similar effect of mutant Vps4 on both intracellular nucleocapsid formation (HBV DNA replication) and extracellular virion secretion was reported independently by Paul Ahlquist's group from the University of Wisconsin.[82] In addition to Vps4, dominant negative mutant of ALIX/AIP1, another MVB protein, was also found to inhibit HBV virion secretion.[82] However, most recently, Prange reported that HBV egress does not require ALIX.[84] The discrepancy in the requirement of ALIX for HBV virion secretion remains to be clarified.

Replication Advantage and Host Factor Independent Phenotype of Mutant I97L

As shown in Fig. 1, core 97L mutation is the most frequent naturally occurring mutation, followed by mutations P5T and P130T.[6] Although we have demonstrated that 97L mutation can produce an immature secretion phenotype, we have no idea whether the immature virions are infectious or not. This question is hampered by two issues: (1) lack of an *in vitro* HBV infection system; (2) difficulty in physical separation of immature from mature virions. If we assume that the immature virions containing SS DNA of mutant 97L are less or not infectious, then it is conceivable that compensatory mutations P5T and P130T would play an important role in restoring the infectivity of mutant I97L virions.[67,69] Similarly, it makes sense that the low secretion phenotype of core mutation P5T can be compensated by core mutation I97L.[62,69] If immature virions are less or not infectious, then why should mutation I97L emerge to predominance in chronic hepatitis patients in the first place? One potential selective advantage of mutation I97L could be immune escape, since amino acid 97 of HBc coincides with a potent T cell epitope.[6,26] However, there should be multiple ways of mutations[85] other than 197L,[86] which can also lead to immune escape. Therefore, what could be the selective advantage of I97L over the other potential immune escape mutations at amino acid 97?

Fat Moon Suk joined my laboratory as a visiting physician scientist from Taipei Medical Hospital with a specialty in hepatology. Suk discovered that viral DNA replication of mutant I97L in Huh7 or J3 hepatoma cells was approximately 4.8-fold better than that of WT HBV. In addition, the level of encapsidated RNA pregenome in mutant I97L was about 5.7-fold higher than that of the WT HBV in Huh7 cells.[86] Unlike Huh7 cells, the difference in viral DNA replication between mutant I97L and WT HBV is greatly diminished in HepG2 cells. We speculate that mutation I97L could have acquired a host factor–independent growth advantage. This advantage is most pronounced in Huh7 cells containing low concentration of such a host factor(s) essential to viral replication. In contrast, this advantage is greatly reduced in HepG2 cells containing much higher concentration of such a host factor. The liver in chronic hepatitis patients is gradually deteriorating, leading to fibrosis, cirrhosis, and hepatocellular carcinoma. It is anticipated that certain host factors could get lost in the de-differentiating hepatocytes. We speculate here that mutant I97L could outcompete WT HBV and emerge from such de-differentiating hepatocytes owing to their host factor–independent growth advantage.[86]

Signal Transduction Via a Critical Hydrophobic Pocket Around Amino Acid 97 of HBc?

It has always been a puzzle to us why a subtle structural change from isoleucine to leucine (I97L) could have such a profound functional effect on viral replication and virion secretion. As mentioned earlier, the genome maturation signal hypothesis postulates that a change in capsid structure may signal genome maturation to the envelopment machinery. We therefore compared the capsid morphology and stability using *E. coli*-expressed capsids between WT HBV and mutant I97L, and found no apparent difference.[87] Ceres *et al.*[88] examined *in vitro* assembly of empty HBV capsids from wild-type and F97L core protein assembly domains. The mutation enhanced both the rate and extent of assembly relative to those for the wild-type protein. Since the structures of the mutant and wild-type capsids are essentially the same and the mutation is not involved in the contact between dimers, it is suggested that the F97L mutation affects the dynamic behavior of dimer and capsid.

The fold of HBV core particles has been reported by Alasdair Stevens' group at National Institute of Health and by Tony Crowther's group at MRC, England, respectively, using cryoEM and X-ray crystallography.[89–91] We examined the topological relationships of the side chain of residue Ile-97 with side chains of neighboring residues within a 6- to ~7-Å distance using the Swiss-PdbViewer and the INSIGHT II Program (accession number 1QGT; http://www.rcsb.org/pdb). We noted that amino acid 97 is in fact surrounded by a hydrophobic pocket, consisting of Val-27 and Leu-31 in α2a helix; Ala-58, Ile-59, Cys-61, and Trp-62 in α3 helix; Met-93, Leu-100, and Leu-101 in α4b helix.[85] Although Lys-96 is in the immediate vicinity of amino acid 97, the side chain of Lys-96 sticks out away from this hydrophobic pocket.

To elucidate the structure-function relationship of HBc at amino acid 97, we systematically replaced the isoleucine residue at amino acid 97 with 18 other amino acids via mutagenesis. Twelve of the 18 mutants exhibited no significant phenotype, while five new mutants displayed strong phenotypes. The I97D mutant had a near lethal phenotype, the I97P mutant exhibited a significantly reduced level of virion secretion, and the I97G mutant lacked the full-length relaxed circular form of viral DNA. Our studies indicate that amino acid 97 of HBc is at a very sensitive and critical position for the structure and function of HBV capsid particles, and only when HBc has a leucine residue at amino acid 97, can it then uniquely lead to an immature secretion phenotype.[85] Using electron cryomicroscopy, Tony Crowther's group compared the structures of the bacterially expressed RNA-containing core particles and the mature RC DNA-containing core particles extracted from virions. They demonstrated significant differences between the structures in RNA- and DNA-containing core particles. One major difference is in the hydrophobic pocket around amino acid 97.[92] Taken together, it is tempting to speculate that this hydrophobic pocket around amino acid 97 could be involved in the signal transduction of genome maturation.

Conclusions and Considerations

Highly frequent mutations accumulate in the core gene of HBV during the progression of liver diseases in natural infection. The most frequent core mutations in patients with chronic active hepatitis or hepatoma include I97L, P5T, and P130T. To date, the mutation I97L remains the most frequent naturally occurring mutation of HBV core gene, followed by mutations P5T and P130T. Although mutation P5T and P130T were first identified from liver samples of hepatoma patients, they are also present in the sera of chronic hepatitis patients without hepatoma. Interestingly, these three most frequent mutations of HBc coincide with three major T cell epitopes in both human and woodchuck systems. These epitopes are restricted by HLA class I and class II. We speculate that they represent natural variants that can escape from immune surveillance during HBV-mediated immunopathogenesis, including acute exacerbation.

In our effort to investigate the functional significance of these naturally occurring mutations, we discovered that mutant I97L exhibited two striking phenotypes: an extracellular phenotype of immature virion secretion, as well as an intracellular phenotype of a host factor–independent replication advantage over the WT HBV in certain hepatocytes. Presumably, as the liver of chronic hepatitis B patients becomes more fibrotic or cirrhotic, the replication of WT HBV becomes more and more restricted, due to the gradual loss of more and more host factors important for HBV replication. Mutant 97L probably emerges from such an unfriendly environment of deteriorating hepatocytes. Such an advantage in replication gained by mutant 97L is probably not free. Instead, it is most likely at the expense of the immature virion secretion. At present, it remains unclear whether the immature virions are less infectious than mature virions. We speculate here that immature virions are likely to have compromised infectivity. If so, then it makes sense that HBV could acquire compensatory mutations P5T or P130T to restore the infectivity and mature virion secretion. The phenomenon of HBV compensatory mutation for the defect in virion secretion can be demonstrated in the contexts of either a wild-type HBV or a naturally occurring variant with multiple mutations. The compensatory phenomenon is not limited to the core protein. We also observed a naturally occurring compensatory mutation W74L of the small envelope protein. This mutation can rescue the virion secretion defect of another small envelope mutation L77R. In the case of the large envelope protein, we demonstrated that an engineered envelope mutation pre-S1-A119F could rescue the immature secretion of core mutant I97L. Therefore, core-envelope interaction must be important for virion secretion. In addition to core and envelope, virion secretion also relies on host factors, such as Vps4. It can be anticipated that more host factors involved in virion secretion will be identified in the future.

The low-secretion phenotype was first observed in the study of the naturally occurring core mutation P5T. This P5T mutation appears to be a "loss of structure/function" mutation, since mutations P5A and P5S also resulted in the same low-secretion phenotype. In contrast, mutations F97L and I97L appear to be a "gain of structure/function" mutation, since none of the other substitutions at amino acid 97 of HBc can result in an immature secretion phenotype. Leucine is the only amino acid at HBc position 97 that can lead to immature secretion.

Such a virion secretion deficiency was also observed in a small envelope mutant L77R. We speculate here that some of these low virion secretion variants could contribute to occult infection in certain groups of patients. In contrast to the core or capsid protein, envelope proteins are well known to be required for virion secretion and viral entry. Our studies here highlight the importance of core protein and core-envelope interaction in virion secretion. Furthermore, many naturally occurring mutations could be compensatory in nature. These kinds of compensatory mutations are most difficult to identify by adequate assays for their functional significance, unless prior knowledge of their primary mutations and phenotypic defects is already available. The lessons that we have learned from the naturally occurring mutations of HBV will hopefully be useful for the understanding of the mutation, selection, and evolution of other infectious agents associated with chronic progressive diseases in humans.

Acknowledgments

The author would like to thank members of the Shih laboratory for helpful discussion and collaboration. Most of the work described here is supported by National Institutes of Health, USA, and University of Texas Medical Branch at Galveston, Texas. The author also acknowledges support from the National Science Council and Academia Sinica, Taiwan, during the manuscript preparation of this book chapter.

References

1. Ganem D, Schneider R. (2001) *Hepadnaviridae*: the viruses and their replication,. In: Knipe DM *et al.* (eds), *Fields virology*, 4th ed. Lippincott Williams & Wilkins, Philadelphia, pp. 2923–2970.
2. Guidotti LG, Chisari FV. (2006) Immunobiology and pathogenesis of viral hepatitis. *Annu Rev Pathol* **1**: 23–61.
3. Ehata T, Omata M, Yolosuka O, *et al.* (1992) Variations in codons 84–101 in the core nucleotide sequence correlate with hepatocellular injury in chronic hepatitis B virus infection. *J Clin Investig* **89**: 332–338.
4. Ehata T, Omata M, Chuang WL, *et al.* (1993) Mutations in core nucleotide sequence of hepatitis B virus correlate with fulminant and severe hepatitis. *J Clin Invest* **91**: 1206–1213.
5. Shih C, Hosono S, Tai PC, *et al.* (1994) The existence of escape mutants of end-stage human hepatitis B virus. In: abstract Book of 1994 International Meeting of Molecular Biology of Hepatitis B Viruses, Pasteur Institute, Paris, p.137.

6. Hosono S, Tai PC, Wang W, *et al.* (1995) Core antigen mutations of human hepatitis B virus in hepatomas accumulate in MHC class II-restricted T cell epitopes. *Virology* **212**: 151–162.
7. Shih C. (2008) Functional significance of naturally occurring hepatitis B virus variants. In: Lai CL and Locarnini S (eds), *Human Virus Guides—Hepatitis B Viruses*. 2nd ed. International Medical Press (IMP), London, Chapter 3.
8. Yuan TT, Sahu GK, Whitehead WE, *et al.* (1999a) The mechanism of an immature secretion phenotype of a highly frequent naturally occurring missense mutation at codon 97 of human hepatitis B virus core antigen. *J Virol* **73**: 5731–5740.
9. Liu CJ, Chen PJ, Lai MY, *et al.* (2003) A prospective study characterizing full-length hepatitis B virus genomes during acute exacerbation. *Gastroenterology* **124**(1): 80–90.
10. Liu CJ, Kao JH, Wang HY, *et al.* (2004) Origin of serum hepatitis B virus in acute exacerbation: comparison with HBV in the liver and from other exacerbation. *Hepatology* **40**: 310–317.
11. Lai MY, Chen DS, Chen PJ, *et al.* (1988) Status of hepatitis B virus DNA in hepatocellular carcinoma: a study based on paired tumor and nontumor liver tissues. *J Med Virol* **25**: 249–258.
12. Shih C, Burke K, Zeldis J, *et al.* (1987) Tight clustering of human hepatitis B virus integration sites in hepatomas near a "triple-stranded" region. *J Virol* **61**: 3491–3498.
13. Chuang WL, Omata M, Ehata T, *et al.* (1993a) Precore mutations and core clustering mutations in chronic hepatitis B virus infection. *Gastroenterology* **104**: 263–271.
14. Chuang WL, Omata M, Ehata T, *et al.* (1993b) Concentrating missense mutations in core gene of hepatitis B virus. Evidence for adaptive mutation in chronic hepatitis B virus infection. *Dig Dis Sci* **38**: 594–600.
15. Shih C, Tai PC, Whitehead W, *et al.* (1996) Hepatitis B and C viruses and liver cancer. In: Bertino JR (ed.), *Encyclopedia of Cancer*, vol. 2. Academic Press, London, pp. 824–834.
16. Akarca US, Lok ASF. (1995) Naturally occurring hepatitis B virus core gene mutations. *Hepatology* **22**: 50–60.
17. Bozkaya H, Ayola B, Lok AS. (1996) High rate of mutations in the hepatitis B core gene during the immune clearance phase of chronic hepatitis B virus infection. *Hepatology* **24**: 32–37.
18. Minami M, Poussin K, Kew M, *et al.* (1996) Precore/core mutations of hepatitis B virus in hepatocellular carcinomas developed on noncirrhotic livers. *Gastroenterology* **111**: 691–700.
19. Zinkernagel RM, Doherty PC. (1974) Restriction of *in vitro* T cell-mediated cytotoxicity in lymphocytic choriomeningitis within a syngeneic or semiallogeneic system. *Nature* **248**(450): 701–702.

20. Pircher H, Moskophidis D, Rohrer U, *et al.* (1990) Viral escape by selection of cytotoxic T cell-resistant virus variants *in vivo. Nature* **346**: 629–633.
21. McMichael AJ, Phillips RE. (1997) Escape of human immunodeficiency virus from immune control. *Annu Rev Immunol* **15**: 271–296.
22. Penna A, Bertoletti A, Cavalli A, *et al.* (1992) Fine specificity of the human T cell response to hepatitis B virus core antigen. *Arch Virol* (Suppl 4): 23–28.
23. Tsai SL, Chen MH, Yeh CT, *et al.* (1996) Purification and characterization of a naturally processed hepatitis B virus peptide recognized by CD8+ cytotoxic T lymphocytes. *J Clin Invest* **97**: 577–584.
24. Lee CJ, Lee PC, Tai JJ, Lee PH. (1990) Human leukocyte antigens in inhabitants of Taiwan. *Taiwan I Hsueh Hui Tsa Chih* **89**: 526–533.
25. Ferrari C, Bertoletti A, Penna A, *et al.* (1991) Identification of immunodominant T cell epitopes of the hepatitis B virus nucleocapsid antigen. *J Clin Invest* **88**(1): 214–222.
26. Menne S, Maschke J, Tolle TK, *et al.* (1997) Characterization of T cell response to woodchuck hepatitis core protein and protection of woodchucks from infection by immunization with peptides containing a T cell epitope. *J Virol* **71**: 65–74.
27. Okamoto H, Tsuda F, Mayumi M. (1987) Defective mutants of hepatitis B virus in the circulation of symptom-free carriers. *Jpn J Exp Med* **57**(4): 217–221.
28. Wakita T, Kakumu S, Shibata M, *et al.* (1991) Detection of pre-C and core region mutants of hepatitis B virus in chronic hepatitis B virus carriers. *J Clin Invest* **88**: 1793–1801.
29. Ackrill AM, Naoumov NV, Eddleston ALWF and Williams R. (1993) Specific deletions in the hepatitis B virus core open reading frame in patients with chronic active hepatitis B. *J Med Virol* **41**: 165–169.
30. Uchida T, Aye TT, Shihata T, *et al.* (1994) Evolution of the hepatitis B virus gene during chronic infection in seven patients. *J Med Virol* 43: 148–154.
31. Gunther S, Fischer L, Pult I, *et al.* (1999) Naturally occurring variants of hepatitis B virus. *Adv Virus Res* **52**: 25–137.
32. Yuan TT, Qiu SM, Shih C. (1998a) Functional characterization of naturally-occurring variants of human hepatitis B virus containing the core internal deletion mutation. *J Virol* **72**: 2168–2176.
33. Yuan TT, Lin MH, Chen DS, Shih C. (1998b) A defective interference-like phenomenon of human hepatitis B virus in chronic carriers. *J Virol* **72**: 578–584.
34. Botta A, Lu M, Zheng X, *et al.* (2000) Naturally occurring woodchuck hepatitis virus (WHV) deletion mutants in chronically WHV-infected woodchucks. *Virology* **277**: 226–234.
35. Sahu GK, Tai PC, Banerjee S, *et al.* (2002) Out-of-frame vs. in-frame core internal deletion (CID) variants of human and woodchuck hepatitis B viruses. *Virology* **292**: 35–43.

36. Rehermann B, Pasquinelli C, Mosier SM, Chisari FV. (1995) Hepatitis B virus (HBV) sequence variation of cytotoxic T lymphocyte epitopes is not common in patients with chronic HBV infection. *J Clin Invest* **96**: 1527–1534.
37. Liu Z, Luo K, He H, Hou J. (2005) Hot-spot mutations in hepatitis B virus core gene: eliciting or evading immune clearance? *J Viral Hepat* **12**: 146–153.
38. Tai PC, Banik D, Lin GI, *et al.* (1997) Novel and frequent mutations of hepatitis B virus coincide with an MHC class I-restricted T cell epitope of the surface antigen. *J Virol* **71**: 4852–4856.
39. Tai PC, Suk FM, Gerlich W, *et al.* (2002) Hypermodification of an internally deleted middle envelope (M) protein of frequent and predominant hepatitis B virus variants. *Virology* **292**: 44–58.
40. Neurath AR, Kent SB, Parker K, *et al.* (1986) Antibodies to a synthetic peptide from the preS 120–145 region of the hepatitis B virus envelope are virus neutralizing. *Vaccine* **4**: 35–37.
41. Guidotti LG, Rochford R, Chung J, *et al.* (1999) Viral clearance without destruction of infected cells during acute HBV infection. *Science* **284**(5415): 825–829.
42. Whalley SA, Murray JM, Brown D, *et al.* (2001) Kinetics of acute hepatitis B virus infection in humans. *J Exp Med* **193**(7): 847–854.
43. Guidotti LG, Ando K, Hobbs MV, *et al.* (1994) Cytotoxic T lymphocytes inhibit hepatitis B virus gene expression by a noncytolytic mechanism in transgenic mice. *Proc Natl Acad Sci USA* **91**: 3764–3768.
44. Liaw YF, Brunetto MR, Hadziyannis S. (2010) The natural history of chronic HBV infection and geographical differences. *Antivir Ther* **15**(Suppl 3): 25–33.
45. Lin CL, Kao JH. (2011) The clinical implications of hepatitis B virus genotype: recent advances. *J Gastroenterol Hepatol* **26** (Suppl 1): 123–130.
46. Zhijian Y, Zhen H, Fan Z, *et al.* (2010) Hepatitis B virus core protein with hot-spot mutations inhibit MxA gene transcription but has no effect on inhibition of virus replication by interferon α. *Virol J* 7: 278–283.
47. Pollicino T, Campo S, Raimondo G. (1995) PreS and core gene heterogeneity in hepatitis B virus (HBV) genomes isolated from patients with long-lasting HBV chronic infection. *Virology* **208**(2): 672–677.
48. Pollicino TS, Zanetti AR, Cacciola I, *et al.* (1997) Pre-S2 defective hepatitis B virus infection in patients with fulminant hepatitis. *Hepatology* **26**: 495–499.
49. Galibert F, Mandart E, Fitoussi F, *et al.* (1979) Nucleotide sequence of the hepatitis B virus genome cloned in *E. coli*. *Nature* **281**: 646–650.
50. Summers J, Mason WS. (1982) Replication of the genome of a hepatitis B-like virus by reverse transcription of an RNA intermediate. *Cell* **29**: 403–415.
51. Gerelsaikhan T, Tavis JE, Bruss V. (1996) Hepatitis B virus nucleocapsid envelopment does not occur without genomic DNA synthesis. *J Virol* **70**: 4269–4274.

52. Wei Y, Tavis JE, Ganem D. (1996) Relationship between viral DNA synthesis and virion envelopment in hepatitis B viruses. *J Virol* **70**: 6455–6458.
53. Mabit H, Schaller H. (2000) Intracellular hepadnavirus nucleocapsids are selected for secretion by envelope protein-independent membrane binding. *J Virol* **74**(24): 11472–11478.
54. Perlman D, Hu J. (2003) Duck hepatitis B virus virion secretion requires a double-stranded DNA genome. *J Virol* 77: 2287–2294.
55. Shih C, Yuan TT. (2004) A cis/trans genetic test for pleiotropic phenotypes associated with a frequent naturally occurring mutation at amino acid 97 of HBV core protein. *Methods Mol Med* **95**: 247–257.
56. Yaginuma K, Shirakata Y, Kobayashi M, Koike K. (1987) Hepatitis B virus (HBV) particles are produced in a cell culture system by transient expression of transfected HBV DNA. *Proc Natl Acad Sci USA* **84**: 2678–2682.
57. Yuan TT, Tai PC, Shih C. (1999b) Subtype-independent immature secretion and subtype-dependent replication deficiency of a highly frequent, naturally occurring mutation of human hepatitis B virus core antigen. *J Virol* **73**: 10122–10128.
58. Schormann W, Kraft A, Ponsel D, Bruss V. (2006) Hepatitis B virus particle formation in the absence of pregenomic RNA and reverse transcriptase. *J Virol* **80**(8): 4187–4190.
59. Tencza MG, Newbold JE. (1997) Heterogeneous response for a mammalian hepadnavirus infection to acyclovir: drug-arrested intermediates of minus-strand viral DNA synthesis are enveloped and secreted from infected cells as virion-like particles. *J Med Virol* **51**: 6–16.
60. Chang SF, Netter HJ, Bruns M, *et al.* (1999) A new avian hepadnavirus infecting snow geese (Anser caerulescens) produces a significant fraction of virions containing single-stranded DNA. *Virology* **262**: 39–54.
61. Le Pogam S, Shih C. (2002) Influence of a putative intermolecular interaction between core and the pre-S1 domain of the large envelope protein on hepatitis B virus secretion. *J Virol* **76**: 6510–6517.
62. Le Pogam S, Yuan TT, Sahu GK, *et al.* (2000) Low-level secretion of human hepatitis B virus virions caused by two independent, naturally occurring mutations (P5T and L60V) in the capsid protein. *J Virol* **74**: 9099–9105.
63. Hui EK, Yi YS, Lo SJ. (1999) Hepatitis B viral core proteins with an N-terminal extension can assemble into core-like particles but cannot be enveloped. *J Gen Virol* **80**(10): 2647–2659.
64. Koschel M, Oed D, Gerelsaikhan T, *et al.* (2000) Hepatitis B virus core gene mutations which block nucleocapsid envelopment. *J Virol* **74**: 1–7.
65. Ponsel D, Bruss V. (2003) Mapping of amino acid side chains on the surface of hepatitis B virus capsids required for envelopment and virion formation. *J Virol* 77: 416–422.
66. Pairan A, Bruss V. (2009) Functional surfaces of the hepatitis B virus capsid. *J Virol* **83**: 11616–11623.

67. Yuan TT, Shih C. (2000) A frequent, naturally occurring mutation (P130T) of human hepatitis B virus core antigen is compensatory for immature secretion phenotype of another frequent variant (I97L). *J Virol* **74**: 4929–4932.
68. Lin X, Ma ZM, Yao X, *et al.* (2002) Replication efficiency and sequence analysis of full-length hepatitis B virus isolates from hepatocellular carcinoma tissues. *Int J Cancer* **102**: 487–491.
69. Chua PK, Wen YM, Shih C. (2003) Coexistence of two distinct secretion mutations (P5T and I97L) in hepatitis B virus core produces a wild-type pattern of secretion. *J Virol* 77: 7673–7676.
70. Chua PK, Wang RY, Lin MH, *et al.* (2005) Reduced secretion of virions and hepatitis B virus (HBV) surface antigen of a naturally occurring HBV variant correlates with the accumulation of the small S envelope protein in the endoplasmic reticulum and Golgi apparatus. *J Virol* **79**: 13483–13496.
71. Kalinina T, Riu A, Fischer L, *et al.* (2001) A dominant hepatitis B virus population defective in virus secretion because of several S-gene mutations from a patient with fulminant hepatitis. *Hepatology* **34**: 385–394.
72. Kalinina T, Iwanski A, Will H, and Sterneck M. (2003) Deficiency in virion secretion and decreased stability of the hepatitis B virus immune escape mutant G145R. *Hepatology* **38**: 1274–1281.
73. Khan N, Guarnieri M, Ahn SH, *et al.* (2004) Modulation of hepatitis B virus secretion by naturally occurring mutations in the S gene. J Virol **78**: 3262–3270.
74. Bruss V, Ganem D. (1991) The role of envelope proteins in hepatitis B virus assembly. *Proc Natl Acad Sci USA* **88**: 1059–1063.
75. Ueda K, Tsurimoto T, Matsubara K. (1991) Three envelope proteins of hepatitis B virus: Large S, middle S, and major S proteins needed for the formation of Dane particles. *J Virol* **65**: 3521–3529.
76. Summers J, Smith PM, Huang MJ, Yu MS. (1991) Morphogenetic and regulatory effects of mutations in the envelope proteins of an avian hepadnavirus. *J Virol* **65**: 1310–1317.
77. Hou J, Wang Z, Cheng J, *et al.* (2001) Prevalence of naturally occurring surface gene variants of hepatitis B virus in nonimmunized surface antigen-negative Chinese carriers. *Hepatology* **34**: 1027–1034.
78. Jeantet D, Chemin I, Mandrand B, *et al.* (2002) Characterization of two hepatitis B virus populations isolated from a hepatitis B surface antigen-negative patient. *Hepatology* **35**: 1215–1224.
79. Garrus JE, von Schwedler UK, Pornillos OW, *et al.* (2001) Tsg101 and the vacuolar protein sorting pathway are essential for HIV-1 budding, *Cell* **107**: 55–65.
80. Kian Chua P, Lin MH, Shih C. (2006) Potent inhibition of human hepatitis B virus replication by a host factor Vps4. *Virology* **354**: 1–6.

81. Lambert C, Döring T, Prange R. (2007) Hepatitis B virus maturation is sensitive to functional inhibition of ESCRT-III, Vps4, and gamma 2-adaptin. *J Virol* **81**: 9050–9060.
82. Watanabe T, Sorensen EM, Naito A, *et al.* (2007) Involvement of host cellular multivesicular body functions in hepatitis B virus budding. *Proc Natl Acad Sci USA* **104**: 10205–10210.
83. Rost M, Mann S, Lambert C, *et al.* (2006) Gamma-adaptin, a novel ubiquitin-interacting adaptor, and Nedd4 ubiquitin ligase control hepatitis B virus maturation. *J Biol Chem* **281**: 29297–29308.
84. Bardens A, Döring T, Stieler J, Prange R. (2011) Alix regulates egress of hepatitis B virus naked capsid particles in an ESCRT-independent manner. *Cell Microbiol* **13**: 602–619.
85. Ning B, Shih C. 2004. Nucleolar localization of human hepatitis B virus capsid protein. *J Virol* **78**: 13653–13668.
86. Suk FM, Lin MH, Newman M, *et al.* (2002) Replication advantage and host factor-independent phenotypes attributable to a common naturally occurring capsid mutation (I97L) in human hepatitis B virus. *J Virol* **76**: 12069–12077.
87. Newman M, Suk FM, Cajimat M, *et al.* (2003) Stability and Morphology Comparisons of Self-assembled Virus-Like Particles from Wild Type and Mutant Human Hepatitis B Virus Capsid Proteins. *J Virol* **77**: 12950–12960.
88. Ceres P, Stray SJ, Zlotnick A. (2004) Hepatitis B virus capsid assembly is enhanced by naturally occurring mutation F97L. *J Virol* **78**: 9538–9543.
89. Conway JF, Cheng N, Zlotnick A, *et al.* (1997) Visualization of a 4-helix bundle in the hepatitis B virus capsid by cryo-electron microscopy. *Nature* **386**: 91–94.
90. Bottcher B, Wynne SA, and Crowther RA. (1997) Determination of the fold of the core protein of hepatitis B virus by electron cryomicroscopy. *Nature* **386**: 88–91.
91. Wynne SA, Crowther RA, and Leslie AG. (1999) The crystal structure of the human hepatitis B virus capsid. *Mol Cell* **3**: 771–780.
92. Roseman AM, Berriman JA, Wynne SA, *et al.* (2005) A structural model for maturation of the hepatitis B virus core. *Proc Natl Acad Sci USA* **102**: 15821–15826.

Chapter 7

Recent Developments in Hepatitis Delta Virus

Camille Sureau*

Introduction

The discovery of the hepatitis delta antigen (HDAg) in 1977 was made upon immunofluorescence examination of liver biopsies obtained from Hepatitis B virus (HBV) chronic carriers who were affected with severe liver injury.[1] It was described at the time as a new antigen–antibody system in HBV-infected individuals. HDAg was detected in liver cell nuclei and later found to be a structural element of viral particles coated with the HBV envelope proteins. At that point it was considered a component of a new virus-like agent whose propagation was dependent upon the presence of HBV.[2–4] After the cloning of the HDV-associated RNA,[5,6] the nucleotide sequence analysis revealed no homology to that of the HBV DNA genome, but the structure of the HDV RNA, a single-stranded, circular molecule of negative polarity, displayed similarities with that of plant viroids,[7] including the ability to fold into a rod-like conformation stabilized by more than 70% intramolecular complementarity.[8] Unlike viroids, the HDV RNA bears an open reading frame (ORF) coding for the

* Laboratoire de Virologie Moléculaire, Institut National de la Transfusion Sanguine, 6 Rue Alexandre Cabanel, 75739 Paris, France, Tel: 33-1-44493056, Fax: 33-1-44493059, e-mail: csureau@ints.fr

HDAg-associated protein, which is detected as two isoforms referred to as small and large (S-HDAg and L-HDAg, respectively).

Soon after the HDV discovery, experiments conducted in tissue culture demonstrated that, when introduced into permissive cells in the absence of HBV, the HDV RNA could replicate and associate with multiple copies of HDV-encoded proteins to form ribonucleoproteins (RNPs).[9] However, the RNPs were unable to exit the cell on their own. Export could only be achieved upon co-expression with the HBV envelope proteins.[10,11] We now know that packaging relies on the ability of the RNP to interact with the HBV envelope proteins and on the unique capacity of the latter for assembly and release of lipoprotein vesicles.[12,13] The resulting HDV particle therefore consists of an inner RNP coated with the HBV envelope proteins, whose role is also to provide a means for entry into human hepatocytes, the target cells of the helper HBV[14] (Fig. 1). Being absolutely dependent on its helper virus, HDV is currently classified as an occasional satellite of HBV; it is the unique member of the *Deltavirus* genus.[8]

Natural HDV infections occur as either a coinfection with HBV or a superinfection of HBV carriers. HDV can cause severe acute and chronic liver diseases in HBV-infected individuals; most of the HBV carriers superinfected with HDV become carriers of both HBV and HDV. It is estimated that 5–10% of the more than 350 million HBV chronic carriers worldwide are coinfected with HDV.[15] Although HDV infection is reportedly declining in some endemic regions, it is not a disappearing disease in several European countries, due to immigration from high endemic countries.[16,17] Acute superinfection leads more often to fulminant hepatitis and liver failure than acute infection with HBV alone. Liver damage is also induced more rapidly in chronically HBV/HDV-infected patients as compared to HBV carriers. HDV is present in many different countries worldwide, and it displays an extended genetic variability classified into eight genotypes.[18] Treatment of chronic HDV infections is limited to alpha-interferon therapy,[19] but the rate of sustained virological response remains unsatisfactory. There is obviously a need for the development of novel antiviral strategies against HDV; hopefully these will emerge from studies currently conducted to decipher the HDV replication cycle.

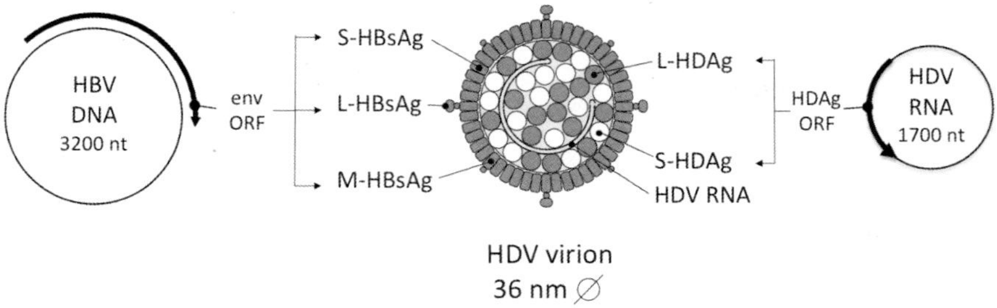

Fig. 1. Schematic representation of the HDV virion. The particle comprises two types of components: (i) the viral envelope of HBV origin, including the HBV envelope proteins S-HBsAg, M-HBsAg, and L-HBsAg, and (ii) the ribonucleoprotein that comprises the circular genomic HDV RNA associated with multiple copies of the HDV-encoded S-HDAg and L-HDAg proteins. The open reading frame for the HBV envelope proteins (env ORF) is represented on the HBV genome (HBV DNA). The ORF for the HDAg proteins (HDAg ORF) is indicated on the HDV genome (HDV RNA).

This review covers the latest developments in the study of the HDV replication cycle, regarding, in particular, the mechanisms of HDV RNA replication, particle assembly, and viral entry.

HDV Assembly

The HDV virions are heterogeneous in size, with an average diameter of 36 nm. The envelope is made of cell-derived lipids associated with HBV envelope proteins, and the inner RNP consists of the RNA genome and approximately 200 copies of the HDAg-protein.[20] The HDV envelope is similar to the one of HBV, including the S-HBsAg, M-HBsAg, and L-HBsAg proteins encoded by a single open reading frame on the HBV genome (Fig. 2). S-HBsAg is 226 amino acid residues in length; M-HBsAg has 55 additional residues (the preS2 domain) at the amino terminus, and L-HBsAg comprises the entire M polypeptide with an additional amino-terminal polypeptide (preS1) of 108–119 residues. S-HBsAg proteins can dimerize and form multimers at the endoplasmic reticulum (ER) membrane through lateral protein–protein interactions, and the resulting aggregates are thought

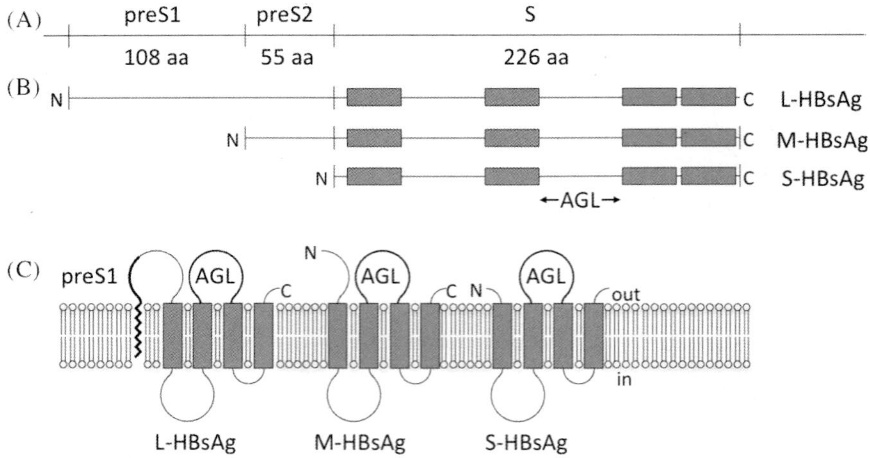

Fig. 2. The HBV envelope proteins. (A) Domains of the HBV envelope proteins open reading frame. (B) The L-HBsAg, M-HBsAg, and S-HBsAg proteins are translated from three in-frame initiation sites located at the amino terminus of the preS1, preS2, and S domains, respectively. Open rectangles indicate transmembrane domains (TMDs). (C) Membrane topology of the HBV envelope proteins. L-, M-, and S-HBsAg proteins adopt a similar topology at the HDV or HBV particles membrane. Broken line indicates the myristate group linked to the amino terminus of L-HBsAg. AGL, antigenic loop bearing the immuno-dominant "a" determinant.

to bud spontaneously into the lumen of a pre-Golgi compartment. Although the S-HBsAg protein provides the driving force in the budding process, it cannot direct HBV virion assembly because the recruitment of the HBV nucleocapsid is mediated by L-HBsAg. Owing to the overproduction of S-HBsAg and to its capacity for auto-assembly, HBV virion formation is a very rare event in comparison with the production of subviral particles (SVPs). An average HBV infectious serum contains approximately 10^{12} to 10^{13} SVPs per mL for only 10^{8} to 10^{9} virions. In comparison, at the onset of an acute HDV infection, an infectious serum has an estimated titer of 10^{9} to 10^{10} HDV virions per mL and only a 10-fold excess of SVPs. Therefore, HDV particles appear to be more efficiently assembled and secreted than are HBV virions.

A peculiarity of the HBV life cycle is that the envelope proteins, irrespective of the presence of the HBV nucleocapsid, drive budding. This characteristic is utilized by HDV. In fact, the sole S-HBsAg protein suffices for envelopment and export of the HDV RNP, demonstrating that it bears a signal for packaging the HDV RNP.[10,11]

The RNP comprises both the S-HDAg and L-HDAg proteins associated with the viral RNA. RNP export from the nucleus relies on a specific nuclear export signal (NES) located in the 19-amino-acid carboxyl terminus of L-HDAg,[21] which also contains a carboxy-terminal CXXQ signal for farnesylation.[84] The farnesyl group covalently bound to the carboxyl terminus of L-HDAg probably serves to anchor the RNP in the ER membrane where the envelope proteins are synthesized.[22] L-HDAg mediates the incorporation of the RNP in the HBV envelope.[23] In fact, it was shown that in tissue culture, the free form of L-HDAg could be packaged in SVPs, but recent data indicate that this would not occur during the natural course of HDV RNA replication when HDV RNPs are produced.[24] The packaging signal on the L-HDAg protein lies within its 19-amino-acid carboxyl terminus, which corresponds to the sequence difference between L- and S-HDAg. Surprisingly, it is not well conserved among the different HDV genotypes, except for the farnesylation signal sequence (CXXQ), for a tryptophan at position 196 and for the presence of at least four proline residues at various positions between W-196 and C-211.[25] In the L-HDAg protein of genotype-1 HDV, Trp-196 was found dispensable for packaging, whereas Pro-201 and -205 were proven important for both the function of the L-HDAg NES and, independently of the NES activity, the activity of the packaging signal.[12,13] Whether L-HDAg binds directly to S-HBsAg is not certain.

Among members closest to HBV in the *Hepadnaviridae* family, the Woodchuck hepatitis virus (WHV) and the Woolly monkey hepatitis B virus (WMHBV) both encode viral envelope proteins that are competent for HDV RNP envelopment.[3,26] This has led to the use of the woodchuck as a practical animal model for HDV, and to the possibility of using primary cultures of spider monkey hepatocytes for testing the infectivity of HDV particles pseudotyped with the WMHBV envelope proteins.[26] In contrast to WHV and WMHBV, the most distantly related *Hepadnavirus*, namely the Duck hepatitis B virus (DHBV), is unable to assist in HDV propagation. This is due in part to the failure of HDV RNA to replicate in avian cells,[27] but also to the inability of the S-DHBsAg envelope protein to package the HDV RNP.[28]

How the HDV RNP recruits the HBV envelope proteins is an interesting question, considering that this process is crucial to the HDV life cycle. Early studies conducted by Chen *et al.*[29] have shown that a truncation of the carboxy-terminal 50 residues of S-HBsAg was sufficient to abolish envelopment and secretion of co-expressed HDAg proteins. Then, O'Malley and Lazinski[28] demonstrated that S-HBsAg, bearing a deletion in the antigenic loop (AGL) (aas 107–147), was defective for HDV RNP envelopment. Interestingly, this mutant remained competent for the incorporation of singly expressed L-HDAg, suggesting that the hindrance observed in HDV maturation was probably reflecting a reduced capacity of the envelope to accommodate the RNP. A deletion of residues 24–28 of the S-HBsAg protein, or the removal of the glycosylation site at position 146 could impair the maturation of HDV particles while having no effect on SVP secretion. In the two latter cases, the defect in HDV assembly was also due to impairment of the ability of the mutant envelope to coat the RNP and not to a lack of interaction with L-HDAg.

More recently, a major determinant of HDV maturation was identified in the carboxyl terminus of S-HBsAg.[12] Within this domain, single substitutions of tryptophan at positions 196, 199, and 201 with alanine or phenylalanine were tolerant of SVP secretion but detrimental for HDV assembly. This was proven to result from the inability of the S-HBsAg mutants to interact with L-HDAg. The Trp-rich domain in S-HBsAg was thus considered a matrix domain for HDV assembly (Fig. 3).[12] An interesting observation was that the Trp-rich motif is strictly conserved in all HBV genotypes as well as in WHV and WMHBV, suggesting that it play an essential function in the HBV life cycle, other than SVP assembly. However, an alanine substitution for Trp-196, -199, and -201 was found permissive for both HBV virion assembly and infectivity,[30] clearly demonstrating that the Trp-rich motif is not essential to the HBV replication cycle *in vitro*. Then why

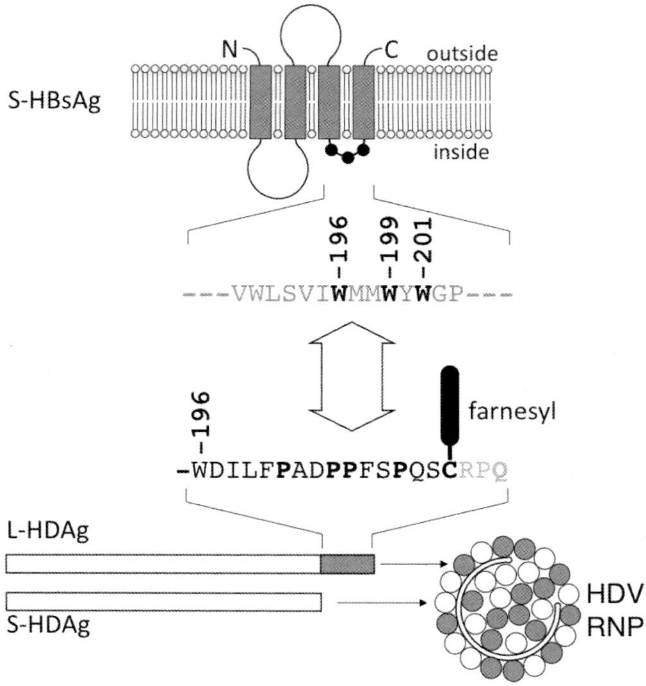

Fig. 3. Elements important for HBV/HDV interaction. Tryptophan residues at positions 196, 199, 201 in the S domain of the HBV envelope proteins constitute the "matrix" domain for assembly of HDV particles. This Trp-rich domain is likely to interact (directly or indirectly) with the carboxyl terminus of the L-HDAg protein of the HDV RNP. Within the carboxy terminal 19 amino acid residues, prolines, and the CXXQ box are important for interaction with HBV envelope proteins. CXXQ is a signal for farnesylation of L-HDAg at cysteine-211, which is required for HDV assembly.

is this motif conserved in all *Orthohepadnaviruses*? It may simply be the result of the overlap between the envelope proteins and polymerase (Pol) genes, which is a characteristic of the HBV genetic organization and a consequence of the small size of the HBV genome. In fact, the envelope proteins ORF is totally included in that of Pol, and the DNA sequence surrounding the codon for Trp-196 in S-HBsAg happens to also encode the YMDD motif of the Pol catalytic domain (Fig. 4). Since there is a strict requirement for a YMDD motif in Pol,

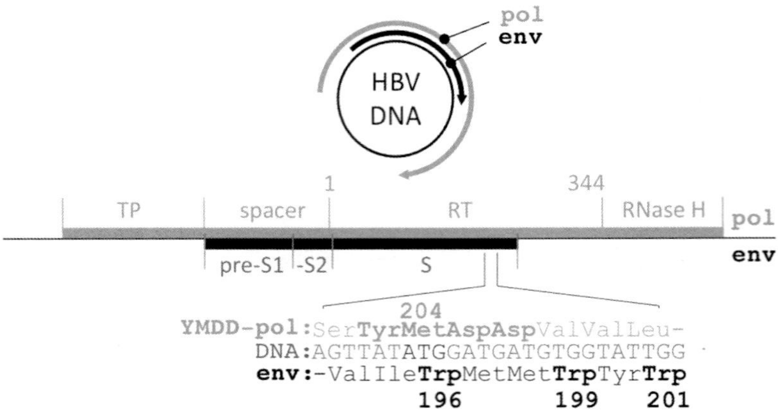

Fig. 4. Sequence of the overlapping HBV polymerase and S-HBsAg genes. Nucleic acid sequence (DNA) coding for the YMDD motif of HBV polymerase (YMDD-pol) and the overlapping 194–201 domain in the carboxyl terminus of S-HBsAg (env). The env and pol genes are indicated on the HBV DNA genome. Pol is divided in the terminal protein domain (TP), the spacer domain, the reverse transcriptase (RT), and RNase H domains. Position 204 in the RT domain corresponds to the methionine residue of the YMDD motif.

there is no possibility other than Trp at position 196 in S-HBsAg. By extrapolation, a few S-HBsAg residues (i.e., Trp-199 and -201), in addition to Trp-196, might as well be conserved solely because they share a DNA coding sequence with Pol. Thus a conserved essential motif in Pol (YMDD) imposes a conserved, useless, Trp-rich motif in S-HBsAg, which is fortuitously used by HDV as a matrix domain for assembly. Note that this HDV matrix domain would be a good target for anti-HDV therapy, because a drug capable of preventing an interaction with HDV RNP by binding to the Trp-rich domain (an envelope protein motif useless to HBV) would not lead to the arising of HBV escape mutants. Overall, one can see the encounter between HBV and HDV as the consequence of (i) the small size of the HBV genome, (ii) the huge overproduction of envelope proteins that can self-assemble into lipoprotein transport vesicles, and (iii) the natural flexibility of the HBV envelope, which can accommodate three types of particles, namely the spherical SVPs of 22 nm in diameter, the empty filaments that are 22 nm in diameter and up to several hundred nanometers in length, and the 42 nm HBV virions. Stoichiometry may also be in favor of S-HBsAg and RNP interaction, since an average infected cell is estimated to contain up to 6×10^6 copies of HDAg proteins and close to 10^5 copies of genomic HDV RNA.[20]

Finally, it is worth mentioning that recent findings from several laboratories suggest that HBV virions and SVPs follow distinct pathways for export: HBV virions would use the late endosomal multivesicular bodies (MVBs) for budding at intracellular membranes, whereas SVPs would be secreted through an MVBs-independent pathway.[31–33] These characteristics may be critically important for HDV, which is more likely to use the SVPs secretion pathway than the MVBs-dependent pathway for the following reasons: (i) the efficiency of HDV virions secretion is close to that of SVPs (at the onset of an acute HDV infection, the titer of HDV virions is only 10-fold lower than that of SVPs), (ii) unlike that of HBV virions, the envelope of SVPs or HDV virions is not enriched in L-HBsAg proteins[2] and does not even require L-HBsAg. So, one might speculate that differences between HBV and HDV in the virion assembly pathway could lead to differences in their respective envelopes and, eventually, to slightly distinct entry pathways.

HDV Entry

The HDV life cycle relies on two HBV elements: (i) the S-HBsAg protein for at least one function, the export of the RNP, and (ii) the L-HBsAg protein for infectivity (Fig. 2). L-HBsAg protein is in fact required for infectivity of both HBV and HDV particles.[14,34] Its amino-terminal preS1 region is assumed to bind to a specific receptor at the surface of human hepatocytes. Note that *in vitro*, M-HBsAg was found dispensable for assembly and infectivity of both HBV and HDV particles.[35,36] Since HDV particles can be assembled by S-HBsAg alone, the L-to-S-HBsAg ratio in the envelope of serum-derived HDV particles might be variable, but the minimum number of L-HBsAg molecules for conferring infectivity is unknown.

Even though the viral entry step remains obscure and the nature of the receptor(s) for HBV/HDV unknown, it is reasonable to assume that HBV and HDV use similar entry pathways, at least in the early phase of attachment. In fact, our current knowledge of the HBV envelope proteins' function at viral entry supports this assumption. One recent study, which extended the experiments of Le Seyec and colleagues,[34] demonstrated that the activity of the preS domain at the entry step of HBV or HDV solely depends on the integrity of its first 75 amino acids.[37] Mutations introduced between amino acids 76 and 113 of the preS domain, overlapping the HBV matrix domain, or deletion of the preS2 domain, did not interfere with infectivity.

preS1 Lipopeptides are Inhibitors of HDV/HBV Entry

A major advance was achieved when synthetic peptides specific for the amino terminus of the preS1 domain were shown to be potent inhibitors of *in vitro* HBV infection. It was demonstrated that, for both HBV and HDV infectivity, there is a requirement for the presence of myristic acid linked to the preS1 amino terminus of L-HBsAg.[38,39] In tissue culture, it has been shown that myristoylated preS1 peptides were efficient in blocking infection at nanomolar concentrations, and the replacement of myristic acid with longer, unbranched fatty acids resulted in even greater activity. HBV infection could be prevented by a short pre-incubation of cells with the lipopeptides, suggesting their rapid and efficient targeting to a receptor at the hepatocyte surface.[40] Infection of susceptible Tupaia hepatocyte cultures with HBV could also be blocked by 2–48 preS1 lipopeptides at nanomolar concentrations. The peptides were found to bind strongly to Tupaia hepatocyte membranes but not to non-hepatic cells or rodent hepatocytes. In this study, the 9–18 amino acids sequence of preS1 was identified as the minimal attachment site.[41] Interestingly, SVPs consisting of S-HBsAg only (lacking M- and L-HBsAg) could not bind to Tupaia hepatocytes, but the addition of L-HBsAg in the SVP envelope restored binding.[42] HBV infection could also be blocked by myristoylated preS1–GST fusion protein, under the condition that the 11–21 preS1 amino acids sequence be intact. In the same study, the essential function of the 11–21 sequence was also directly demonstrated when HBV particles bearing lesions in this domain were shown to lose infectivity. The inhibition potency of the preS1 lipopeptides was also proven in the HDV model, further supporting the assumption that HBV and HDV use at least one common step for entry, which involves the highly conserved amino terminus of the preS1 sequence.[43]

More importantly, the antiviral potential of preS1 lipopeptides was comforted by the recent demonstration of their activity *in vivo*. This was achieved in immuno-deficient transgenic mice repopulated with human hepatocytes.[44] In this model, the peptides could be delivered subcutaneously at low doses, and they were able to accumulate in the liver and to prevent HBV infection of the human hepatocytes. Surprisingly, the preS1 lipopeptides would not only bind to transplanted human hepatocytes, but also to mice hepatocytes — the latter observation was in contradiction with the abovementioned study of Glebe and colleagues, who found no affinity of preS1-bearing SVPs for rodent hepatocytes. [42] The preS1 determinant would thus be organ- and not species-specific. Whatever mechanism is used by the preS1 lipopeptides to block HBV entry, these compounds actually constitute a very promising therapeutic approach to prevent, or cure, HBV or HDV infections.[44]

Role of Glycosaminoglycans in HDV Entry?

Glycosaminoglycan (GAG) side chains of cell surface–associated proteoglycans are used by various viruses as primary docking sites, and, according to Schulze *et al.*,[45] cell surface-associated heparan sulfate proteoglycans are also implicated in HBV entry. HBV virions were shown to bind to GAGs at the surface of susceptible HepaRG cells, or to immobilized heparin, in an L-HBsAg-dependent manner. Furthermore, infection could be blocked by treatment of virions with heparin (or highly sulfated dextran sulfate). The treatment of the HepaRG cell surface with heparinase, or the inhibition of GAG synthesis by sodium chlorate, also prevented HBV binding and infection. The requirement for cell-surface GAGs as low-affinity receptor for

HBV was also demonstrated in primary cultures of Tupaia hepatocytes.[46] In this system, HBV infection was inhibited by soluble heparin (or other higher-sulfated polymers), and pre-treatment of susceptible cells with heparinase decreased viral binding and infection. Interestingly, infection of primary hepatocyte cultures could be blocked with preS1-lipopeptides but not with heparin, when provided after virus attachment. The interpretation of these data was that HBV would initially bind with low-affinity-to-cell-surface GAGs before preS1 could initiate binding to a high-affinity receptor. Whether this scenario pertains also to HDV remains to be established.

Role of the "a" Determinant in HDV/HBV Entry

Besides the central preS1 infectivity determinant, a second determinant was recently identified in the AGL borne by the three HBV envelope proteins (Fig. 2). Its activity was first established in the HDV model,[47] using recombinant particles bearing lesions in the AGL sequence. Interestingly, mutations of the AGL cysteine residues were found to block viral entry but also to alter the antigenicity of the conserved "a" determinant. The immuno-dominant "a" determinant consists in conformational epitopes born by the AGL sequence, and it is present at the surface of HBV particles of all genotypes. Furthermore, antibodies directed to the "a" determinant are highly neutralizing, and this characteristic is at the origin of the HBV vaccine.

Infectivity, and the "a" determinant, were also lost when virions were treated with membrane-impermeable inhibitors of thiol/disulfide isomerization prior to or during the virus–cell interaction period. These findings not only demonstrated the essential role of the AGL cysteines at viral entry, but they established a correlation between the AGL cysteine disulfide network, the conformation of the "a" determinant, and infectivity.[48] This was the first report associating the "a" determinant with an essential function in the HBV life cycle, providing at the same time an explanation of its strict conservation among all HBV genotypes. The AGL thus bears a second entry determinant in the HBV envelope proteins, which may function independently of preS1, as a ligand to a specific receptor, and/or as an essential element of the viral envelope disassembly process.

HDV/HBV Entry Through the Endocytic Pathway?

The intracellular uptake of HBV or HDV virions has not been characterized in detail. For both viruses, entry may proceed via receptor-mediated endocytosis,[49] but once released in the cytoplasm, the HDV RNP and the HBV nucleocapsid most likely follow separate pathways to the cell nucleus. In support of an endocytic route for HBV entry, a recent study demonstrated that extraction of cholesterol from HBV virions reduced infectivity without affecting particle integrity, antigenicity, and the ability to bind to hepatocytes.[50] There was no evidence for a role of lipid rafts in the HBV entry process, and the conclusion of this study was that the cholesterol content of the viral envelope was dispensable for viral binding but required at a post-binding step in the entry process.[50] Similar results were obtained in the DHBV model, in which cholesterol was seen as instrumental in the endosomal escape of endocytosed DHBV particles.[51] It will be interesting to determine if, as with HBV virions, the infectivity of HDV particles is dependent on the cholesterol content of the viral envelope.

Obviously, many questions remain about the HBV/HDV entry pathway: What is the extent of the similarities between the HBV and HDV entry mechanisms? Do these two viruses use exactly the same set of receptors? Do they follow a classical endocytic pathway for entry? And, if so, is there a need for acidification of an endosomal compartment? Assuming that the viral envelope proteins are cross-linked, what is the mechanism of envelope disassembly? Then how do the HDV and HBV nucleocapsids differ from each other for nuclear import?[52]

Replication of HDV RNA

After viral entry, the HDV RNA is delivered to the nucleus, where it is used as a template for rolling-circle transcription that generates multimers of antigenomic HDV RNA. These multimers are cleaved into monomers by a ribozyme activity borne by the antigenomic RNA. The monomers are then ligated to form circular antigenomic molecules. In turn, the antigenomic RNA becomes the template for a second rolling-circle replication mechanism to produce circular genomic RNA. The HDV RNA replication cycle includes also the transcription of a messenger RNA for the HDAg protein and an edition of the RNA, which leads to the production of two isoforms of the HDAg protein. These two isoforms (S- and L-HDAg) have multiple functions in the HDV life cycle, but no RNA polymerase activity. Thus, somehow, HDV must replicate its RNA genome using cellular RNA polymerase(s), a characteristic of the viroid replication cycle in plants. RNA-directed transcription had long been considered nonexistent in mammalian cells; however, there is now evidence for such a function, but its evolutionary origin remains largely unknown. It is worth mentioning that RNA-directed transcription in mammalian cells was clearly suggested by the identification of mirror-spliced antisense transcripts, which are reverse complement RNAs of spliced mRNAs, likely generated by direct transcription.[53]

Cellular RNA Polymerase II is Involved in HDV RNA Transcription

In general, RNA viruses manage to replicate their genome in mammalian cells by encoding either a reverse transcriptase to generate a DNA intermediate (retroviruses) or an RNA-dependent RNA polymerase (RdRP) for direct transcription. In the case of HDV, RNA polymerase II (Pol II), which normally uses a DNA template, seems to be the key enzyme for transcription of the viral RNA.[54,55] But whether Pol II is involved in the synthesis of all HDV RNA species is still a subject of debate. Experiments conducted in cell culture using a cDNA-free transfection system (i.e., using synthetic HDV RNA molecules only) have shown that synthesis of full-length antigenomic HDV RNA was completely resistant to 100 μg/mL of alpha-amanitin, whereas synthesis of genomic RNA was totally inhibited at 2.5 μg/mL.[56] These findings suggested that RNA polymerase I, an enzyme involved in the transcription of ribosomal RNAs, rather than Pol II, would be responsible for antigenomic RNA synthesis.[57]

Studies that identified Pol II as responsible for the synthesis of all HDV RNAs were conducted using an experimental system based on a cell line in which the production of S-HDAg, a protein essential for HDV RNA replication, is inducible. To initiate HDV RNA replication, cells are first transfected with full-length HDV RNA molecules that are deficient for HDAg expression; then, the induction of S-HDAg expression launches the replication of HDV RNA. Using this model, Chang and colleagues[58] have shown that alpha-amanitin would block the accumulation of genomic, antigenomic, and messenger HDV RNAs, suggesting that only Pol II was involved in the synthesis of all types of HDV transcripts. The same conclusion was reached when nuclear run-on assays were performed to demonstrate that synthesis, rather than accumulation or stability of the HDV RNAs, was inhibited by alpha-amanitin.[59] In live cells, there is also indirect evidence that Pol II could carry out RNA synthesis from an HDV RNA template, including template switching, non-templated nucleotide incorporation, and RNA recombination.[60,61]

Syntheses of Genomic and Antigenomic HDV RNAs Occur in Different Nuclear Compartments

HDV RNA replication was associated with an intracellular co-localization of HDAg with Pol II and the splicing factor SC35 in structures of the nucleoplasm referred to as speckles or transcription factories (Fig. 5).[59] In the absence of HDV replication, HDAg would instead localize in the nucleoli. According to Huang *et al.*,[62] when HDAg was artificially targeted to the nucleoli, initiation of genomic RNA synthesis from antigenomic RNA was abolished, whereas antigenomic RNA synthesis from genomic template was not affected. Then, forcing the release of S-HDAg from the nucleoli correlated with a restoration of genomic RNA synthesis and an apparent interaction between S-HDAg and Pol II. The data suggested that synthesis of genomic and antigenomic RNAs would segregate to different subnuclear compartments and use different cell machineries: synthesis of genomic RNA and HDAg

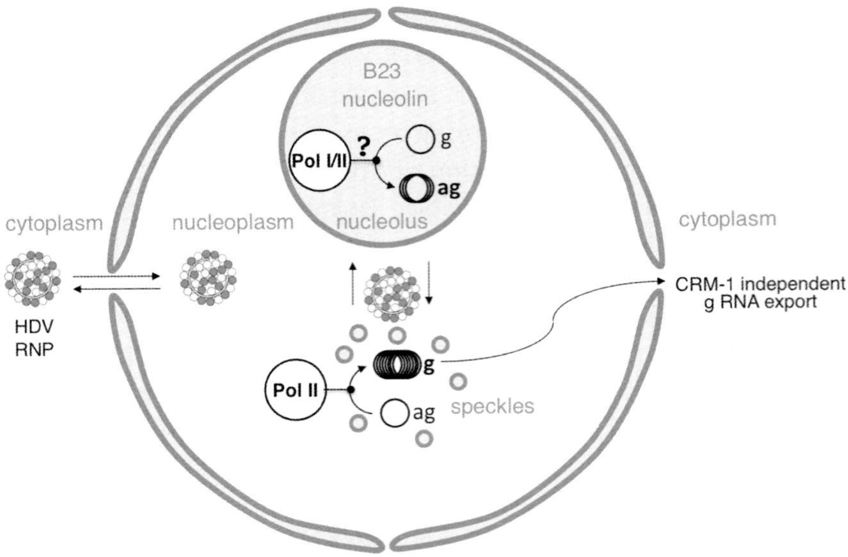

Fig. 5. Replication of HDV RNA in the cellular nucleus. Transcription of genomic RNA (g) from antigenomic (ag) is thought to occur in structures referred to as speckles in the nucleoplasm, and to be carried out by cellular RNA-polymerase II (Pol II). Transcription of antigenomic from genomic templates might co-localize with nucleolin and B23 in the nucleolus. Export of genomic HDV RNA from the nucleus was reported to be independent of HDAg and CRM-1.[77]

mRNA would occur in the nucleoplasm and be carried out by Pol II, whereas synthesis of antigenomic RNA would take place in the nucleoli but the origin of the RNA polymerase activity (Pol I or Pol II) is still uncertain (Fig. 5).

HDV RNA Replication is Controlled by HDAg Post-Translational Modifications

The transcription of the different HDV RNA species appears to be regulated by post-translational modifications of S-HDAg.[63] In fact, S-HDAg is subjected to important post-translational modifications, including Arg-13 methylation, Ser-177 phosphorylation, and Lys-72 acetylation, which are essential to HDV mRNA transcription and genomic RNA synthesis but dispensable for antigenomic RNA synthesis. Syntheses of the three RNA species were shown to have different sensitivities to acetylation/deacetylation inhibitors, and this was taken as further evidence that antigenomic RNA, on the one hand, and genomic and messenger RNAs, on the other, are carried out by independent mechanisms. Furthermore, it was observed that both S- and L-HDAg could interact with the Yin Yang 1(YY1) transcription factor and its associated acetyltransferases, CREB-binding protein (CBP), and p300, to modulate the replication of HDV RNA.[64] Exogenous expression of YY1, CBP, or p300 could activate HDV RNA replication, and the p300 acetyltransferase activity was important in this process. Moreover, S-HDAg could be acetylated *in vivo*, and the treatment with cellular deacetylase inhibitor increased S-HDAg acetylation and HDV RNA replication.

In Vitro Transcription of HDV RNA by Cellular RNA Polymerase II

Several studies have been conducted *in vitro* using various transcription assays to characterize the RNA-directed transcription activity of Pol II and its implication in HDV RNA replication. In one study, it was shown that S-HDAg could stimulate Pol II transcription elongation.[65] The rationale for this experiment was that elongation by Pol II was known to be negatively regulated by the DRB-sensitivity inducing factor (DSIF) or the negative elongation factor (NELF), a subunit of which displays limited sequence similarity to S-HDAg. In the *in vitro* assay, S-HDAg could bind directly to Pol II and stimulate transcription by displacing NELF. S-HDAg could also regulate Pol II-directed elongation for both cellular messenger RNA synthesis and HDV RNA transcription.[65] Using reconstituted elongation complexes, S-HDAg was shown to interact with the clamp of Pol II, a mobile structure that usually holds DNA and RNA in place, affecting transcription fidelity by accelerating forward the translocation of Pol II.[66] This was consistent with the capacity of S-HDAg to assist Pol II in RNA-dependent RNA transcription.

The intrinsic RdRP activity of Pol II was also shown *in vitro* using pure polymerase, NTPs, and an RNA template-product scaffold.[67] In these experimental conditions, RdRP activity appeared slower and less processive than DNA-dependent activity. Moreover, RdRP activity was obtained using part of the HDV antigenome as a template. A complex made of transcription factor IIS (TFIIS) associated with Pol II could cleave an HDV RNA strand, create a reactive stem-loop, and extend the new RNA 3′-end. Pol II could extend different forms of short RNA stem-loops with a 5′ extension.

According to Abrahem and colleagues,[68] an RNA-Pol II preinitiation complex could form in nuclear extracts on an RNA promoter derived from genomic HDV RNA, to initiate a transcription reaction that involves the TATA-binding protein. Transcription initiated by direct binding of purified Pol II to the RNA promoter could be inhibited by an antibody specific to the largest subunit of Pol II. The initiation complex was shown to contain the core subunit of Pol II and the main transcription factors.[68]

A Mechanism for Initiation of HDV RNA Transcription by Pol II

The promoter activity for synthesis of antigenomic HDV RNA was proposed to map to a 29-nucleotide region (nucleotides 1650–1679) and the one for genomic RNA to a 160-nucleotide region around position 1679 that overlaps with the antigenomic promoter region on the opposite strand.[69] These sequences both map to one tip of the rod-like structures (referred to here as the pode) of the genomic and the antigenomic HDV RNAs, in a highly conserved region (Fig. 6). Pol II would recognize the quasi-double-stranded structures of these hairpins as promoters. Then, in a recent study, it was shown that the pode hairpins (genomic and antigenomic) were critical for the initiation of transcription.[70] In cells undergoing active HDV RNA replication, two 5′-capped HDV RNA species of approximately 18–25 nucleotides were identified: one antigenomic oligo-RNA corresponded to the 5′-end of HDAg mRNA and could interact with HDAg and Pol II; the second oligo-RNA mapped to a structurally analogous

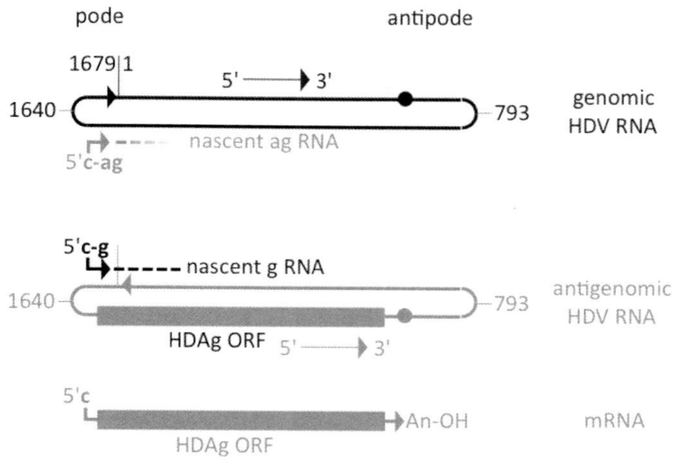

Fig. 6. Initiation of HDV RNA transcription. Genomic, antigenomic, and messenger HDV RNAs are represented. Initiation of transcription is to occur at the pode of the genomic and antigenomic HDV RNAs, and to be mediated by small 5′-capped RNAs (5′-c-ag and 5′-c-g) at the indicated positions of the antigenomic and genomic templates, respectively. The 5′-end of the antigenomic small RNA (5′-c-ag) corresponds to that of the HDAg mRNA at position 1630/1631. The genomic small RNA starts at around position 1645. Arrows on the RNA strands indicate the 5′-to 3′-direction.

region on the genomic strand (Fig. 6). These genomic and antigenomic small RNAs would act in initiating transcription of antigenomic and genomic RNAs, respectively. In the same study,[70] it was shown that HDAg could interact with the RNA helicase MOV10, and cells depleted for MOV10 would not support HDV RNA replication while permissive for translation from HDAg mRNA, indicating an involvement of MOV10 in RNA-directed transcription.

Taken together, the above data further suggest an RdRP activity of Pol II and its central role in the HDV RNA replication mechanism.

HDV Ribozymes

The rolling-circle mechanism of HDV RNA replication includes the generation of longer-than-unit-length products that are cleaved in cis by a ribozyme present in both the genomic and antigenomic RNA strands to produce monomers. The linear monomers are then circularized and serve as templates for further rounds of replication by the rolling-circle mechanism. The genomic and antigenomic HDV ribozymes are similar in sequence and structure,[71] but they differ from the sequence of hammerhead or hairpin ribozymes. The enzymatic activity is a trans-esterification reaction that converts a 3′,5′-phosphodiester bond to a 2′,3′-cyclic monophosphate group and a 5′-hydroxyl group. The reaction can be observed *in vitro* in the absence of cellular proteins, and it is modulated by divalent metal ions. The minimal size of the HDV ribozymes for self-cleavage is approximately 85 nucleotides, and a single nucleotide, 5′ to the cleavage site, is sufficient for cleavage. Considerable advances have been made recently that provide details about the mechanism of HDV ribozymes activity (for review, see Ref. 72). *In vivo*, self-cleavage and ligation are probably regulated to best integrate into the HDV life cycle. For instance, since the genomic RNA serves as a template for synthesis of antigenomic RNA and also for transcription of the HDAg mRNA, one might logically expect that on the nascent antigenomic transcript, ribozyme activity could interfere with polyadenylation reaction, because the self-cleavage site is located just 33 nt downstream of the polyadenylation site for HDAg mRNA. An efficient polyadenylation may thus require that ribozyme activity be suppressed. This issue was addressed in a study conducted *in vitro* using an HDV antigenomic ribozyme precursor RNA.[73] It was shown that non-ribozyme sequences located upstream of the cleavage site and including the polyadenylation site sequence could affect the ribozyme activity. Relative to a ribozyme precursor without the extra upstream sequences, the kinetic profile for self-cleavage of the longer precursor was altered. During the HDV RNA replication process, a modulation of ribozyme activity could eventually be exerted through competing alternative structures to control the balance between maturation of the mRNA and processing of replication intermediates. However, as indicated above, mRNA transcription and synthesis of antigenomic RNA likely occur in separate nuclear compartments, and these two mechanisms may thus be totally independent of each other.[56]

An HDV-like Ribozyme Encoded by the Human Genome

The structure of the HDV ribozymes, which, until recently, was thought to be unique among other known ribozymes, has raised questions about its phylogenetic origin. Interestingly, a recent publication reports the identification of a conserved mammalian sequence coding for an active HDV-like ribozyme that resides in an intron of the gene for cytoplasmic polyadenylation element binding protein 3 (CPEB3).[74] The so-called CPEB3 ribozyme was shown to be structurally and biochemically related to HDV ribozymes, suggesting that HDV and CPEB3 ribozymes are evolutionarily related. Prior to this report, a cellular homolog of HDAg, termed delta-interacting protein A (DIPA), had been identified, which suggested that HDV had evolved from a primitive viroid-like RNA through capture of a cellular transcript.[75] Since HDV is found only in humans, these findings may indicate that the origin of HDV may be recent, with both its ribozyme and HDAg coding sequence arising from the human transcriptome (Fig. 7).[74]

Export of Genomic HDV RNA from the Nucleus

The trafficking events required for HDV entry are largely unknown. It is generally assumed that the HDV RNP can shuttle between nucleus and cytoplasm,[76] as expected if one considers that RNA replication occurs in the nucleus and assembly at the ER membrane. All forms of HDV RNA are synthesized and accumulated in the nucleus. Since extracellular HDV virions contain exclusively the genomic HDV RNA strand, there must be a process for selection of this molecule to assemble RNPs that are ultimately enveloped and released as virions from the infected cell. As reported by Macnaughton *et al.*,[77] soon after synthesis, a significant amount of newly synthesized genomic HDV RNA could be detected not only in the nucleus where replication occurs, but also in the cytoplasm. In contrast, most of the antigenomic HDV RNA was retained in the

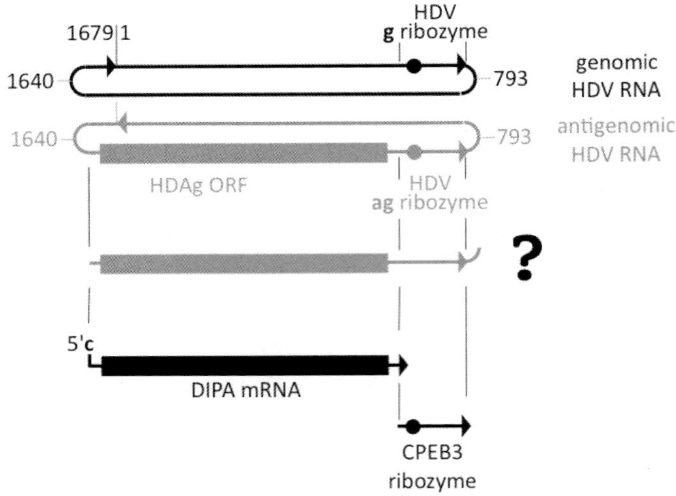

Fig. 7. A possible scenario for the origin of HDV. HDV may have evolved from a cellular HDV-like ribozyme, residing in an intron of the CPEB3 gene (CPEB3 ribozyme) fused to the mRNA of a cellular homolog of HDAg (DIPA).

nucleus, and the proportion of genomic HDV RNA in the nucleus compared to cytoplasm remained relatively constant over time, indicating that export of genomic HDV RNA occurred continuously. Surprisingly, L-HDAg, which is responsible for virus packaging, was not required for nuclear export of genomic RNA. The mechanism for export of genomic RNA was resistant to leptomycin B, indicating that a cell region maintenance-1 (Crm1)-independent pathway was involved. Genomic HDV RNA would then be exported by the same mechanism as that for the splicing-dependent export of cellular mRNAs, consistent with genomic RNA being synthesized by Pol II.

The HDAg Proteins

As indicated above, a single ORF on the HDV genome encodes the S- and L-HDAg proteins, which differ from each other by only 19 additional amino acid residues at the carboxyl terminus of L-HDAg.

The origin of these two isoforms was revealed by the analysis of HDV RNA sequences that emerge during the natural course of infection: a heterogeneity at codon 196 in the HDAg ORF was observed, changing from a UAG stop codon to UGG for tryptophan, and corresponding to S-HDAg (195 aas in length) and L-HDAg (214 aas), respectively. The sequence modification was found to occur during the course of HDV RNA replication, leading to the production of S-HDAg in the early phase of active HDV RNA replication, and L-HDAg in the late phase of assembly. It was later found to result from an editing event on the full-length antigenomic HDV RNA, catalyzed by cellular adenosine deaminase that acts on RNA (ADAR). ADARs are RNA-editing enzymes that catalyze a change from adenosine to inosine, a nucleoside that is then recognized as guanosine by the translational machinery. The small form of ADAR-1 is assumed to be the enzyme in charge of editing HDV RNA.[78] Hence, this enzyme plays a critical role in the regulation of the HDV life replication cycle by controlling the switch between HDV RNA amplification and virion assembly. Interestingly, there are significant differences in the editing efficiency between HDV genotypes I and III.[79]

The Functions of the HDAg Proteins

A number of functional domains have been identified within the 195 amino acid sequence common to both HDAg proteins, which are involved at different stages of the replication cycle (Fig. 8). An RNA chaperon domain is located in the lysine-rich coiled-coil domain of both S- and L-HDAg spanning amino acids 13–60, which also promotes protein–protein interactions. A bipartite nuclear localization signal (NLS) had been initially identified in the central region of HDAg (aas 69–88), but more recently, a single short sequence (aas 66–75) was shown to be sufficient for nuclear localization activity.[80] In addition, two nucleolin binding sites critical for nucleolar targeting of HDAg and for HDV RNA replication, spanning aas 35–50 and 51–65, respectively, were identified.[81] An RNA-binding domain required for HDV RNA binding and replication, comprising two arginine-rich motifs, maps to the carboxy terminal third of the

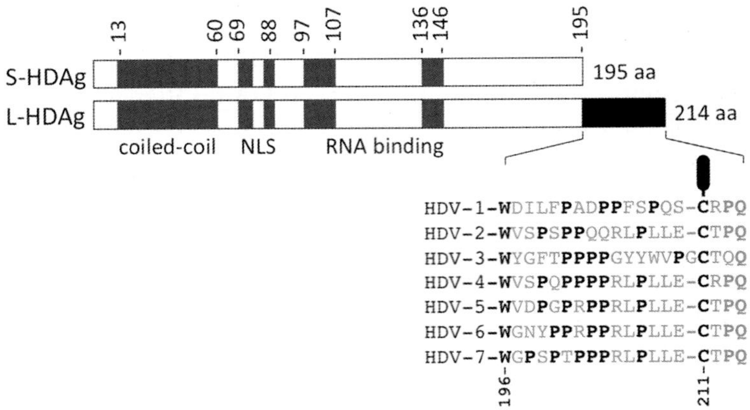

Fig. 8. Functional domains of the HDAg proteins. NLS, nuclear localization signal. The alignment of the L-HDAg carboxy-terminal sequences of seven HDV genotypes shown. Conserved residues are tryptophan at position 196 and the carboxy-terminal CXXQ box. At least 4 proline residues, indicated in bold, are present. Farnesylation of cysteine 211 is indicated.

proteins (aas 97–146). Both S- and L-HDAg bind HDV RNA, but only S-HDAg is essential to genome replication.[82] According to Chao and colleagues,[82] not only is L-HDAg unable to assist RNA replication, it can also act as a negative dominant. However, for Macnaughton and Lai,[83] L-HDAg is not an inhibitor of replication when expressed at the beginning of the replication cycle. What is clearly established is that S-HDAg is essential for replication and L-HDAg for assembly.

As mentioned above, the 19 carboxy-terminal amino acids of L-HDAg, which make the sequence difference with S-HDAg, contain a signal for farnesylation,[84] and this post-translational modification is essential for the ability of L-HDAg to drive HDV assembly.[85] It is in fact necessary and sufficient for L-HDAg to interact with S-HBsAg. This short sequence includes a CXXQ motif that directs the farnesylation of the protein at cysteine residue 211, a modification that could promote membrane-binding activity necessary for the interaction with the envelope proteins (Fig. 8). The packaging signal on L-HDAg relies also on the presence of proline residues diversely spaced within the carboxy-terminal 19 amino acids, according to genotype.

Within this short sequence lies a chromosome region maintenance 1-independent nuclear export signal (NES) that could also play a role in HDV assembly.[21] It was shown to mediate an interaction between L-HDAg and a protein designated NESI (for NES interacting protein). NESI consists of 467 amino acid residues, including a putative actin-binding site and an NLS, and its down-regulation was shown to abolish L-HDAg-mediated nuclear export and virion assembly.[86]

More recently, a putative clathrin box, a motif conserved in clathrin adaptors, was identified in the L-HDAg carboxyl terminus, and a direct interaction between the L-HDAg clathrin box and the amino terminus of the clathrin heavy chain could be characterized. Inhibitors of clathrin-mediated exocytosis and endosomal trafficking were shown to block HDV assembly but not to affect SVP assembly, raising the possibility that L-HDAg interaction with clathrin might be relevant to HDV assembly.[87]

Additional HDAg post-translational modifications have been documented, such as phosphorylation, acetylation, and methylation, that impact the HDV RNA replication mechanism. [88] For instance, the methylation reaction that occurs at arginine-13 in the RNA-binding domain of S-HDAg is essential for synthesis of genomic RNA strand. A methylation-defective mutant (R13A) could not support HDV RNA replication, was unable to form a speckled structure in the cell nucleus, and, unlike wt S-HDAg, permeated into the cytoplasm.[89] S-HDAg is also phosphorylated at Serine-177, described as critical for S-HDAg interaction with Pol II and for synthesis of genomic HDV RNA.[90] As already indicated, acetylation of S-HDAg at Lys-72 is required for genomic RNA synthesis and HDV mRNA transcription, but dispensable for antigenomic RNA synthesis. The acetylation status of S-HDAg could thus constitute a molecular switch for the synthesis of genomic or antigenomic RNAs.[63]

Overall, it is remarkable that HDV can manage to fulfill numerous functions at different steps of the HDV life cycle using the translation product of a single ORF on the HDV genome. It is achieved by combining an RNA-editing strategy with numerous post-translational modifications that in turn promote interaction with several proteins, including enzymes such as Pol II.

Conclusions

The recent advances in the study of the HDV replication cycle further point to the uniqueness of this infectious agent. Interestingly, if one considers the following findings — (i) the existence of a cellular homolog of the HDAg protein,[75] (ii) the presence of an HDV-like ribozyme in a cellular gene intron,[74] and (iii) the demonstration of an intrinsic RdRP activity of cellular RNA polymerase II and its apparent role in the replication mechanism of HDV RNA — the hypothesis of the HDV arising from the human transcriptome becomes plausible. Furthermore, as is true for viruses in general, experiments performed on HDV have provided important information on general aspects of biology, including the abovementioned RNA-editing process, the HDV-related cellular ribozymes, and the Pol II RdRP activity. The interaction between HDV and its helper HBV is also a unique phenomenon in virology that needs to be further explored because it may lead to the development of new HDV-specific antivirals that could, for instance, target the HDV matrix domain of the HBV envelope proteins.

Acknowledgments

The work in the author's laboratory is supported by ANRS grants.

References

1. Rizzetto M, Canese MG, Arico S, *et al.* (1977) Immunofluorescence detection of new antigen-antibody system (delta/anti-delta) associated to hepatitis B virus in liver and in serum of HBsAg carriers. *Gut* **18**: 997–1003.
2. Bonino F, Heermann KH, Rizzetto M, Gerlich WH. (1986) Hepatitis delta virus: protein composition of delta antigen and its hepatitis B virus-derived envelope. *J Virol* **58**: 945–950.
3. Ponzetto A, Cote PJ, Popper H, *et al.* (1984) Transmission of the hepatitis B virus-associated delta agent to the eastern woodchuck. *Proc Natl Acad Sci USA* **81**: 2208–2212.
4. Rizzetto M, Canese MG, Gerin JL, *et al.* (1980) Transmission of the hepatitis B virus-associated delta antigen to chimpanzees. *J Infect Dis* **141**: 590–602.
5. Denniston KJ, Hoyer BH, Smedile A, *et al.* (1986) Cloned fragment of the hepatitis delta virus RNA genome: sequence and diagnostic application. *Science* **232**: 873–875.

6. Wang KS, Choo OL, Weiner AJ, *et al.* (1987) The viroid-like structure of the hepatitis delta (delta) genome: synthesis of a viral antigen in recombinant bacteria. *Prog Clin Biol Res* **234**: 71–82.
7. Chen PJ, Kalpana G, Goldberg J, *et al.* (1986) Structure and replication of the genome of the hepatitis delta virus. *Proc Natl Acad Sci USA* **83**: 8774–8778.
8. Taylor JM (2006) Hepatitis delta virus. *Virology* **344**: 71–76.
9. Ryu WS, Netter HJ, Bayer M, Taylor J. (1993) Ribonucleoprotein complexes of hepatitis delta virus. *J Virol* **67**: 3281–3287.
10. Sureau C, Moriarty AM, Thornton GB, Lanford RE. (1992) Production of infectious hepatitis delta virus *in vitro* and neutralization with antibodies directed against hepatitis B virus pre-S antigens. *J Virol* **66**: 1241–1245.
11. Wang CJ, Chen PJ, Wu JC, *et al.* (1991) Small-form hepatitis B surface antigen is sufficient to help in the assembly of hepatitis delta virus-like particles. *J Virol* **65**: 6630–6636.
12. Komla-Soukha I, Sureau C. (2006) A tryptophan-rich motif in the carboxyl terminus of the small envelope protein of hepatitis B virus is central to the assembly of hepatitis delta virus particles. *J Virol* **80**: 4648–4655.
13. O'Malley B, Lazinski DW. (2005) Roles of carboxyl-terminal and farnesylated residues in the functions of the large hepatitis delta antigen. *J Virol* **79**: 1142–1153.
14. Sureau C, Guerra B, Lanford RE. (1993) Role of the large hepatitis B virus envelope protein in infectivity of the hepatitis delta virion. *J Virol* **67**: 366–372.
15. Koytak ES, Yurdaydin C, Glenn JS. (2007) Hepatitis d. *Curr Treat Options Gastroenterol* **10**: 456–463.
16. Cross TJ, Rizzi P, Horner M, *et al.* (2008) The increasing prevalence of hepatitis delta virus (HDV) infection in South London. *J Med Virol* **80**: 277–282.
17. Wedemeyer H, Heidrich B, Manns MP. (2007) Hepatitis D virus infection--not a vanishing disease in Europe! *Hepatology* **45**: 1331–1332; author reply 1332–1333.
18. Deny P. (2006) Hepatitis delta virus genetic variability: from genotypes I, II, III to eight major clades? *Curr Top Microbiol Immunol* **307**: 151–171.
19. Farci P, Chessa L, Balestrieri C, Serra G, Lai ME. (2007) Treatment of chronic hepatitis D. *J Viral Hepat* **14** Suppl **1:** 58–63.
20. Gudima S, Chang J, Moraleda G, *et al.* (2002) Parameters of human hepatitis delta virus genome replication: the quantity, quality, and intracellular distribution of viral proteins and RNA. *J Virol* **76**: 3709–3719.
21. Lee CH, Chang SC, Wu CH, Chang MF. (2001) A novel chromosome region maintenance 1-independent nuclear export signal of the large form of hepatitis delta antigen that is required for the viral assembly. *J Biol Chem* **276**: 8142–8148.
22. Otto JC, Casey PJ. (1996) The hepatitis delta virus large antigen is farnesylated both *in vitro* and in animal cells. *J Biol Chem* **271**: 4569–4572.

23. Chang FL, Chen PJ, Tu SJ, *et al.* (1991) The large form of hepatitis delta antigen is crucial for assembly of hepatitis delta virus. *Proc Natl Acad Sci USA* **88**: 8490–8494.
24. Chai N, Chang HE, Nicolas E, *et al.* (2008) Properties of subviral particles of hepatitis B virus. *J Virol* **82**: 7812–7817.
25. Radjef N, Gordien E, Ivaniushina V, *et al.* (2004) Molecular phylogenetic analyses indicate a wide and ancient radiation of African hepatitis delta virus, suggesting a deltavirus genus of at least seven major clades. *J Virol* **78**: 2537–2544.
26. Barrera A, Guerra B, Lee H, Lanford RE. (2004) Analysis of host range phenotypes of primate hepadnaviruses by *in vitro* infections of hepatitis D virus pseudotypes. *J Virol* **78**: 5233–5243.
27. Liu YT, Brazas R, Ganem D. (2001) Efficient hepatitis delta virus RNA replication in avian cells requires a permissive factor(s) from mammalian cells. *J Virol* **75**: 7489–7493.
28. O'Malley B, Lazinski D. (2002) A hepatitis B surface antigen mutant that lacks the antigenic loop region can self-assemble and interact with the large hepatitis delta antigen. *J Virol* **76**: 10060–10063.
29. Chen PJ, Lai WJ, Wang CJ, Chen DS. (1993) Hepatitis B surface antigen and large-form hepatitis delta antigen in HDV assembly: a further study. *Prog Clin Biol Res* **382**: 29–34.
30. Blanchet M, Sureau C. (2006) Analysis of the cytosolic domains of the hepatitis B virus envelope proteins for their function in viral particle assembly and infectivity. *J Virol* **80**: 11935–11945.
31. Kian Chua P, Lin MH, Shih C. (2006) Potent inhibition of human Hepatitis B virus replication by a host factor Vps4. *Virology* **354**: 1–6.
32. Lambert C, Doring T, Prange R. (2007) Hepatitis B virus maturation is sensitive to functional inhibition of ESCRT-III, Vps4, and gamma 2-adaptin. *J Virol* **81**: 9050–9060.
33. Watanabe T, Sorensen EM, Naito A, *et al.* (2007) Involvement of host cellular multivesicular body functions in hepatitis B virus budding. *Proc Natl Acad Sci USA* **104**: 10205–10210.
34. Le Seyec J, Chouteau P, Cannie I, *et al.* (1999) Infection process of the hepatitis B virus depends on the presence of a defined sequence in the pre-S1 domain. *J Virol* **73**: 2052–2057.
35. Fernholz D, Galle PR, Stemler M, *et al.* (1993) Infectious hepatitis B virus variant defective in pre-S2 protein expression in a chronic carrier. *Virology* **194**: 137–148.
36. Sureau C, Guerra B, Lee H. (1994) The middle hepatitis B virus envelope protein is not necessary for infectivity of hepatitis delta virus. *J Virol* **68**: 4063–4066.
37. Blanchet M, Sureau C. (2007) Infectivity determinants of the hepatitis B virus pre-S domain are confined to the *N*-terminal 75 amino acid residues. *J Virol* **81**: 5841–5849.

38. Abou-Jaoude G, Molina S, Maurel P, Sureau C. (2007) Myristoylation signal transfer from the large to the middle or the small HBV envelope protein leads to a loss of HDV particles infectivity. *Virology* **365**: 204–209.
39. Gripon P, Le Seyec J, Rumin S, Guguen-Guillouzo C. (1995) Myristylation of the hepatitis B virus large surface protein is essential for viral infectivity. *Virology* **213**: 292–299.
40. Gripon P, Cannie I, Urban S. (2005) Efficient inhibition of hepatitis B virus infection by acylated peptides derived from the large viral surface protein. *J Virol* **79**: 1613–1622.
41. Glebe D. (2006) Attachment sites and neutralising epitopes of hepatitis B virus. *Minerva Gastroenterol Dietol* **52**: 3–21.
42. Glebe D, Urban S, Knoop EV, *et al.* (2005) Mapping of the hepatitis B virus attachment site by use of infection-inhibiting preS1 lipopeptides and tupaia hepatocytes. *Gastroenterology* **129**: 234–245.
43. Engelke M, Mills K, Seitz S, *et al.* (2006) Characterization of a hepatitis B and hepatitis delta virus receptor binding site. *Hepatology* 43: 750–760.
44. Petersen J, Dandri M, Mier W, *et al.* (2008) Prevention of hepatitis B virus infection *in vivo* by entry inhibitors derived from the large envelope protein. *Nat Biotechnol* **26**: 335–341.
45. Schulze A, Gripon P, Urban S. (2007) Hepatitis B virus infection initiates with a large surface protein-dependent binding to heparan sulfate proteoglycans. *Hepatology* **46**: 1759–1768.
46. Leistner CM, Gruen-Bernhard S, Glebe D. (2008) Role of glycosaminoglycans for binding and infection of hepatitis B virus. *Cell Microbiol* **10**: 122–133.
47. Jaoude GA, Sureau C. (2005) Role of the antigenic loop of the hepatitis B virus envelope proteins in infectivity of hepatitis delta virus. *J Virol* **79**: 10460–10466.
48. Abou-Jaoude G, Sureau C. (2007) Entry of hepatitis delta virus requires the conserved cysteine residues of the hepatitis B virus envelope protein antigenic loop and is blocked by inhibitors of thiol-disulfide exchange. *J Virol* **81**: 13057–13066.
49. Kock J, Borst EM, Schlicht HJ. (1996) Uptake of duck hepatitis B virus into hepatocytes occurs by endocytosis but does not require passage of the virus through an acidic intracellular compartment. *J Virol* **70**: 5827–5831.
50. Bremer CM, Bung C, Kott N, *et al.* (2009) Hepatitis B virus infection is dependent on cholesterol in the viral envelope. *Cell Microbiol* **11**: 249–260.
51. Funk A, Mhamdi M, Hohenberg H, *et al.* (2008) Duck hepatitis B virus requires cholesterol for endosomal escape during virus entry. *J Virol* **82**: 10532–10542.
52. Kann M, Schmitz A, Rabe B. (2007) Intracellular transport of hepatitis B virus. *World J Gastroenterol* **13**: 39–47.
53. Cheng J, Kapranov P, Drenkow J, *et al.* (2005) Transcriptional maps of 10 human chromosomes at 5-nucleotide resolution. *Science* **308**: 1149–1154.

54. Gudima S, Wu SY, Chiang CM, *et al.* (2000) Origin of hepatitis delta virus mRNA. *J Virol* **74**: 7204–7210.
55. Modahl LE, Macnaughton TB, Zhu N, *et al.* (2000) RNA-Dependent replication and transcription of hepatitis delta virus RNA involve distinct cellular RNA polymerases. *Mol Cell Biol* **20**: 6030–6039.
56. Modahl LE, Lai MM. (1998) Transcription of hepatitis delta antigen mRNA continues throughout hepatitis delta virus (HDV) replication: A new model of HDV RNA transcription and replication. *J Virol* **72**: 5449–5456.
57. Macnaughton TB, Shi ST, Modahl LE, Lai MM. (2002) Rolling circle replication of hepatitis delta virus RNA is carried out by two different cellular RNA polymerases. *J Virol* **76**: 3920–3927.
58. Chang J, Nie X, Gudima S, Taylor J. (2006) Action of inhibitors on accumulation of processed hepatitis delta virus RNAs. *J Virol* **80**: 3205–3214.
59. Chang J, Nie X, Chang HE, Han Z, Taylor J. (2008) Transcription of hepatitis delta virus RNA by RNA polymerase II. *J Virol* **82**: 1118–1127.
60. Chang J, Taylor J. (2002) *In vivo* RNA-directed transcription, with template switching, by a mammalian RNA polymerase. *EMBO J* **21**: 157–164.
61. Gudima SO, Chang J, Taylor JM. (2006) Restoration *in vivo* of defective hepatitis delta virus RNA genomes. *RNA* **12**: 1061–1073.
62. Huang WH, Chen YS, Chen PJ. (2008) Nucleolar targeting of hepatitis delta antigen abolishes its ability to initiate viral antigenomic RNA replication. *J Virol* **82**: 692–699.
63. Tseng CH, Jeng KS, Lai MM. (2008) Transcription of subgenomic mRNA of hepatitis delta virus requires a modified hepatitis delta antigen that is distinct from antigenomic RNA synthesis. *J Virol* **82**: 9409–9416.
64. Huang WH, Mai RT, Lee YH. (2008) Transcription factor YY1 and its associated acetyltransferases CBP and p300 interact with hepatitis delta antigens and modulate hepatitis delta virus RNA replication. *J Virol* **82**: 7313–7324.
65. Yamaguchi Y, Filipovska J, Yano K, *et al.* (2001) Stimulation of RNA polymerase II elongation by hepatitis delta antigen. *Science* **293**: 124–127.
66. Yamaguchi Y, Mura T, Chanarat S, *et al.* (2007) Hepatitis delta antigen binds to the clamp of RNA polymerase II and affects transcriptional fidelity. *Genes Cells* **12**: 863–875.
67. Lehmann E, Brueckner F, Cramer P. (2007) Molecular basis of RNA-dependent RNA polymerase II activity. *Nature* **450**: 445–449.
68. Abrahem A, Pelchat M. (2008) Formation of an RNA polymerase II preinitiation complex on an RNA promoter derived from the hepatitis delta virus RNA genome. *Nucleic Acids Res* **36**: 5201–5211.
69. Macnaughton TB, Beard MR, Chao M, *et al.* (1993) Endogenous promoters can direct the transcription of hepatitis delta virus RNA from a recircularized cDNA template. *Virology* **196**: 629–636.

70. Haussecker D, Cao D, Huang Y, *et al.* (2008) Capped small RNAs and MOV10 in human hepatitis delta virus replication. *Nat Struct Mol Biol* **15**: 714–721.
71. Ferre-D'Amare AR, Zhou K, Doudna JA. (1998) Crystal structure of a hepatitis delta virus ribozyme. *Nature* **395**: 567–574.
72. Been MD. (2006) HDV ribozymes. *Curr Top Microbiol Immunol* **307**: 47–65.
73. Brown AL, Perrotta AT, Wadkins TS, Been MD. (2008) The poly(A) site sequence in HDV RNA alters both extent and rate of self-cleavage of the antigenomic ribozyme. *Nucleic Acids Res* **36**: 2990–3000.
74. Salehi-Ashtiani K, Luptak A, Litovchick A, Szostak JW. (2006) A genome wide search for ribozymes reveals an HDV-like sequence in the human CPEB3 gene. *Science* **313**: 1788–1792.
75. Brazas R, Ganem D. (1996) A cellular homolog of hepatitis delta antigen: implications for viral replication and evolution. *Science* **274**: 90–94.
76. Tavanez JP, Cunha C, Silva MC, *et al.* (2002) Hepatitis delta virus ribonucleoproteins shuttle between the nucleus and the cytoplasm. *RNA* **8**: 637–646.
77. Macnaughton TB, Lai MM. (2002) Genomic but not antigenomic hepatitis delta virus RNA is preferentially exported from the nucleus immediately after synthesis and processing. *J Virol* **76**: 3928–3935.
78. Wong SK, Lazinski DW. (2002) Replicating hepatitis delta virus RNA is edited in the nucleus by the small form of ADAR1. *Proc Natl Acad Sci USA* **99**: 15118–15123.
79. Casey JL, Tennant BC, Gerin JL (2006) Genetic changes in hepatitis delta virus from acutely and chronically infected woodchucks. *J Virol* **80**: 6469–6477.
80. Alves C, Freitas N, Cunha C (2008) Characterization of the nuclear localization signal of the hepatitis delta virus antigen. *Virology* **370**: 12–21.
81. Lee CH, Chang SC, Chen CJ, Chang MF. (1998) The nucleolin binding activity of hepatitis delta antigen is associated with nucleolus targeting. *J Biol Chem* **273**: 7650–7656.
82. Chao M, Hsieh SY, Taylor J. (1990) Role of two forms of hepatitis delta virus antigen: evidence for a mechanism of self-limiting genome replication. *J Virol* **64**: 5066–5069.
83. Macnaughton TB, Lai MM. (2002) Large hepatitis delta antigen is not a suppressor of hepatitis delta virus RNA synthesis once RNA replication is established. *J Virol* **76**: 9910–9919.
84. Glenn JS, Watson JA, Havel CM, White JM. (1992) Identification of a prenylation site in delta virus large antigen. *Science* **256**: 1331–1333.
85. Bordier BB, Ohkanda J, Liu P, *et al.* (2003) *In vivo* antiviral efficacy of prenylation inhibitors against hepatitis delta virus. *J Clin Invest* **112**: 407–414.
86. Wang YH, Chang SC, Huang C, *et al.* (2005) Novel nuclear export signal-interacting protein, NESI, critical for the assembly of hepatitis delta virus. *J Virol* **79**: 8113–8120.

87. Huang C, Chang SC, Yu IC, *et al.* (2007) Large hepatitis delta antigen is a novel clathrin adaptor-like protein. *J Virol* **81**: 5985–5994.
88. Huang WH, Chen CW, Wu HL, Chen PJ. (2006) Post-translational modification of delta antigen of hepatitis D virus. *Curr Top Microbiol Immunol* **307**: 91–112.
89. Li YJ, Stallcup MR, Lai MM. (2004) Hepatitis delta virus antigen is methylated at arginine residues, and methylation regulates subcellular localization and RNA replication. *J Virol* **78**: 13325–13334.
90. Chen YS, Huang WH, Hong SY, *et al.* (2008) ERK1/2-mediated phosphorylation of small hepatitis delta antigen at serine 177 enhances hepatitis delta virus antigenomic RNA replication. *J Virol* **82**: 9345–9358.

HEPATITIS C VIRUS

Chapter 8

Development of an HCV Cell Culture System

Takanobu Kato and Takaji Wakita*

Introduction

Since the discovery of the hepatitis C virus (HCV) in 1989, it has been recognized as a serious medical and public health problem.[1] An estimated 170 million people are currently infected with HCV, and many trials attempting to eradicate this virus from chronically infected patients have been undertaken. However, the establishment of new therapy and the evaluation of anti-HCV activity of new compounds have been hampered by the lack of a reliable and reproducible cell culture system for HCV. Because HCV was identified by molecular cloning without visualization of virus particles, growing HCV in cultured cells remains difficult. Several in vitro replication models of HCV were reported using established cell lines or primary hepatocytes from humans or other susceptible animals, but the viral yields were insufficient for the desired uses.[2–10] The HCV infection and

* Corresponding Author: Takaji Wakita, M.D., Ph.D. Department of Virology II, National Institute of Infectious Diseases, Toyama 1-23-1, Shinjuku, Tokyo 162-8640, Japan, Tel: +81-3-5285-1111, Fax: +81-3-5285-1161, e-mail: wakita@nih.go.jp

replication observed in these models were limited, and highly sensitive reverse transcription polymerase chain reaction (RT-PCR) were required to detect the replicating HCV.

In 1999, the establishment of the subgenomic HCV RNA replicon system was reported — a major breakthrough allowing the study of HCV replication in cultured cells.[11] This HCV subgenomic replicon system consists of the two parts of the cistron — the HCV internal ribosome entry site (HCV-IRES) that directs expression of the neomycin resistance gene and the encephalomyocarditis virus (EMCV)-IRES that directs expression of the HCV nonstructural (NS) proteins NS3, NS4A, NS4B, NS5A, and NS5B (Fig. 1a). By transfecting the *in vitro*-transcribed replicon RNA into the human hepatoma cell line Huh7 and selecting cells using neomycin sulfate (G418), continuous high-level replication of replicons was achieved in transfected cells. Only the replicon-replicating cells (which produce neomycin phosphotransferase) can survive G418 selection and form visible colonies. Although this system is a powerful tool for the study of HCV replication mechanisms and the search for potential antiviral agents, it does not support virus particle formation and cannot recapitulate the entire viral life cycle in cultured cells. In our previous study, we reported a unique HCV genotype 2a strain called JFH-1, and showed that this strain can replicate efficiently in cultured cells. Our data also revealed that cell culture-generated HCV particles can infect both cultured cells and a chimpanzee.[12,13] This strain allowed us to obtain efficient HCV infection and replication in vitro.[13–15]

Isolation of the JFH-1 Strain

The HCV genotype 2a strain JFH-1 was isolated from a case of fulminant hepatitis.[12] The patient was a 32-year-old man who was admitted to hospital with liver dysfunction. This patient's serum aspartate aminotransferase and alanine aminotransferase concentrations were 9160 IU/L and 6970 IU/L, respectively, and the minimum prothrombin time was 16%. Stage II encephalopathy developed five days after admission. HCV RNA was detected in the acute phase serum by RT-PCR, and showed a high level of HCV viremia (10^5 copies/mL).

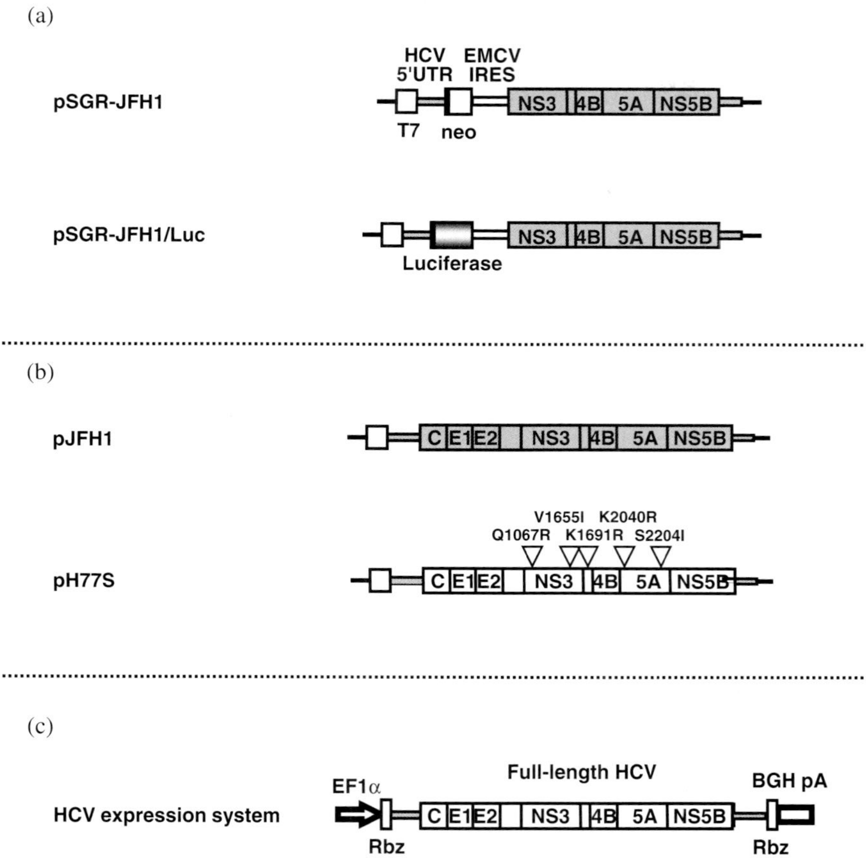

Fig. 1. Schematic structures of various constructs produced for evaluation of HCV replication and infection in cell culture. (**a**) Structures of JFH-1 replicons. pSGR-JFH1, subgenomic replicon JFH-1; pSGR-JFH-1/Luc, subgenomic luciferase reporter replicon JFH-1. (**b**) Structures of molecular clones for the cell culture–generated HCV strains JFH-1 and H77S. Introduced adaptive mutations of H77S are indicated at the top by triangles. (**c**) Structure of the HCV expression system with ribozymes.

Abbreviations: 5′-UTR, 5′-untranslated region; EMCV, encephalomyocarditis virus; IRES, internal ribosome entry site; T7, T7 promoter; neo, neomycin resistance gene; NS3, nonstructural protein 3; 4B, nonstructural protein 4B; 5A, nonstructural protein 5A; NS5B, nonstructural protein 5B; C, core protein; E1, envelope protein 1; E2, envelope protein 2; EF1α, EF1a promoter; BGH pA, polyadenylation signal of the bovine growth hormone gene; Rbz, hammerhead ribozyme.

Tests to detect anti-HCV antibody were negative at admission, but became positive six weeks after admission. All other hepatitis virus markers were negative. Thus, the patient was diagnosed with HCV-associated fulminant hepatitis. The source of the HCV infection was unclear. There was no evidence of prior liver disease, and the patient had no history of illegal drug use. He had not received any blood transfusions or intravenous drugs, undergone acupuncture, or had sexual contact with an HCV carrier in the previous six months. The predominant HCV strain in this patient was isolated from the acute phase serum, and the entire genome sequence of this strain was determined. Based on homology and phylogenetic analysis, this HCV strain (JFH-1) was shown to belong to genotype 2a, although it deviated slightly from other genotype 2a strains isolated from chronic hepatitis patients.

Subgenomic Replicon System with JFH-1

To investigate the replication capacity of JFH-1, we constructed a subgenomic JFH-1 replicon[16] (Fig. 1a). After transfecting *in vitro*-transcribed JFH-1 replicon RNA into Huh7 cells, colony formation was observed after selection with neomycin sulfate. The efficiency of colony formation of JFH-1 was approximately 500-fold higher than that of the prototypic HCV genotype 1b replicon Con1, and 50-fold higher than that of a Con-1 replicon containing cell culture-adaptive mutations. In a transient replication assay, this JFH-1 clone showed marked replication efficiency without G418 selection. This JFH-1 replicon also showed replication not only in Huh7 cells but also in the hepatocyte-derived cell lines HepG2 and IMY-N9 and in the non-hepatocyte-derived cell lines HeLa and HEK293.[17,18] Importantly, the JFH-1 replicon does not require an adaptive mutation to replicate in cultured cells. To develop a sensitive and accurate assay system for anti-HCV effects, we also constructed a JFH-1 reporter replicon system in which the neomycin resistance gene was replaced with a firefly luciferase gene[19] (Fig. 1a). The relative luciferase activity of this JFH-1 reporter replicon increased exponentially during the course of the experiment. Therefore, we reasoned that this clone replicated autonomously in cultured cells. Using this system, we were able to easily and accurately assess the anti-HCV effects of IFN and ribavirin. Based on these analyses, JFH-1 was shown to be able to replicate efficiently without adaptive mutations. The full-length RNA of the HCV genotype 1b strain with multiple cell culture-adaptive mutations did not produce an active HCV infection. Thus, it appears that high replication capacity without a need for adaptive mutations is an important factor in the development of an HCV infection system.

Establishment of JFH-1 Infection and the Replication System

To take advantage of this replication efficiency, we constructed full-length JFH-1 cDNA for use in assaying replication of HCV in cultured cells[13] (Fig. 1b). We constructed full-length JFH-1 cDNA that was identical to the consensus sequence of infected virus from the fulminant hepatitis patient. To produce the full-length JFH-1 RNA, we inserted the T7 promoter sequence immediately upstream of the full-length JFH-1 cDNA sequence, and then used T7 RNA polymerase to transcribe the RNA. When synthesized full-length JFH-1 RNA was transfected into naive Huh7 cells, viral RNA replication and viral protein expression were observed in the transfected cells. To determine whether viral particles were secreted from cells transfected with full-length JFH-1 RNA, we first assayed for HCV core protein in the culture medium from transfected cells using a highly sensitive immunoassay.[20] The cells transfected with full-length JFH-1 RNA secreted HCV core protein. Next, culture supernatant from cells transfected with full-length JFH-1 RNA was analyzed with a sucrose density gradient assay. Viral RNA and all structural proteins were detected in fractions with a density of approximately 1.15–1.17 g/mL, suggesting formation and secretion of complete viral particles. Viral particles secreted into the culture medium were visualized by immuno-electron microscopy. Viral particles were spherical, with an outer diameter of approximately 55 nm. These findings clearly indicate that HCV viral particles were produced by full-length JFH-1 RNA-transfected Huh7 cells and secreted into the culture medium. To test the infectivity of secreted viral particles, culture medium from full-length JFH-1 RNA-transfected cells was inoculated into naive Huh7 cells. Using indirect immuno-staining with anti-HCV antibodies, HCV proteins were detected in very few cells. We then collected the culture medium of RNA-transfected Huh7 cells, cleared it by low-speed centrifugation, and filtered it through a disk filter. Cleared virus in this medium was concentrated by ultrafiltration. By inoculating this concentrated culture medium into naive Huh7 cells, not only was expression of HCV proteins observed, but an increase in intracellular HCV RNA was also detected in infected cells. These data clearly indicate that the viral particles generated by transfection of full-genome JFH-1 RNA are infectious to Huh7 cells, although at low efficiency.

To confirm the infectivity of the viral particles, we tested for neutralization of the infectivity. CD81 has been identified as an HCV receptor molecule. Pre-treatment with anti-CD81 antibody decreased the infectivity of the culture supernatant. This is consistent with the previous finding that pseudotype virus infection was inhibited by anti-CD81 antibodies, and indicates that infectivity of the secreted viral particles depends on a CD81-specific pathway.

To confirm the *in vivo* infectivity of cell culture-generated JFH-1 virus particles, culture supernatant from Huh7 cells transfected with full-length JFH-1 RNA was inoculated into a chimpanzee. The chimpanzee was first inoculated with JFH-1 RNA-untransfected culture medium, and no infection was observed. Then the chimpanzee was inoculated with 10^4 times diluted culture medium, and no infection developed. Next, the chimpanzee was inoculated with 10^3 times diluted culture medium containing 8×10^3 copies of JFH-1 virus, which induced viremia. Thus, the cell culture-generated JFH-1 virus is infectious *in vivo*, and infection of cultured cells with this virus mimics the infection observed in the patient's liver. Curiously, the infection of this strain in the chimpanzee was transient and showed low-level viremia, no elevation of ALT, and no evidence of hepatitis on liver biopsy, although this JFH-1 strain was isolated from a patient with fulminant hepatitis. Thus, the pathogenicity of this virus *in vivo* may be different from its replication ability *in vitro*.

Robust HCV Infection Using Permissive Cell Lines

JFH-1 was the first HCV strain found to replicate efficiently and to produce infectious virus particles in cultured cells. However, the infection efficiency of JFH-1 was quite low in the first studies in which it was used, with only a small percentage of inoculated cells becoming infected. This limitation was overcome using permissive cell lines, such as Huh7.5 or Huh7.5.1 cells.[14,15] Huh7.5, which is a cured cell line established from a replicon cell line, supports high levels of subgenomic replication of HCV strains Con1 and H77, and is known to have a defective point mutation in RIG-I, resulting in defective intracellular interferon signaling against HCV RNA replication.[21] Huh7.5.1 is a subline derived from Huh7.5. Using these cell lines, robust expansion of JFH-1 and J6/JFH-1 viruses (described below) could be observed. Nearly 100% of Huh7.5 and Huh7.5.1 cells became HCV positive when inoculated with JFH-1 and J6/JFH-1 viruses.

HCV Infection and Replication System for Other Genotypes

To attempt to establish an *in vitro* infection and replication model for other HCV genotypes, many clones (including those isolated from other patients with fulminant hepatitis) have been used,[22,23] but most of these attempts did not yield sufficient replication *in vitro*. Only one strain, H77S, was able to produce infectious virus by transfection of the synthesized full-length RNA[24] (Fig. 1b). This H77S strain is a derivative of the genotype 1a Hutchinson strain and contains five cell culture-adaptive mutations. As an alternative approach to assessing the *in vitro* infection of other genotype strains, chimeric HCV constructs were produced by combining genomic sequences of JFH-1 with other HCV strains. Initially, an intra-genotypic (genotype 2a) chimeric construct containing J6CF structural genes (J6/JFH-1) was reported to produce infectious viruses more efficiently than the JFH-1 wild-type.[15,25] For other genotype strains, efficient virus production was obtained when genomic fragments were fused to a site located immediately after the first transmembrane domain of NS2.[24] Experiments with these chimeric viruses will be important for assessing the cross-neutralization of anti-HCV antibodies.

To produce the infectious viruses of various HCV genotypes *in vitro*, a DNA expression system using HCV cDNA has been reported[26,27] (Fig. 1c). This system contains genome-length cDNA of various HCV genotypes (1a, 1b, and 2a) with self-cleaving ribozymes for efficient HCV particle production.[27,28] Although efficient production of HCV particles was achieved using this system, *in vitro* infection was not observed when the viruses produced, except for the JFH-1 strain, were inoculated into cells. The absence of infection of strains other than JFH1 is probably a result of the lower replication efficiencies of these strains and the sensitivity limit of the detection method. Because *in vivo* infection could be detected by inoculating the cell culture-generated genotype 1a and 1b viruses into chimpanzees, this system certainly can produce infectious viruses in vitro and may be a valuable tool for studying the production of various HCV genotypes and to screen for antiviral agents that target this step.

Applications of HCV Cell Culture System

The HCV cell culture system enabled us to investigate the entire life cycle of this virus *in vitro*. However, the system depends on JFH-1, and it is still unclear why other strains do not replicate in these cell lines. This issue may be clarified by constructing chimeric viruses containing elements of JFH-1 and other strains. It has been reported that the structural and NS2 regions of JFH-1 can be replaced by the corresponding regions of the J6CF strain without affecting RNA replication or virus particle formation.[15,25] Thus, it seems clear that the region responsible for efficient replication of JFH-1 is located in the nonstructural region downstream of NS3.[29] By constructing various chimeras of the HCV J6CF and JFH-1 strains and investigating RNA replication and virus particle production in Huh7 cells, both the NS5B-to-3'X and the NS3 helicase regions have been shown to be responsible for efficient replication of the JFH-1 strain. Further analysis will be needed to determine the precise region responsible for the phenotype of JFH-1.

Using this system, several host factors have been identified that affect HCV infection and replication, including receptor candidates for HCV, such as CD81, SR-BI, claudin-1, and occludin.[30–32] Because JFH-1 can replicate in mouse cells,[33] these molecules will be important for the development of small-animal models of HCV infection. Molecules of intracellular antiviral signaling pathways are known to play a pivotal role in innate immunity from HCV infection, and some of them have been shown to be disrupted by HCV infection. The NS3/4a serine protease disrupts the signal or cleaves the critical adaptor molecules in these pathways, including IRF-3, RIG-I, Cardif/ IPS-1/ MAVS/VISA, and TRIF/TCAM-1.[34–40] It will be important to identify additional factors involved in viral infection and replication. This information may enable us to develop more permissive cell lines and to assess the infectivity of other HCV strains or viruses from clinical samples.

This system has also been useful for the detailed analysis of HCV biology. Recently, by investigating the intracellular distribution of HCV proteins, it has been reported that the association of core proteins and lipid droplets is important for production of infectious virus particles, and that the interaction of the NS5a-HCV RNA complex with core proteins is also involved in regulating HCV particle formation.[41,42] These observations are important for understanding the life cycle of HCV, and may be novel targets for antiviral compounds.

Conclusions

After many attempts, we were able to obtain an HCV replication and infection system *in vitro*. Use of this system will continue to shed light on HCV virology and will likely aid in the development of antiviral compounds. However, there are still some limitations. Only two clones, JFH-1 and H77S, or intra- and inter-genotype chimeric clones can replicate in cultured cells. We are still unable to observe infection of HCV from patient serum. We have not yet generated a convenient small-animal model for HCV infection. Thus, continued efforts and refinements will be required for understanding the biology of this virus.

Acknowledgments

This work was partially supported by a grant-in-aid for Scientific Research from the Japan Society for the Promotion of Science, from the Ministry of Health, Labour and Welfare of Japan and from the Ministry of Education, Culture, Sports, Science and Technology, and by the Research on Health Sciences Focusing on Drug Innovation from the Japan Health Sciences Foundation.

References

1. Choo QL, Kuo G, Weiner AJ, *et al.* (1989) Isolation of a cDNA clone derived from a blood-borne non-A, non-B viral hepatitis genome. *Science* **244**: 359–362.
2. Shimizu YK, Iwamoto A, Hijikata M, *et al.* (1992) Evidence for *in vitro* replication of hepatitis C virus genome in a human T-cell line. *Proc Natl Acad Sci USA* **89**: 5477–5481.
3. Bertolini L, Iacovacci S, Ponzetto A, *et al.* (1993) The human bone-marrow-derived B-cell line CE, susceptible to hepatitis C virus infection. *Res Virol* **144**: 281–285.
4. Ito T, Mukaigawa J, Zuo J, *et al.* (1996) Cultivation of hepatitis C virus in primary hepatocyte culture from patients with chronic hepatitis C results in release of high titre infectious virus. *J Gen Virol* **77**(Pt 5): 1043–1054.
5. Mizutani T, Kato N, Saito S, *et al.* (1996) Characterization of hepatitis C virus replication in cloned cells obtained from a human T-cell leukemia virus type 1-infected cell line, MT-2. *J Virol* **70**: 7219–7223.
6. Iacovacci S, Manzin A, Barca S, *et al.* (1997) Molecular characterization and dynamics of hepatitis C virus replication in human fetal hepatocytes infected *in vitro* . *Hepatology* **26**: 1328–1337.
7. Fournier C, Sureau C, Coste J, *et al.* (1998) *In vitro* infection of adult normal human hepatocytes in primary culture by hepatitis C virus. *J Gen Virol* **79 (Pt 10)**: 2367–2374.
8. Rumin S, Berthillon P, Tanaka E, *et al.* (1999) Dynamic analysis of hepatitis C virus replication and quasispecies selection in long-term cultures of adult human hepatocytes infected *in vitro. J Gen Virol* **80**(Pt 11): 3007–3018.

9. Ito T, Yasui K, Mukaigawa J, *et al.* (2001) Acquisition of susceptibility to hepatitis C virus replication in HepG2 cells by fusion with primary human hepatocytes: establishment of a quantitative assay for hepatitis C virus infectivity in a cell culture system. *Hepatology* **34**: 566–572.
10. Zhao X, Tang ZY, Klumpp B, *et al.* (2002) Primary hepatocytes of Tupaia belangeri as a potential model for hepatitis C virus infection. *J Clin Invest* **109**: 221–232.
11. Lohmann V, Korner F, Koch J, *et al.* (1999) Replication of subgenomic hepatitis C virus RNAs in a hepatoma cell line. *Science* **285**: 110–113.
12. Kato T, Furusaka A, Miyamoto M, *et al.* (2001) Sequence analysis of hepatitis C virus isolated from a fulminant hepatitis patient. *J Med Virol* **64**: 334–339.
13. Wakita T, Pietschmann T, Kato T, *et al.* (2005) Production of infectious hepatitis C virus in tissue culture from a cloned viral genome. *Nat Med* **11**: 791–796.
14. Zhong J, Gastaminza P, Cheng G, *et al.* (2005) Robust hepatitis C virus infection *in vitro*. *Proc Natl Acad Sci USA* **102**: 9294–9299.
15. Lindenbach BD, Evans MJ, Syder AJ, *et al.* (2005) Complete replication of hepatitis C virus in cell culture. *Science* **309**: 623–626.
16. Kato T, Date T, Miyamoto M, *et al.* (2003) Efficient replication of the genotype 2a hepatitis C virus subgenomic replicon. *Gastroenterology* **125**: 1808–1817.
17. Date T, Kato T, Miyamoto M, *et al.* (2004) Genotype 2a hepatitis C virus subgenomic replicon can replicate in HepG2 and IMY-N9 cells. *J Biol Chem* **279**: 22371–22376.
18. Kato T, Date T, Miyamoto M, *et al.* (2005) Nonhepatic Cell Lines HeLa and 293 Support Efficient Replication of the Hepatitis C Virus Genotype 2a Subgenomic Replicon. *J Virol* **79**: 592–596.
19. Kato T, Date T, Miyamoto M, *et al.* (2005) Detection of anti-hepatitis C virus effects of interferon and ribavirin by a sensitive replicon system. *J Clin Microbiol* **43**: 5679–5684.
20. Aoyagi K, Ohue C, Iida K, *et al.* (1999) Development of a simple and highly sensitive enzyme immunoassay for hepatitis C virus core antigen. *J Clin Microbiol* **37**: 1802–1808.
21. Sumpter R Jr, Loo YM, Foy E, *et al.* (2005) Regulating intracellular antiviral defense and permissiveness to hepatitis C virus RNA replication through a cellular RNA helicase, RIG-I. *J Virol* **79**: 2689–2699.
22. Sakai A, Takikawa S, Thimme R, *et al.* (2007) *In vivo* study of the HC-TN strain of hepatitis C virus recovered from a patient with fulminant hepatitis: RNA transcripts of a molecular clone (pHC-TN) are infectious in chimpanzees but not in Huh7.5 cells. *J Virol* **81**: 7208–7219.
23. Kimura T, Imamura M, Hiraga N, *et al.* (2008) Establishment of an infectious genotype 1b hepatitis C virus clone in human hepatocyte chimeric mice. *J Gen Virol* **89**: 2108–2113.

24. Yi M, Villanueva RA, Thomas DL, *et al.* (2006) Production of infectious genotype 1a hepatitis C virus (Hutchinson strain) in cultured human hepatoma cells. *Proc Natl Acad Sci USA* **103**: 2310–2315.
25. Pietschmann T, Kaul A, Koutsoudakis G, *et al.* (2006) Construction and characterization of infectious intragenotypic and intergenotypic hepatitis C virus chimeras. *Proc Natl Acad Sci USA* **103**: 7408–7413.
26. Cai Z, Zhang C, Chang KS, *et al.* (2005) Robust production of infectious hepatitis C virus (HCV) from stably HCV cDNA-transfected human hepatoma cells. *J Virol* **79**: 13963–13973.
27. Kato T, Matsumura T, Heller T, *et al.* (2007) Production of infectious hepatitis C virus of various genotypes in cell cultures. *J Virol* **81**: 4405–4411.
28. Heller T, Saito S, Auerbach J, *et al.* (2005) An *in vitro* model of hepatitis C virion production. *Proc Natl Acad Sci USA* **102**: 2579–2583.
29. Murayama A, Date T, Morikawa K, *et al.* (2007) The NS3 helicase and NS5B-to-3′X regions are important for efficient hepatitis C virus strain JFH-1 replication in Huh7 cells. *J Virol* **81**: 8030–8040.
30. Evans MJ, von Hahn T, Tscherne DM, *et al.* (2007) Claudin-1 is a hepatitis C virus co-receptor required for a late step in entry. *Nature* **446**: 801–805.
31. Liu S, Yang W, Shen L, *et al.* (2009) Tight junction proteins claudin-1 and occludin control hepatitis C virus entry and are downregulated during infection to prevent superinfection. *J Virol* **83**: 2011–2014.
32. Ploss A, Evans MJ, Gaysinskaya VA, *et al.* (2009) Human occludin is a hepatitis C virus entry factor required for infection of mouse cells. *Nature* **457**: 882–886.
33. Uprichard SL, Chung J, Chisari FV, Wakita T. (2006) Replication of a hepatitis C virus replicon clone in mouse cells. *Virol J* **3**: 89.
34. Foy E, Li K, Wang C, Sumpter R Jr, *et al.* (2003) Regulation of interferon regulatory factor-3 by the hepatitis C virus serine protease. *Science* **300**: 1145–1148.
35. Foy E, Li K, Sumpter R Jr, *et al.* (2005) Control of antiviral defenses through hepatitis C virus disruption of retinoic acid-inducible gene-I signaling. *Proc Natl Acad Sci U S A* **102**: 2986–2991.
36. Li K, Foy E, Ferreon JC, *et al.* (2005) Immune evasion by hepatitis C virus NS3/4A protease-mediated cleavage of the Toll-like receptor 3 adaptor protein TRIF. *Proc Natl Acad Sci USA* **102**: 2992–2997.
37. Meylan E, Curran J, Hofmann K, *et al.* (2005) Cardif is an adaptor protein in the RIG-I antiviral pathway and is targeted by hepatitis C virus. *Nature* **437**: 1167–1172.
38. Li XD, Sun L, Seth RB, *et al.* (2005) Hepatitis C virus protease NS3/4A cleaves mitochondrial antiviral signaling protein off the mitochondria to evade innate immunity. *Proc Natl Acad Sci USA* **102**: 17717–17722.

39. Loo YM, Owen DM, Li K, *et al.* (2006) Viral and therapeutic control of IFN-beta promoter stimulator 1 during hepatitis C virus infection. *Proc Natl Acad Sci USA* **103**: 6001–6006.
40. Lin R, Lacoste J, Nakhaei P, *et al.* (2006) Dissociation of a MAVS/IPS-1/VISA/Cardif-IKKepsilon molecular complex from the mitochondrial outer membrane by hepatitis C virus NS3-4A proteolytic cleavage. *J Virol* **80**: 6072–6083.
41. Miyanari Y, Atsuzawa K, Usuda N, *et al.* (2007) The lipid droplet is an important organelle for hepatitis C virus production. *Nat Cell Biol* **9**: 1089–1097.
42. Masaki T, Suzuki R, Murakami K, *et al.* (2008) Interaction of hepatitis C virus nonstructural protein 5A with core protein is critical for the production of infectious virus particles. *J Virol* **82**: 7964–7976.

Chapter 9

Regulation of Innate Immunity and Interferon Defenses by Hepatitis C Virus

Stacy M. Horner and Michael Gale Jr*

Introduction

As described earlier in this book, hepatitis C virus (HCV) is a hepatotropic RNA virus that continues to be a major public health problem. HCV persistently infects nearly 200 million people worldwide. A further problem is that approximately 3–4 million people are newly infected annually. While 20–30% of those acutely infected with HCV may clear the virus, the majority of those infected can count on developing a chronic infection, leading to liver inflammation, fibrosis, and cirrhosis. As such, HCV is a primary indicator for liver transplantation and a leading etiology for hepatocellular carcinoma.[1,2] The disease associated with HCV places a substantial global burden on the health care system and this cost is set to increase dramatically as new cases of HCV-associated illness are presented in the clinic in the coming decades.[3] There is no vaccine for HCV, and the current HCV

* Corresponding Author: Michael Gale, Jr., Department of Immunology, University of Washington School of Medicine, 1959 N.E. Pacific Street, Box 357650, Seattle, WA 98195, USA, Tel: 206-685-7953, Fax: 206-543-1013, e-mail: mgale@u.washington.edu

therapy of pegylated interferon (IFN)-α in combination with ribavirin leads to a sustained virologic response (SVR) in only half of those chronically infected.[4] In recent years, HCV-related liver disease has become the leading cause of death in those people coinfected with both HCV and HIV.[5] As the pandemics of HCV and HIV progress, it is believed that the two viruses will expand their dual infection to more completely track together in at-risk populations. Indeed, the risk factors for HCV or HIV transmission each include exposure to blood products, intravenous drug and/or mucosal-associated drug use, sexual contact, and other shared risk factors. Thus, defining the molecular mechanisms by which HCV regulates innate immunity will provide novel insights into how to approach the management of both HCV mono-infection and HCV/HIV dual infection.

The innate intracellular immune response to virus infection is activated when conserved pattern associated molecular patterns (PAMPs) generated during infection are recognized in cells by proteins known as pattern recognition receptors (PRRs). Among PRRs, three major classes of these proteins include the Toll-like receptors (TLRs), RIG-I-like receptors (RLRs), and nucleotide oligomerization domain (NOD)-like receptors (NLRs). Viral engagement of TLRs and RLRs leads to downstream signaling that results in the activation of latent transcription factors, including the IFN regulatory factors (IRFs) and nuclear factor-κB (NF-κB), and culminates in the induction of IRF-3 target genes, type I IFN, and proinflammatory cytokines. Type I IFN coordinates immunity to prevent virus infection of new cells and limit virus spread. Importantly, type I IFN signals in an autocrine and paracrine manner through the type I IFN receptor to induce hundreds of IFN stimulated genes (ISGs) that establish a tissue-wide antiviral state, and it sensitizes infected cells to apoptosis.[6,7] Furthermore, type I IFN modulates several aspects of adaptive immunity, including establishment of cytotoxic T cell responses, generation of natural killer cells, and B cell differentiation, for the elimination of virus-infected cells.[7,8] NLR-mediated PAMP detection activates caspase-1, leading to the secretion of proinflammatory cytokines, such as interleukin-1β,[9] which directs an antiviral inflammatory response.

Recent studies have made progress in our understanding of how HCV is detected by the host cell for the activation of innate immune response. In spite of this detection and ability to stimulate innate immune defenses and IFN production, HCV is successful at mediating a chronic course of infection in the majority of those infected. This review will discuss how HCV is sensed by the innate immune response and how the virus subsequently antagonizes the antiviral effects of type I IFN.

Hepatocytes Detect HCV Infection Through RIG-I

HCV replicates *in vivo* in human patients to produce an estimated 10^{12} virions per day,[10] and this high level of HCV transiently activates the innate immune system early during infection. Acute HCV infection in chimpanzees induces an ISG expression profile in the liver consistent with a type I IFN response.[11] Furthermore, HCV RNA is potent inducer of type I IFN,[12,13] mediated by HCV engagement of PRRs.

Actions of PRRs

A key foundation of immune defense initiation is the discrimination by the infected host of non-self versus self. That is, the host cell must know it is infected by a microbial agent in order to initiate immune defenses against that agent. Identification of a microbe as non-self is facilitated by the various PRRs expressed in a given cell. In the case of viruses, viral PAMPs produced during infection of a given cell are recognized as non-self by three distinct classes of sensors that initiate intracellular signaling, the TLRs, NLRs, and RLRs.[14] The function of these antiviral sensors is regulated in part by their compartmentalization within the cell. RLRs and NLRs sense PAMPs in the cytoplasm of most cells, including hepatocytes, while TLRs are expressed in endosomes and on the cell surface and primarily detect viral PAMPs in various immune cells, such as macrophages, dendritic cells, B cells, and some types of T cells.[15] The role of NLRs in sensing RNA viruses is still unclear. NLRs have recently been implicated in triggering virus-responsive programs of inflammasome signaling, and they are also activated by bacterial PAMPs or intracellular stress signals leading to inflammation.[9,16]

The TLR family members that detect viral PAMPs include TLR2, TLR3, TLR4, TLR7/8, and TLR9, and their activation leads to production of proinflammatory cytokines and IFN-α. TLR2 and TLR4 are expressed on the plasma membrane and are involved in detecting viral proteins at the cell surface. Both the HCV core and NS3 proteins have been shown to activate TLR2 signaling, leading to host inflammation through actions of responding myeloid cells, but the biology of these events in HCV infection is not fully understood.[17,18] TLR3, TLR7/8, and TLR9 detect viral nucleic acids. TLR7/8 and TLR9 are primarily expressed in endosomes, while TLR3 can be found in endosomes and at the cell surface, depending on the cell type.[15] TLR9 serves as a PRR for unmethylated CpG motifs in DNA, and therefore a role in the detection of HCV, an RNA virus, is not expected. TLR7/8 sense uridine and guanosine-rich single-stranded RNA (ssRNA)[19,20] in myeloid cells, including conventional dendritic cells and plasmacytoid dendritic cells, the major producers of IFN-α in response to virus infection. TLR3 is expressed in conventional dendritic cells, macrophages, and several non-immune cell types, such as fibroblasts and epithelial cells, and has been proposed to sense double-stranded RNA (dsRNA). The physiological viral ligand and antiviral function of TLR3 and TLR7/8 are undefined,[21,22] although it appears that TLR3 is involved in the induction of type II IFN following picornavirus infection.[23] Because TLR3 and TLR7/8, which are transmembrane proteins, primarily sense nucleic acids in an intracellular location, detection of HCV by these TLRs would likely require phagocytosis of virus-infected cells. Hepatocytes, the primary site of HCV infection, are deficient in TLR signaling.[24] To date, no HCV RNA ligand has been shown to directly activate TLR pathways. Therefore, TLRs likely mediate a secondary response to HCV infection, perhaps sensing infection in phagocytosed cells,[25] resulting in amplification of intra-hepatic inflammatory signals.

The RLR family consists of retinoic acid-inducible gene-I, RIG-I; melanoma differentiation-associated gene 5, MDA5; and laboratory of genetics and physiology 2, LGP2. Because these RLRs are localized to the cytoplasmic compartment in cells, including hepatocytes, they are candidates to be the primary intracellular sensors of HCV infection.[26] Both RIG-I and MDA5 contain two amino terminal caspase activation and recruitment domains (CARD). All the RLRs have a DExD/H RNA helicase domain and bind to RNA ligands. Studies using RIG-I −/− and MDA5 −/− mice revealed that RIG-I and MDA5 recognize distinct classes of viruses.[26] RIG-I has been shown to be essential for the detection of a specific set of ssRNA viruses, including paramyxoviruses, flaviviruses, orthomyxoviruses, and rhabdoviruses, while MDA5 is essential for detection of picornaviruses, such as encephalomyocarditis virus.[13,26–28] These studies also showed that production of IFN-β following stimulation with poly I:C, the synthetic double-stranded (ds) RNA analogue, was mediated by MDA5, and not RIG-I.[26,29] The proposed ligand for MDA5 is long dsRNA greater than three kilobases.[30]

Transfection of HCV RNA into human hepatoma cells leads to transient activation of IRF-3 and IFN-β.[13,31] This activation is defective in cells that lack functional RIG-I, but can be complemented with wild-type RIG-I, demonstrating that RIG-I is required for activation of IRF-3 by HCV RNA.[13] HCV infection is also more efficient in cells that lack functional RIG-I signaling.[32] Taken together, these studies indicate that RIG-I is the primary intracellular sensor of HCV. Biochemical studies revealed that RIG-I, but not MDA5, recognizes the 5′- and 3′-non-coding regions (NTRs) of HCV RNA.[31] The HCV NTRs are highly structured and contain partial double-stranded regions (Fig. 1). Recent work has defined the detailed mechanism of HCV RNA ligand recognition by RIG-I for the induction of antiviral signaling.[12,33]

HCV is Recognized as Non-Self by RIG-I

Cytoplasmic ssRNA containing a 5′-triphosphate (ppp), short dsRNA, and uridine- or adenosine-rich viral RNA motifs are recognized by RIG-I.[12,30,33–36] The RNA ligand base composition, sugar backbone, and length of the RNA sequence also provide important signals required for RIG-I activation.[12,33] The HCV genome contains several

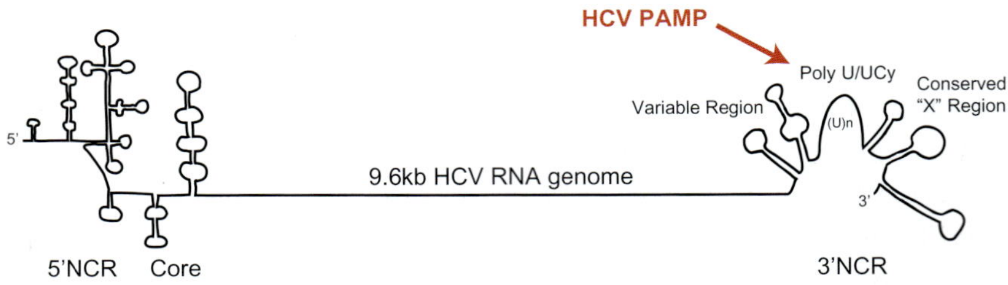

HCV PAMP sequence: 5'ppp.....$U_6CCCU_{11}CU_{34}CUCCU_9CCUCU_7CCU_4CUUUCCUUU$...

Fig. 1. RNA genome structure of hepatitis C virus (HCV) and the nature of the HCV PAMP. **Upper:** The 9.6 kilobase positive-strand RNA genome of HCV is illustrated. The RNA secondary structure of the 5′ non-translated region (NTR) and 3′-NTR are depicted. The HCV 3′-NTR contains three domains: a variable region with two RNA stem loops, a single-stranded poly U/UC region, and a conserved "X" region with three RNA stem loops. The minimal HCV pattern-associated molecular pattern (PAMP) that activates RIG-I signaling is indicated by the arrow. **Lower:** The nature of the HCV PAMP region, known as the poly-U/UC domain, is depicted to show the poly-uridine motifs and interspersed C nucleotides. This motif requires the presence of a free 5′-triphosphate on the viral RNA. The viral negative-strand product of this motif can also be recognized as a PAMP by RIG-I but also requires a free 5′ triphosphate.

motifs that facilitate its recognition by RIG-I. HCV RNA is not capped and therefore its RNA contains the 5′-ppp required for detection by RIG-I. The 3′-NTR of the HCV genome is the primary HCV PAMP that activates RIG-I signaling.[13,31] Importantly, this region is highly conserved among HCV genotypes and is critical for HCV replication.[37] The 3′-NTR consists of three parts: a variable region containing two stem loops, a poly U-rich region that is single-stranded, and a conserved "X" region, which contains three stem loops (Fig. 1). It was expected that dsRNA or RNA with secondary structure located within the 3′-NTR would be the primary HCV PAMP for RIG-I through interactions with the RIG-I helicase domain. In fact, the highly-structured X region of the HCV genome does not activate RIG-I signaling, but surprisingly, the poly U/UC region is a potent activator of RIG-I signaling. A 5′-ppp is necessary for this activation, but it is not sufficient, as the X region containing a 5′-ppp does not activate RIG-I.[12]

How does RIG-I detect these two distinct signatures (5′ ppp and poly U/UC sequence) in one RNA molecule, when they are separated by over 9,600 nucleotides in the HCV genome? There are three possibilities. First, well-established long-range RNA interactions between the 5′ and 3′ ends of the HCV genome that are important for efficient HCV translation could bring the 5′ ppp and poly U/UC into close proximity for stable interaction with and activation of RIG-I.[38,39] Alternatively, it was recently shown that the ability of HCV poly U/UC RNA to activate RIG-I does not require immediate adjacency of the 5′-ppp, suggesting that RIG-I recognizes two distinct features of the PAMP RNA, both the 5′-ppp and the poly U/UC sequence.[33] In fact, structural studies suggest that RIG-I has two distinct RNA sensing domains.[36,40] RIG-I binds 5′-ppp ssRNA through a positively charged groove in its C-terminal domain, which includes the RD. As this groove can only accommodate three nucleotides, RIG-I must use an additional domain, most likely the helicase domain, to discriminate specific RNA sequences, such as the poly U/UC present in the HCV PAMP. Finally, the corresponding replication intermediate sequence of the poly U/UC region, poly A/AG, also activates RIG-I signaling. In this case, during HCV replication, the 5′-ppp of the negative strand would be in close proximity to the poly A/AG region, and together these motifs could bind to RIG-I. In fact, it has been proposed that RIG-I may bind to substrate RNA and then translocate on the RNA to scan for activating sequences.[41]

Though it is not known where in the cell and when in the HCV replication program the PAMP that activates RIG-I is generated, we do know that HCV replicates in association with cellular membranes. HCV induces web-like membrane structures that compartmentalize HCV replication complexes and probably protect HCV RNA from RIG-I detection.[42,43] However, this does not exclude the possibility of the presence of some non-membranous HCV RNA in the cytoplasm. In addition to these large perinuclear replication complexes on web-like cellular membranes, smaller HCV replication complexes are distributed on membranes located throughout the cytoplasm.[44,45] These smaller replication complexes, distinct from the large perinuclear replication complexes, appear to be formed during early times after infection (<24 h),[44,45] and therefore could be initial detection sites of the HCV genome by RIG-I. In support of this idea, HCV activates IRF-3 at early times during infection before extensive viral protein synthesis has occurred,[32,46] suggesting that the genome amplification process of HCV replication is essential for PAMP recognition by RIG-I. However, further studies are needed to determine the intracellular localization and activation of RIG-I during HCV infection.

The HCV PAMP (5′-ppp poly U/UC ssRNA) triggers hepatic immune responses *in vivo*, and these responses are mediated by RIG-I.[12] The PAMPs in the 3′-NTR of three different HCV strains (Con1, J4L6, JFH-1) have been tested so far, and all three activate RIG-I signaling.[12,33] The PAMP from the HCV genotype 2a strain JFH-1, which replicates and produces infectious particles in cell culture, is a relatively weak activator of RIG-I signaling, suggesting that the ability of the JFH-1 strain to replicate in cell culture could be due in part to its low immuno-stimulatory activity.[33] Other viruses that activate RIG-I, including rabies virus, ebola virus, and measles virus, all contain PAMP regions with similar sequences to the HCV PAMP that potently activate RIG-I signaling, thereby demonstrating a conserved mechanism of PAMP recognition by RIG-I.[12]

The features present in the HCV PAMP distinguish it from cellular RNAs and illustrate the exquisite discrimination that RIG-I uses to distinguish self or endogenous RNA from non-self or viral RNA. While HCV RNA is uncapped and contains a 5′-ppp, endogenous RNA has 5′ end modifications (such as a methylguanosine cap or a monophosphate), interacts with ribonucleoproteins, or has extensive base modifications, all of which would mask detection by RIG-I. In fact, some viruses have evolved mechanisms to avoid detection by RIG-I, such as removal of the 5′-ppp from their viral genomes or attachment of viral proteins to the 5′ ends of their genomes.[47,48] The poly U/UN (where N denotes an interspersed non-U nucleotide) sequence seems to be a general immune mechanism for non-self recognition, as TLR7/8 are also activated by ssRNA PAMPs containing this sequence.[19,20]

Following RNA virus infection or dsRNA treatment of cells, the antiviral endoribonuclease RNaseL serves an important role in amplifying the PAMP-driven signaling events that drive high-level production of IFN-β. Upon activation by 2′,5′-linked oligoadenylate, which is produced by the 2′–5′-oligoadenylate synthetase (2–5 OAS), RNaseL cleaves RNA to generate small RNA cleavage products that signal through the RIG-I/MDA5 pathway.[49] RNaseL can cleave HCV RNA at single-stranded UA and UU dinucleotides *in vitro*[50,51]; however, it is unclear how RNaseL-cleaved RNA is recognized by RIG-I. It is possible that cleavage of HCV RNA by RNaseL could generate higher levels of HCV PAMP leading to amplification of RIG-I signaling during virus infection.

Molecular Mechanism of RIG-I Activation

In the absence of ligand, RIG-I is held in an inactive confirmation by intramolecular interactions between the N-terminal CARD and the C-terminal RD, which suppress ATPase activity of RIG-I.[31,52] Structural studies revealed that both short dsRNA and 5′-ppp RNA bind to the C-terminal domain of RIG-I.[36] RIG-I interactions with these RNA ligands, including the HCV PAMP, induces a conformational change in RIG-I that facilitates oligomerization through the RD, activation of the ATPase domain, and exposure of the CARD for interaction with the CARD domain of the adaptor protein IFN promoter stimulating factor-1 (IPS-1, also known as Cardif, MAVS, and VISA) and activation of downstream signaling molecules (see below).[31,36,40,53] Recently published structural studies of RIG-1 in complex with a dsRNA ligand are beginning to reveal the details of RIG-1 activation, as well as the role of the RIG-1 helicase domain in RIG-1 activation.[1–3] However, further structural studies with a PAMP-activating RNA ligand are required to understand the full mechanism of RIG-1 activation.

In addition to internal regulation, RIG-I activation is subject to external regulation. Ubiquitination of RIG-I on the CARD by TRIM25 or on the C-terminal domain by Riplet/RNF135, promotes the interaction of RIG-I with IPS-1 and activation of downstream signaling.[57,58] The RLR LGP2 also plays a role in regulating RIG-I signaling and controlling virus infection. LGP2, which lacks a CARD, has been reported to be a negative regulator of the RIG-I pathway.[31,59] Several mechanisms for this negative regulation have been suggested, including sequestration of RNA ligands, heterodimerization with RIG-I, and prevention of the kinase IκB kinase-ε (IKKε) from binding to IPS-1.[31,56,60] While the C-terminal domain of LGP2 can bind to dsRNA, including HCV RNA, this binding activity does not appear to be required for its ability to regulate RIG-I signaling.[31,61,62] Studies in LGP2$^{-/-}$ mice demonstrate that, depending on the type of virus, LGP2 is either a positive or negative regulator of antiviral signaling.[63] Future studies are needed to define the mechanism of LGP2 regulation of RIG-I signaling during HCV infection.

IFN Signaling During HCV Infection

During HCV infection, RIG-I detects the 3′ NTR of the HCV RNA leading to activation of IRF-3.[13,31] Activated IRF-3 directly stimulates the expression of specific set of genes within the infected cell, including type I IFN.[64,65] Secreted IFN-β transmits local and tissue-wide signals for the induction of hundreds of ISGs by interacting in an autocrine and paracrine manner with the α/β IFN receptor located on the cell surface. This leads to activation of the Janus kinase (Jak)-Signal transducer and activator of transcription (STAT) pathway. STAT1 and STAT2 are phosphorylated by JAK-1 and protein tyrosine kinase-2 and, together with IRF9, form the ISG factor 3 (ISGF3) transcription factor complex. ISGF3 binds to IFN-stimulated response elements within the promoters of ISGs to activate gene transcription.

Type I IFN successfully suppresses HCV replication in cell culture[66,67] and induces expression of many ISGs.[6] In addition to having roles in antiviral activities, ISGs are involved in such processes as lipid metabolism, apoptosis, protein degradation, and inflammatory responses.[6] Microarray analysis of liver samples from HCV-infected chimpanzees or chronically infected humans demonstrate that many ISGs are expressed in the HCV-infected liver.[11,46,68–72] Characterization of these ISGs has revealed that several have anti-HCV activity, including protein kinase R (PKR), ISG56, ISG20, ADAR1, and viperin. Both PKR and ISG56 block HCV replication at the level of translational inhibition. PKR phosphorylates the alpha subunit of the eukaryotic initiation factor (eIF) 2 and the IRF-3-activated gene ISG56 binds to eIF3; both of these processes prevent initiation of viral protein translation.[73] ADAR1 is an RNA editing enzyme that deaminates adenosines in dsRNA, resulting in destabilization of RNA and accumulation of mutations.[74] Viperin has also been characterized as an anti-HCV effector. Its expression induces the formation of crystalloid ER and interferes with protein secretion.[75] Deletion of the ER localization domain of viperin abrogates its anti-HCV activity.[76] Therefore, it is possible that viperin inhibits HCV replication by interfering with the formation or activity of the HCV replication complex on ER membranes.

The current HCV therapy of pegylated IFN-α and ribavirin is successful for only approximately 50% of those treated, despite the fact that type I IFN can block HCV replication and stimulate the expression of ISGs that have anti-HCV activity. HCV has several mechanisms for disrupting the IFN pathway (see below) that may contribute to the low success rate of HCV therapy. This disruption of the IFN pathway by HCV may prevent the activation of a specific subset of ISGs important for HCV clearance. Examination of paired liver biopsies from patients with chronic HCV before and after peg-IFN-α therapy found that non-responders to therapy had similarly high levels of ISGs before and after therapy.[77] Additionally, this pre-activation of ISGs was higher in patients infected with the more difficult to treat HCV genotypes 1 and 4 than in genotypes 2 and 3, which are easier to treat. While the exact mechanism for this pre-activation of ISGs and non-response to therapy is not known, this study, along with similar studies in chimpanzees and human patients,[65,66,78] suggests that non-responders may not be activating a specific subset of ISGs that are important for viral clearance or that negative feedback loops, such as those mediated by ISG15 protease USP18,[79] due to the high levels of IFN signaling in these patients prevent the induction of anti-HCV ISGs following peg-IFN-a therapy. Indeed, consensus IFN, which has increased anti-HCV potency

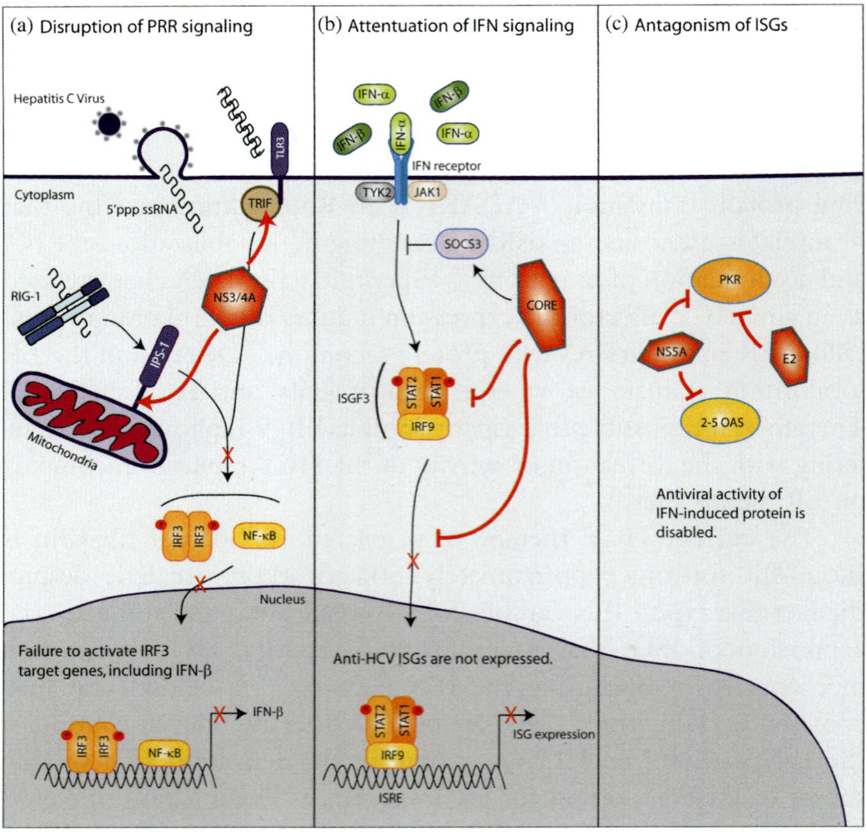

Fig. 2. Hepatitis C virus (HCV) evades the innate immune response of infected hepatocytes to dysregulate hepatic innate immunity. The intracellular innate immune response to HCV is activated when RIG-I (retinoic inducible gene-I) binds to the HCV PAMP RNA motif a 5′-triphosphate (5′-ppp). HCV proteins can regulate the signaling pathways that activate this response. **(a)** The HCV NS3/4A protease cleaves the host cell IPS-1 protein to disrupt the RIG-I pathway. NS3/4A has also been shown to target TLR3 (Toll-like receptor-3) signaling by proteolysis of the essential TLR3 adaptor protein, Toll/interleukin-1 receptor/resistance domain-containing adaptor-inducing IFN (TRIF), though this regulation has not been demonstrated *in vivo*. Cleavage of these proteins by NS3/4A prevents viral PAMP-driven IRF-3 signaling and IFN-β expression within the infected cell. **(b)** The HCV core protein has been shown to block several aspects of Jak-STAT signaling in cultured cells to alter the expression of ISGs. HCV core protein induces suppressor of cytokine signaling-3 (SOCS3), a negative regulator of Jak/STAT signaling. HCV core also prevents STAT1 phosphorylation and nuclear import *in vitro*. **(c)** HCV proteins can

over peg-IFN-α, induces a subset of ISGs distinct from those induced by peg-IFN-α.[80] Perhaps one of these ISGs or a specific IFN-induced micro-RNA will be implicated as a factor required for successful HCV clearance.[81]

Control of Innate Immune Signaling by HCV

Despite the fact that RIG-I signaling and ISGs are induced by HCV, 80% of those infected with HCV become chronically infected. HCV successfully evades several aspects of the host immune response, including expression of IFN-β, IFN signaling, and antiviral activities of IFN-induced proteins (Fig. 2). These activities blunt the intracellular innate immune response and also most likely cause defects in the adaptive immune response, contributing to HCV persistence.

HCV Disruption of IFN-β Induction

RIG-I signaling is activated upon interaction with the HCV PAMP, which promotes a conformational change in RIG-I to facilitate its interaction with IPS-1. This interaction and subsequent activation of IPS-1 leads to the assembly of a complex on the mitochondria that signals downstream to activate NF-κB and the kinases IKKε and Tank-binding kinase 1 (TBK1), which phosphorylate the transcription factors IRF-3 and IRF7. STING (also called MITA[82] and MPYS[83]) has been identified as a new member of this pathway.[84] It appears to be localized in both the ER and the mitochondria, and, upon virus infection, it recruits TBK1 to the mitochondria for the phosphorylation and activation of IRF-3. The exact mechanism by which interaction of RIG-I with IPS-1 activates signal transduction through IPS-1 remains to be determined, especially since overexpression of IPS-1 alone leads to downstream signaling. Both NLRX1 and

---

Fig. 2. (*Continued*) antagonize the antiviral activity of IFN-induced proteins in cultured cells. Both the HCV E2 and NS5A proteins have been shown to bind to protein kinase R (PKR) and inhibit its activation. NS5A also has been shown to block the antiviral activity of 2′–5′ oligoadenylate synthetase (2–5 OAS) in cultured cells. TYK2, protein tyrosine kinase-2; PRR, pattern recognition receptor.

the C1q receptor gC1qR have been identified as negative regulators of IPS-1 signaling.[85–87] When these proteins are overexpressed during virus infection, they interact with IPS-1 on the mitochondria and prevent IPS-1 signaling. Therefore, activation of RIG-I by PAMP binding and subsequent ubiquitination of RIG-I may promote strong interactions with IPS-1 to displace these negative regulators from IPS-1 for the activation of downstream antiviral signaling. Interestingly, extracellular HCV core protein interacts with gC1qR on the surface of T cells and inhibits T cell activation.[88,89] It will be of interest to determine if HCV core protein interacts with intracellular gC1qR or if any of the HCV proteins antagonize the function of any of these new members in the RIG-I pathway.

Regulation of the RIG-I Pathway

HCV has several mechanisms for disrupting the RIG-I signaling pathway that it triggers to prevent and control activation of type I IFN.[90] The NS3/4A protein is the primary HCV protein responsible for evasion of these antiviral signaling pathways. NS3/4A is the major HCV protease and is essential for viral replication. NS3 interacts with its cofactor, the NS4A peptide, to anchor the NS3/4A complex to intracellular membranes and to facilitate complete activation of the NS3 protease domain.[91,92] During HCV replication, NS3/4A uses its serine protease domain to cleave the HCV polyprotein and release the mature nonstructural viral proteins. To block signaling by RIG-1, NS3/4A also cleaves the innate immune adaptor protein IPS-1 on the mitochondrial-associated membrane, a specialized ER subdomain.[4] This cleavage abolishes RIG-I mediated signal transduction, including IRF-3 activation, in the infected cell.[32,93–96] Recent studies have demonstrated that IPS-1 oligomerization is required for activation of antiviral signaling, and it appears that NS3/4A cleavage of IPS-1 also prevents signaling by disrupting IPS-1 oligomerization.[97,98] NS3/4A has also been shown to cleave TRIF (Toll/interleukin-1 receptor/resistance domain-containing adaptor-inducing IFN), the signaling adaptor molecule for TLR3, to prevent TLR3-mediated antiviral signaling.[94] However, the role of TLR3/TRIF in innate immunity to

HCV has not been defined and other studies have not reproduced this finding.[99,100] NS3/4A protease inhibitors are currently under development as antiviral therapies for HCV.[4] These protease inhibitors, which block the ability of NS3/4A to cleave the HCV polyprotein and IPS-1, restore innate immune signaling in HCV infected cells.[101,102] In addition to NS3/4A-mediated cleavage of IPS-1 and TRIF, NS3 has also been reported to physically bind to TBK1 to block the association of TBK1 with IRF-3 and the activation of IRF-3.[103]

The HCV core protein interacts with the DEAD box protein 3 (DDX3), a member of the DEAD box helicase family.[104–106] It has recently been shown that, during virus infection, DDX3 interacts with IKKε to prevent IRF activation.[107] Further studies are required to determine if the interaction of HCV core with DDX3, which is required for HCV replication,[108] has an effect on IRF activation during HCV infection. In one study, HCV NS4B was shown to block RIG-I and IPS-1 mediated signal transduction, although other studies have not observed an effect of NS4B expression on IRF-3 activation.[90,109]

HCV Regulation of IFN Signaling and Antiviral Activities of IFN-Induced Proteins

The Jak/STAT pathway is another target of HCV for virus-directed regulation of innate immune defenses. Expression of the HCV polyprotein may attenuate Jak/STAT signal transduction by inducing expression of protein phosphatase 2A (PP2A)[110] in chronic HCV infection, resulting in a decreased activity of STAT1 and loss of ISG activation by the transcription factor ISGF3.[111,112] The HCV core protein expressed alone can block Jak/STAT signal transduction. It directly binds to STAT1 to prevent its phosphorylation and subsequent activation of downstream anti-HCV ISGs, including PKR, MxA, and 2–5 OAS.[113–115] Additionally, the HCV core protein induces expression of the suppressor of cytokine signaling-3 (SOCS3), a negative regulator of the Jak/STAT signaling pathway, resulting in decreased STAT1 activation in response to type I IFN.[116] Therefore, antagonism of the Jak/STAT pathway by HCV core protein prevents ISG induction in response to virus infection.

The HCV NS5A and E2 proteins directly interfere with the antiviral actions of ISGs, including PKR. The HCV glycoprotein E2 acts as competitive substrate with eIF2α for PKR binding, resulting in inhibition of PKR kinase activity.[117] The HCV NS5A protein, which is involved in RNA replication and HCV particle assembly,[118–120] prevents PKR activation through direct interactions that prevent PKR-mediated translation control, resulting in increased HCV replication and subversion of the IFN response.[121,122] NS5A also blocks the antiviral function of 2–5 OAS through a direct interaction.[123] Expression of NS5A seems to induce secretion of IL-8, resulting in a reduced expression of ISGs.[124] In fact, IL-8 levels are increased in HCV-infected patients who do not respond to therapy, as compared to those that do respond to therapy, suggesting that NS5A may have direct effects on responses to IFN therapy.[125] Finally, NS5A binds to MyD88, the major signaling adaptor protein for TLRs (except for TLR3), in macrophages, and prevents cytokine induction in response to TLR ligands, although the relevance of this is unclear because HCV has not been reported to infect macrophages.[126]

Conclusion and considerations

The HCV/host interactions that regulate innate antiviral defenses play important roles in directing the outcome of infection and immunity. The recent discovery and characterization of the actual PAMP that is recognized as non-self to initiate the immune response to HCV infection has provided us with a more complete understanding of how RIG-I engages its ligands for activation of downstream signaling pathways that mediate the production of type I IFN and ISGs. In spite of the fact that HCV induces RIG-I signaling, 80% of those infected do not clear the virus and become chronically infected. This high frequency of virus persistence underscores the observations that HCV has a diverse set of strategies for evading the actions of RIG-I signaling, type I interferon actions, and the function of specific antiviral ISGs, thus creating an environment suitable for persistent infection. These innate immune evasion mechanisms also associate with defective adaptive immune responses to HCV, reflecting the importance of innate immune programming of adaptive immunity. A detailed understanding of the interactions of HCV with the innate immune response is necessary for developing novel immuno-therapeutics and effective vaccine strategies against HCV infection.

Summary

Hepatitis C virus persistence now impacts nearly 200 million people with chronic infection. A key to viral persistence is the ability of HCV to control and evade innate immune defenses that would otherwise suppress infection. HCV persistence is linked to viral control of the host cell RIG-I pathway and regulation of interferon actions within infected hepatocytes. This chapter provides a contemporary overview of the virus and host processes that regulate innate immune defenses during HCV infection. Understanding of the HCV-host interactions that govern innate immunity is essential for improving the current interferon-based therapeutic regimen for HCV infection.

Acknowledgments

The authors would like to thank members of the Gale laboratory and our colleagues for helpful discussion, collaboration, and consultation. SMH is a fellow of the Irvington Institute Fellowship Program of the Cancer Research Institute. Work in the Gale laboratory is supported by funds from the State of Washington, National Institutes of Health (AI060389, AI40035, DA024563, and AI057568), and by the Burroughs-Wellcome Fund.

References

1. Bosch FX, Ribes J, Borras J. (1999) Epidemiology of primary liver cancer. *Semin Liver Dis* **19**: 271–285.
2. Sharma P, Lok A. (2006) Viral hepatitis and liver transplantation. *Semin Liver Dis* **26**: 285–297.
3. Lavanchy D. (2009) The global burden of hepatitis C. *Liver Int* **29**(Suppl 1): 74–81.

4. Soriano V, Peters MG, Zeuzem S. (2009) New therapies for hepatitis C virus infection. *Clin Infect Dis* **48**: 313–320.
5. Pineda JA, Garcia-Garcia JA, Aguilar-Guisado M, *et al.* (2007) Clinical progression of hepatitis C virus-related chronic liver disease in human immunodeficiency virus-infected patients undergoing highly active antiretroviral therapy. *Hepatology* **46**: 622–630.
6. De Veer MJ, Holko M, Frevel M, *et al.* (2001) Functional classification of interferon-stimulated genes identified using microarrays. *J Leukoc Biol* **69**: 912–920.
7. Stetson DB, Medzhitov R. (2006) Type I interferons in host defense. Immunity **25**: 373–381.
8. Purtha WE, Chachu KA, Virgin HWt, Diamond MS. (2008) Early B-cell activation after West Nile virus infection requires alpha/beta interferon but not antigen receptor signaling. *J Virol* **82**: 10964–10974.
9. Petrilli V, Dostert C, Muruve DA, Tschopp J. (2007) The inflammasome: A danger sensing complex triggering innate immunity. *Curr Opin Immunol* **19**: 615–622.
10. Neumann AU, Lam NP, Dahari H, *et al.* (1998) Hepatitis C viral dynamics *in vivo* and the antiviral efficacy of interferon-alpha therapy. *Science* **282**: 103–107.
11. Su AI, Pezacki JP, Wodicka L, *et al.* (2002) Genomic analysis of the host response to hepatitis C virus infection. *Proc Natl Acad Sci USA* **99**: 15669–15674.
12. Saito T, Owen DM, Jiang F, *et al.* (2008) Innate immunity induced by composition-dependent RIG-I recognition of hepatitis C virus RNA. *Nature* **454**: 523–527.
13. Sumpter R, Jr., Loo YM, Foy E, *et al.* (2005) Regulating intracellular antiviral defense and permissiveness to hepatitis C virus RNA replication through a cellular RNA helicase, RIG-I. *J Virol* **79**: 2689–2699.
14. Saito T, Gale M Jr. (2007) Principles of intracellular viral recognition. *Curr Opin Immunol* **19**: 17–23.
15. Eisenacher K, Steinberg C, Reindl W, Krug A. (2007) The role of viral nucleic acid recognition in dendritic cells for innate and adaptive antiviral immunity. *Immunobiology* **212**: 701–714.
16. Kanneganti TD, Body-Malapel M, Amer A, *et al.* (2006) Critical role for Cryopyrin/Nalp3 in activation of caspase-1 in response to viral infection and double-stranded RNA. *J Biol Chem* **281**: 36560–36568.
17. Chang S, Dolganiuc A, Szabo G. (2007) Toll-like receptors 1 and 6 are involved in TLR2-mediated macrophage activation by hepatitis C virus core and NS3 proteins. *J Leukoc Biol* **82**: 479–487.
18. Dolganiuc A, Oak S, Kodys K, *et al.* (2004) Hepatitis C core and nonstructural 3 proteins trigger toll-like receptor 2-mediated pathways and inflammatory activation. *Gastroenterology* **127**: 1513–1524.

19. Diebold SS, Kaisho T, Hemmi H, *et al.* (2004) Innate antiviral responses by means of TLR7-mediated recognition of single-stranded RNA. *Science* **303**: 1529–1531.
20. Heil F, Hemmi H, Hochrein H, *et al.* (2004) Species-specific recognition of single-stranded RNA via toll-like receptor 7 and 8. *Science* **303**: 1526–1529.
21. Schroder M, Bowie AG. (2005) TLR3 in antiviral immunity: key player or bystander? *Trends Immunol* **26**: 462–468.
22. Vercammen E, Staal J, Beyaert R. (2008) Sensing of viral infection and activation of innate immunity by toll-like receptor 3. *Clin Microbiol Rev* **21**: 13–25.
23. Negishi H, Osawa T, Ogami K, *et al.* (2008) A critical link between Toll-like receptor 3 and type II interferon signaling pathways in antiviral innate immunity. *Proc Natl Acad Sci USA* **105**: 20446–20451.
24. Seki E, Brenner DA. (2008) Toll-like receptors and adaptor molecules in liver disease: update. *Hepatology* **48**: 322–335.
25. Schulz O, Diebold SS, Chen M, *et al.* (2005) Toll-like receptor 3 promotes cross-priming to virus-infected cells. *Nature* **433**: 887–892.
26. Kato H, Takeuchi O, Sato S, *et al.* (2006) Differential roles of MDA5 and RIG-I helicases in the recognition of RNA viruses. *Nature* **441**: 101–105.
27. Loo YM, Fornek J, Crochet N, *et al.* (2008) Distinct RIG-I and MDA5 signaling by RNA viruses in innate immunity. *J Virol* **82**: 335–345.
28. Yoneyama M, Kikuchi M, Natsukawa T, *et al.* (2004) The RNA helicase RIG-I has an essential function in double-stranded RNA-induced innate antiviral responses. *Nat Immunol* **5**: 730–737.
29. Gitlin L, Barchet W, Gilfillan S, *et al.* (2006) Essential role of mda-5 in type I IFN responses to polyriboinosinic: polyribocytidylic acid and encephalomyocarditis picornavirus. *Proc Natl Acad Sci USA* **103**: 8459–8464.
30. Kato H, Takeuchi O, Mikamo-Satoh E, *et al.* (2008) Length-dependent recognition of double-stranded ribonucleic acids by retinoic acid-inducible gene-I and melanoma differentiation-associated gene 5. *J Exp Med* **205**: 1601–1610.
31. Saito T, Hirai R, Loo YM, *et al.* (2007) Regulation of innate antiviral defenses through a shared repressor domain in RIG-I and LGP2. *Proc Natl Acad Sci USA* **104**: 582–587.
32. Loo YM, Owen DM, Li K, *et al.* (2006) Viral and therapeutic control of IFN-beta promoter stimulator 1 during hepatitis C virus infection. *Proc Natl Acad Sci USA* **103**: 6001–6006.
33. Uzri D, Gehrke L. (2009) Nucleotide sequences and modifications that determine RIG-I/RNA binding and signaling activities. *J Virol* **83**: 4174–4184.
34. Hornung V, Ellegast J, Kim S, *et al.* (2006) 5′-Triphosphate RNA is the ligand for RIG-I. *Science* **314**: 994–997.
35. Pichlmair A, Schulz O, Tan CP, *et al.* (2006) RIG-I-mediated antiviral responses to single-stranded RNA bearing 5′-phosphates. *Science* **314**: 997–1001.

36. Takahasi K, Yoneyama M, Nishihori T, *et al.* (2008) Nonself RNA-sensing mechanism of RIG-I helicase and activation of antiviral immune responses. *Mol Cell* **29**: 428–440.
37. You S, Rice CM. (2008) 3′ RNA elements in hepatitis C virus replication: kissing partners and long poly(U). *J Virol* **82**: 184–195.
38. Ito T, Tahara SM, Lai MM. (1998) The 3′-untranslated region of hepatitis C virus RNA enhances translation from an internal ribosomal entry site. *J Virol* **72**: 8789–8796.
39. Song Y, Friebe P, Tzima E, *et al.* (2006) The hepatitis C virus RNA 3′-untranslated region strongly enhances translation directed by the internal ribosome entry site. *J Virol* **80**: 11579–11588.
40. Cui S, Eisenacher K, Kirchhofer A, *et al.* (2008) The C-terminal regulatory domain is the RNA 5′-triphosphate sensor of RIG-I. *Mol Cell* **29**: 169–179.
41. Myong S, Cui S, Cornish PV, *et al.* (2009) Cytosolic viral sensor RIG-I is a 5′-triphosphate-dependent translocase on double-stranded RNA. *Science* **323**: 1070–1074.
42. Egger D, Wolk B, Gosert R, *et al.* (2002) Expression of hepatitis C virus proteins induces distinct membrane alterations including a candidate viral replication complex. *J Virol* **76**: 5974–5984.
43. Gosert R, Egger D, Lohmann V, *et al.* (2003) Identification of the hepatitis C virus RNA replication complex in Huh-7 cells harboring subgenomic replicons. *J Virol* **77**: 5487–5492.
44. Targett-Adams P, Boulant S, McLauchlan J. (2008) Visualization of double-stranded RNA in cells supporting hepatitis C virus RNA replication. *J Virol* **82**: 2182–2195.
45. Wolk B, Buchele B, Moradpour D, Rice CM. (2008) A dynamic view of hepatitis C virus replication complexes. *J Virol* **82**: 10519–10531.
46. Lau DT, Fish PM, Sinha M, *et al.* (2008) Interferon regulatory factor-3 activation, hepatic interferon-stimulated gene expression, and immune cell infiltration in hepatitis C virus patients. *Hepatology* **47**: 799–809.
47. Ferrer-Orta C, Arias A, Escarmis C, Verdaguer N. (2006) A comparison of viral RNA-dependent RNA polymerases. *Curr Opin Struct Biol* **16**: 27–34.
48. Habjan M, Andersson I, Klingstrom J, *et al.* (2008) Processing of genome 5′ termini as a strategy of negative-strand RNA viruses to avoid RIG-I-dependent interferon induction. *PLoS ONE* **3**: e2032.
49. Malathi K, Dong B, Gale M Jr., Silverman RH. (2007) Small self-RNA generated by RNase L amplifies antiviral innate immunity. *Nature* **448**: 816–819.
50. Han JQ, Barton DJ. (2002) Activation and evasion of the antiviral 2′–5′ oligoadenylate synthetase/ribonuclease L pathway by hepatitis C virus mRNA. *RNA* **8**: 512–525.
51. Washenberger CL, Han JQ, Kechris KJ, *et al.* (2007) Hepatitis C virus RNA: dinucleotide frequencies and cleavage by RNase L. *Virus Res* **130**: 85–95.

52. Gee P, Chua PK, Gevorkyan J, *et al.* (2008) Essential role of the N-terminal domain in the regulation of RIG-I ATPase activity. *J Biol Chem* **283**: 9488–9496.
53. Saito T, Gale M Jr. (2008) Regulation of innate immunity against hepatitis C virus infection. *Hepatol Res* **38**: 115–122.
54. Luo D, Ding SC, Vela A, *et al.* (2011) Structural insights into RNA recognition by RIG-I. *Cell* **147**(2): 409–422.
55. Kowalinski E, Lunardi T, McCarthy AA, *et al.* (2011) Structural basis for the activation of innate immune pattern-recognition receptor RIG-1 by viral RNA. **147**(2): 423–435.
56. Jiang F, Ramanathan A, Miller MT, *et al.* (2011). Structural basis of RNA recognition and activation by innate immune receptor RIG-1. *Nature* **479**(7373): 423–427.
57. Gack MU, Shin YC, Joo CH, *et al.* (2007) TRIM25 RING-finger E3 ubiquitin ligase is essential for RIG-I-mediated antiviral activity. *Nature* **446**: 916–920.
58. Oshiumi H, Matsumoto M, Hatakeyama S, Seya T. (2009) Riplet/RNF135, a RING finger protein, ubiquitinates RIG-I to promote interferon-beta induction during the early phase of viral infection. *J Biol Chem* **284**: 807–817.
59. Yoneyama M, Kikuchi M, Matsumoto K, *et al.* (2005) Shared and unique functions of the DExD/H-box helicases RIG-I, MDA5, and LGP2 in antiviral innate immunity. *J Immunol* **175**: 2851–2858.
60. Komuro A, Horvath CM. (2006) RNA- and virus-independent inhibition of antiviral signaling by RNA helicase LGP2. *J Virol* **80**: 12332–12342.
61. Bamming D, Horvath CM. (2009) Regulation of signal transduction by enzymatically inactive antiviral RNA helicase proteins MDA5, RIG-I and LGP2. *J Biol Chem* **284**: 9700–9712.
62. Li X, Ranjith-Kumar CT, Brooks MT, *et al.* (2009) The RIG-I like receptor LGP2 recognizes the termini of double-stranded RNA. *J Biol Chem* **284**: 13881–13891.
63. Venkataraman T, Valdes M, Elsby R, *et al.* (2007) Loss of DExD/H box RNA helicase LGP2 manifests disparate antiviral responses. *J Immunol* **178**: 6444–6455.
64. Elco CP, Guenther JM, Williams BR, Sen GC. (2005) Analysis of genes induced by Sendai virus infection of mutant cell lines reveals essential roles of interferon regulatory factor 3, NF-kappaB, and interferon but not toll-like receptor 3. *J Virol* **79**: 3920–3929.
65. Grandvaux N, Servant MJ, tenOever B, *et al.* (2002) Transcriptional profiling of interferon regulatory factor 3 target genes: direct involvement in the regulation of interferon-stimulated genes. *J Virol* **76**: 5532–5539.
66. Guo JT, Bichko VV, Seeger C. (2001) Effect of alpha interferon on the hepatitis C virus replicon. *J Virol* **75**: 8516–8523.

67. Pawlotsky JM. (2003) The nature of interferon-alpha resistance in hepatitis C virus infection. *Curr Opin Infect Dis* **16**: 587–592.
68. Bigger CB, Brasky KM, Lanford RE. (2001) DNA microarray analysis of chimpanzee liver during acute resolving hepatitis C virus infection. *J Virol* **75**: 7059–7066.
69. Bigger CB, Guerra B, Brasky KM, *et al.* (2004) Intrahepatic gene expression during chronic hepatitis C virus infection in chimpanzees. *J Virol* **78**: 13779–13792.
70. Helbig KJ, Lau DT, Semendric L, *et al.* (2005) Analysis of ISG expression in chronic hepatitis C identifies viperin as a potential antiviral effector. *Hepatology* **42**: 702–710.
71. Lanford RE, Guerra B, Lee H, *et al.* (2006) Genomic response to interferon-alpha in chimpanzees: implications of rapid downregulation for hepatitis C kinetics. *Hepatology* **43**: 961–972.
72. Smith MW, Yue ZN, Korth MJ, *et al.* (2003) Hepatitis C virus and liver disease: global transcriptional profiling and identification of potential markers. *Hepatology* **38**: 1458–1467.
73. Wang C, Pflugheber J, Sumpter R Jr., *et al.* (2003) Alpha interferon induces distinct translational control programs to suppress hepatitis C virus RNA replication. *J Virol* **77**: 3898–3912.
74. Taylor DR, Puig M, Darnell ME, *et al.* (2005) New antiviral pathway that mediates hepatitis C virus replicon interferon sensitivity through ADAR1. *J Virol* **79**: 6291–6298.
75. Hinson ER, Cresswell P. (2009) The N-terminal amphipathic alpha-helix of viperin mediates localization to the cytosolic face of the endoplasmic reticulum and inhibits protein secretion. *J Biol Chem* **284**: 4705–4712.
76. Jiang D, Guo H, Xu C, *et al.* (2008) Identification of three interferon-inducible cellular enzymes that inhibit the replication of hepatitis C virus. *J Virol* **82**: 1665–1678.
77. Sarasin-Filipowicz M, Oakeley EJ, Duong FH, *et al.* (2008) Interferon signaling and treatment outcome in chronic hepatitis C. *Proc Natl Acad Sci USA* **105**: 7034–7039.
78. Chen L, Borozan I, Feld J, *et al.* (2005) Hepatic gene expression discriminates responders and nonresponders in treatment of chronic hepatitis C viral infection. *Gastroenterology* **128**: 1437–1444.
79. Randall G, Chen L, Panis M, *et al.* (2006) Silencing of USP18 potentiates the antiviral activity of interferon against hepatitis C virus infection. *Gastroenterology* **131**: 1584–1591.
80. Erickson AK, Seiwert S, Gale M Jr. (2008) Antiviral potency analysis and functional comparison of consensus interferon, interferon-alpha2a and pegylated interferon-alpha2b against hepatitis C virus infection. *Antivir Ther* **13**: 851–862.
81. Pedersen IM, Cheng G, Wieland S, *et al.* (2007) Interferon modulation of cellular microRNAs as an antiviral mechanism. *Nature* **449**: 919–922.

82. Zhong B, Yang Y, Li S, *et al.* (2008) The adaptor protein MITA links virus-sensing receptors to IRF3 transcription factor activation. *Immunity* **29**: 538–550.
83. Jin L, Waterman PM, Jonscher KR, *et al.* (2008) MPYS, a novel membrane tetraspanner, is associated with major histocompatibility complex class II and mediates transduction of apoptotic signals. *Mol Cell Biol* **28**: 5014–5026.
84. Ishikawa H, Barber GN. (2008) STING is an endoplasmic reticulum adaptor that facilitates innate immune signalling. *Nature* **455**: 674–678.
85. Moore CB, Bergstralh DT, Duncan JA, *et al.* (2008) NLRX1 is a regulator of mitochondrial antiviral immunity. *Nature* **451**: 573–577.
86. Tattoli I, Carneiro LA, Jehanno M, *et al.* (2008) NLRX1 is a mitochondrial NOD-like receptor that amplifies NF-kappaB and JNK pathways by inducing reactive oxygen species production. *EMBO Rep* **9**: 293–300.
87. Xu L, Xiao N, Liu F, *et al.* (2009) Inhibition of RIG-I and MDA5-dependent antiviral response by gC1qR at mitochondria. *Proc Natl Acad Sci USA* **106**: 1530–1535.
88. Kittlesen DJ, Chianese-Bullock KA, Yao ZQ, *et al.* (2000) Interaction between complement receptor gC1qR and hepatitis C virus core protein inhibits T-lymphocyte proliferation. *J Clin Invest* **106**: 1239–1249.
89. Yao ZQ, Eisen-Vandervelde A, Waggoner SN, *et al.* (2004) Direct binding of hepatitis C virus core to gC1qR on CD4+ and CD8+ T cells leads to impaired activation of Lck and Akt. *J Virol* **78**: 6409–6419.
90. Gale M, Jr., Foy EM. (2005) Evasion of intracellular host defence by hepatitis C virus. *Nature* **436**: 939–945.
91. Brass V, Berke JM, Montserret R, *et al.* (2008) Structural determinants for membrane association and dynamic organization of the hepatitis C virus NS3–4A complex. *Proc Natl Acad Sci USA* **105**: 14545–14550.
92. Wolk B, Sansonno D, Krausslich HG, *et al.* (2000) Subcellular localization, stability, and trans-cleavage competence of the hepatitis C virus NS3-NS4A complex expressed in tetracycline-regulated cell lines. *J Virol* **74**: 2293–2304.
93. Foy E, Li K, Wang C, *et al.* (2003) Regulation of interferon regulatory factor-3 by the hepatitis C virus serine protease. *Science* **300**: 1145–1148.
94. Li XD, Sun L, Seth RB, *et al.* (2005) Hepatitis C virus protease NS3/4A cleaves mitochondrial antiviral signaling protein off the mitochondria to evade innate immunity. *Proc Natl Acad Sci USA* **102**: 17717–17722.
95. Meylan E, Curran J, Hofmann K, *et al.* (2005) Cardif is an adaptor protein in the RIG-I antiviral pathway and is targeted by hepatitis C virus. *Nature* **437**: 1167–1172.
96. Horner SM, Liu HM, Park HS, *et al.* (2011) Mitochondrial-associated endoplasmic reticulum membranes (MAM) form innate immune synapses and are targeted by hepatitis C virus. *Proc Natl Acad Sci USA* 2011 Aug 15.
97. Baril M, Racine ME, Penin F, Lamarre D. (2009) MAVS dimer is a crucial signaling component of innate immunity and the target of hepatitis C virus NS3/4A protease. *J Virol* **83**: 1299–1311.

98. Tang ED, Wang CY. (2009) MAVS self-association mediates antiviral innate immune signaling. *J Virol* **83**: 3420–3428.
99. Dansako H, Ikeda M, Kato N. (2007) Limited suppression of the interferon-beta production by hepatitis C virus serine protease in cultured human hepatocytes. *FEBS J* **274**: 4161–4176.
100. Dansako H, Ikeda M, Ariumi Y, *et al.* (2009) Double-stranded RNA-induced interferon-beta and inflammatory cytokine production modulated by hepatitis C virus serine proteases derived from patients with hepatic diseases. *Arch Virol.*
101. Johnson CL, Owen DM, Gale M Jr. (2007) Functional and therapeutic analysis of hepatitis C virus NS3.4A protease control of antiviral immune defense. *J Biol Chem* **282**: 10792–10803.
102. Liang Y, Ishida H, Lenz O, *et al.* (2008) Antiviral suppression vs restoration of RIG-I signaling by hepatitis C protease and polymerase inhibitors. *Gastroenterology* **135**: 1710–1718 e1712.
103. Otsuka M, Kato N, Moriyama M, *et al.* (2005) Interaction between the HCV NS3 protein and the host TBK1 protein leads to inhibition of cellular antiviral responses. *Hepatology* **41**: 1004–1012.
104. Mamiya N, Worman HJ. (1999) Hepatitis C virus core protein binds to a DEAD box RNA helicase. *J Biol Chem* **274**: 15751–15756.
105. Owsianka AM, Patel AH. (1999) Hepatitis C virus core protein interacts with a human DEAD box protein DDX3. *Virology* **257**: 330–340.
106. You LR, Chen CM, Yeh TS, *et al.* (1999) Hepatitis C virus core protein interacts with cellular putative RNA helicase. *J Virol* **73**: 2841–2853.
107. Schroder M, Baran M, Bowie AG. (2008) Viral targeting of DEAD box protein 3 reveals its role in TBK1/IKKepsilon-mediated IRF activation. *EMBO J* **27**: 2147–2157.
108. Ariumi Y, Kuroki M, Abe K, *et al.* (2007) DDX3 DEAD-box RNA helicase is required for hepatitis C virus RNA replication. *J Virol* **81**: 13922–13926.
109. Tasaka M, Sakamoto N, Itakura Y, *et al.* (2007) Hepatitis C virus non-structural proteins responsible for suppression of the RIG-I/Cardif-induced interferon response. *J Gen Virol* **88**: 3323–3333.
110. Heim MH, Moradpour D, Blum HE. (1999) Expression of hepatitis C virus proteins inhibits signal transduction through the Jak-STAT pathway. *J Virol* **73**: 8469–8475.
111. Blindenbacher A, Duong FH, Hunziker L, *et al.* (2003) Expression of hepatitis c virus proteins inhibits interferon alpha signaling in the liver of transgenic mice. *Gastroenterology* **124**: 1465–1475.
112. Duong FH, Filipowicz M, Tripodi M, *et al.* (2004) Hepatitis C virus inhibits interferon signaling through up-regulation of protein phosphatase 2A. *Gastroenterology* **126**: 263–277.

113. De Lucas S, Bartolome J, Carreno V. (2005) Hepatitis C virus core protein down-regulates transcription of interferon-induced antiviral genes. *J Infect Dis* **191**: 93–99.
114. Lin W, Kim SS, Yeung E, *et al.* (2006) Hepatitis C virus core protein blocks interferon signaling by interaction with the STAT1 SH2 domain. *J Virol* **80**: 9226–9235.
115. Melen K, Fagerlund R, Nyqvist M, *et al.* (2004) Expression of hepatitis C virus core protein inhibits interferon-induced nuclear import of STATs. *J Med Virol* **73**: 536–547.
116. Bode JG, Ludwig S, Ehrhardt C, *et al.* (2003) IFN-alpha antagonistic activity of HCV core protein involves induction of suppressor of cytokine signaling-3. *FASEB J* **17**: 488–490.
117. Taylor DR, Shi ST, Romano PR, *et al.* (1999) Inhibition of the interferon-inducible protein kinase PKR by HCV E2 protein. *Science* **285**: 107–110.
118. Appel N, Zayas M, Miller S, *et al.* (2008) Essential role of domain III of non-structural protein 5A for hepatitis C virus infectious particle assembly. *PLoSPathog* 4: e1000035.
119. Huang Y, Staschke K, De Francesco R, Tan SL. (2007) Phosphorylation of hepatitis C virus NS5A nonstructural protein: a new paradigm for phosphorylation-dependent viral RNA replication? *Virology* **364**: 1–9.
120. Tellinghuisen TL, Foss KL, Treadaway J. (2008) Regulation of hepatitis C virion production via phosphorylation of the NS5A protein. *PLoSPathog* **4**: e1000032.
121. Gale M Jr., Blakely CM, Kwieciszewski B, *et al.* (1998) Control of PKR protein kinase by hepatitis C virus nonstructural 5A protein: molecular mechanisms of kinase regulation. *Mol Cell Biol* **18**: 5208–5218.
122. Pflugheber J, Fredericksen B, Sumpter R Jr., *et al.* (2002) Regulation of PKR and IRF-1 during hepatitis C virus RNA replication. *Proc Natl Acad Sci USA* **99**: 4650–4655.
123. Taguchi T, Nagano-Fujii M, Akutsu M, *et al.* (2004) Hepatitis C virus NS5A protein interacts with 2′,5′-oligoadenylate synthetase and inhibits antiviral activity of IFN in an IFN sensitivity-determining region-independent manner. *J Gen Virol* **85**: 959–969.
124. Polyak SJ, Khabar KS, Paschal DM, *et al.* (2001) Hepatitis C virus nonstructural 5A protein induces interleukin-8, leading to partial inhibition of the interferon-induced antiviral response. *J Virol* **75**: 6095–6106.
125. Polyak SJ, Khabar KS, Rezeiq M, Gretch DR. (2001) Elevated levels of interleukin-8 in serum are associated with hepatitis C virus infection and resistance to interferon therapy. *J Virol* **75**: 6209–6211.
126. Abe T, Kaname Y, Hamamoto I, *et al.* (2007) Hepatitis C virus nonstructural protein 5A modulates the toll-like receptor-MyD88-dependent signaling pathway in macrophage cell lines. *J Virol* **81**: 8953–8966.

Chapter 10

Expression and Functions of Hepatitis C Virus F Protein

Kamile Yuksek and Jing-hsiung James Ou*

Introduction

Hepatitis C virus (HCV) is a positive-stranded RNA virus with a genome size of about 9.6 Kb. This virus is the sole member of the *Hepacivirus* genus of the *Flaviviridae* family. The genomic sequences of different HCV isolates are highly heterogeneous. Based on phylogenetic analysis, HCV has been grouped into six major genotypes and many subtypes.[1] The HCV genome codes for a polyprotein, which is cleaved by cellular and viral proteases into 10 different proteins (Fig. 1). The structural proteins, which are the core protein and E1 and E2 envelope proteins, are located at the N-terminus of the polyprotein sequence. These structural proteins are followed by the p7 protein, which possesses the ion channel activity, and then by NS2, NS3, NS4A, NS4B, NS5A, and NS5B non-structural proteins, which are involved in viral replication and maturation. The translation of the HCV polyprotein is cap-independent and mediated by the internal ribosome entry site (IRES), a highly structured RNA sequence encompassing most of the 5′-noncoding region and the first few codons of the polyprotein

* Department of Molecular Microbiology and Immunology, University of Southern California, Keck School of Medicine, Los Angeles, California. Email: jamesou@usc.edu

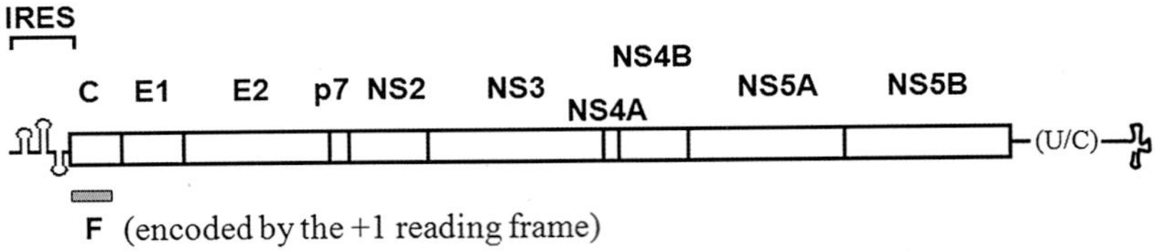

Fig. 1. The HCV genomic organization. The gray bar indicates the F protein coding sequence. IRES, internal ribosomal entry site; (U/C), U/C-rich sequence.

coding sequence. In addition to the polyprotein, an additional protein termed F protein, alternative reading frame (ARF) protein, or core+1 has also been reported.[2–4] The coding sequence of this protein overlaps with that of the core protein sequence in the +1/−2 reading frame and is conserved in more than 90% of the HCV sequences compiled in the database (http://s2as02.genes.nig.ac.jp/) (Fig. 1). Its length, however, varies from approximately 120 to 160 amino acids, depending on the genotypes.[3] The conservation of the F protein reading frame is unlikely fortuitous or simply caused by the presence of two conserved RNA stem-loop structures (SLV and SLVI) in the core protein coding sequence,[5,6] as the +2/−1 reading frame is disrupted by multiple stop codons,[2] and the F protein coding sequence is also conserved in the genome of the distantly related GB virus-B (GBV-B).[3] In this article, we will review the expression and the possible biological functions of this F protein.

The Discovery of the HCV F Protein

The expression of a 16 kDa protein from the HCV core protein sequence was first observed in the *in vitro* translation experiments using the core protein RNA of the HCV-1 isolate,[7] a genotype 1a virus. This 16 kDa protein was later also observed in mammalian cells when the core protein sequence of HCV-1 and another genotype 1a virus HCV-RH was expressed, although the expression efficiency for this protein was significantly lower for the HCV-RH strain.[8,9] The difference in expression efficiency of the 16 kDa protein between these two strains is due to the presence of a continuous stretch of

10 adenosine residues at codons 8–11 of the HCV-1 core protein sequence. This continuous 10–A sequence is absent in the HCV-RH strain due to the replacement of the fourth A residue with a G in codon 9 and the tenth A residue with a C in codon 11. This HCV-RH sequence will be referred to as the wild-type sequence in this article as it is the sequence present in the great majority of the HCV sequences that have been reported. Further analysis of the HCV core protein coding sequence revealed an additional coding sequence in the +1 reading frame. The possibility that this 16 kDa protein might be encoded by this alternative reading frame was confirmed by our laboratory when a premature termination codon introduced into this +1 reading frame of the HCV-1 RNA was found to reduce the size of this 16 kDa protein without affecting the size of the core protein on a protein gel.[3] How this 16 kDa protein might be expressed from this alternative reading frame was unclear due to the lack of an initiator AUG codon that may be used to direct its protein synthesis. The possibility that this protein is translated from an internal non-AUG codon was eliminated because fusing the coding sequence of an HA tag to the 5′-end of the HCV-1 core protein coding sequence increased the size of both the core protein and this 16 kDa protein. This particular experiment also indicated that this 16 kDa protein uses the core protein AUG codon for its translation initiation and, since this protein terminates at the stop codon of the +1 reading frame, translational ribosomal frameshift is likely involved in its synthesis. Indeed, radio-sequencing of the core protein and the F protein synthesized using rabbit reticulocyte lysates confirmed that a −2/+1 frameshift occurred at codons 8–11 of the core protein sequence, resulting in the production of the 16 kDa protein sharing the first 10 amino acids with the core protein. This protein was hence named the F protein for "frameshifting."[3]

The F protein was also independently discovered by Walewski *et al.* They aligned eight diverse full-length HCV sequences from GenBank to search for additional reading frames or cryptic elements that might be important for the HCV life cycle.[2] They identified two regions that were excessively conserved. One of these two regions is the core protein coding sequence, which contains 22 highly conserved codons. As this high degree of conservation cannot be attributed to one single coding function of the viral RNA, they conducted further analysis on the core protein coding sequence in the HCV database and also came to the identification of the overlapping F protein coding sequence, which they termed ARF, for alternative reading frame. Walewski *et al.* confirmed the expression of the F/ARF protein during natural HCV infection by using synthetic peptides derived from the F/ARF protein sequence to test HCV- or HBV-infected patient sera. They found that there was a correlation between HCV infection and the antiserum reactivity to F/ARF peptides.[2]

The F protein was also named core+1, to indicate the position of the new reading frame, by Varaklioti *et al.* who showed that the expression of the F protein is dependent on the translation of the core protein.[4] They studied the expression of the HCV-1 core protein and found that the mutation of the core protein initiation codon to a stop codon would abolish the production of the F protein *in vitro*.[4] Their data also support that the F protein is synthesized by translational ribosomal frameshift.

Expression of the F Protein During Natural Infection

Although the HCV F protein has not been detected in HCV-infected hepatocytes in patients, serological evidence provided irrefutable evidence that the F protein is expressed during natural HCV infection in patients.[2,3,10–14] As mentioned above, Walewski *et al.* demonstrated the presence of antibodies in patients that reacted with the F protein–derived peptides.[2] By using the F protein synthesized either *in vitro* using rabbit reticulocyte lysates or in *E. coli*, we also demonstrated the presence of F-reactive antibodies in HCV patients.[3] Subsequent studies by several laboratories indicated that the F protein reactive antibodies were present in 40–90% of HCV patients, depending on the studies.[3,10–15] The anti-F response is independent of HCV genotypes, coinfection by human immunodeficiency virus-1 (HIV-1), the stage of liver disease, or viral load.[10–13] Moreover, there is no correlation between the anti-F response and the response to antiviral treatments.[12] The current anti-HCV treatment is a combination therapy using both pegylated interferon (peg-IFN) and ribavirin. Patients who recovered from HCV infection after the combination therapy had reduced anti-F antibody titers compared to non-responders or chronic HCV patients.[14,15]

Another line of evidence that indicates the expression of the F protein during natural HCV infection is the presence of T lymphocytes that are reactive to the F protein in HCV patients.[11,13] Based on the F protein sequence shared among the HCV genotypes, Bain *et al.* predicted several T cell epitopes and synthesized these class I restricted peptides from genotype 1b.[13] To measure the T cell response against the F protein, they analyzed peripheral blood from chronically or previously HCV-infected patients by measuring IFN-γ and/or IL-10 production from T cells in response to these synthetic peptides, which is indicative of T cell activation. In 20% of the HCV seropositive patients, the F-specific T cell response was observed.[13] Troesch *et al.* also analyzed the cytotoxic T lymphocyte (CTL) activity in the peripheral blood mononuclear cells (PBMC) of 11 HCV-infected subjects.[11] They found F protein-specific CTL activities in 9 of these 11 subjects. The detection of F-specific T cell responses again demonstrates the production of the F protein during natural HCV infection.

Molecular Mechanisms for the Expression of the HCV F Protein

Expression by Programmed Ribosomal Frameshift

Multiple mechanisms for the expression of the HCV F protein have been proposed. As discussed above, the initial mechanism proposed for the expression of the F protein is +1/–2 translational ribosomal frameshift at codons 8–11 of the core protein coding sequence. This would result in the shifting of the ribosome into the F reading frame and the production of a protein that contains a 10-amino acid leader from the core protein sequence at the N-terminus. Programmed ribosomal frameshift is a mechanism used mostly by viruses to increase the coding capacity of their genomes.[16] It is also used by certain cellular genes.[16] Ribosomes can shift either backward (−1) or forward (+1) during translation, leading to the production of different proteins.[17] Members of *Retroviridae, Astroviridae, Coronaviridae, Nidovirales,* etc., make use of the ribosomal frameshift to produce viral proteins.[18] The consensus slippery sequence for −1 frameshifting is X XXY YYZ, where X is any nucleotide, Y is A or U, and Z is any nucleotide but G. This slippery sequence will allow the two tRNAs occupying the P site and the A site of the ribosome to maintain two base pairs with the codons after −1 ribosomal frameshift.

The initial studies on the expression of the HCV F protein were conducted using the HCV-1 isolate.[3,4] This isolate contains a stretch of 10 continuous adenosine residues at codons 8–11. This sequence will allow the ribosomes to slip either forward or backward at codons 9, 10, or 11 by one or two nucleotides and still allow the tRNAs to maintain at least two correct base pairings with the codons. It is interesting that there is at least one copy of the consensus −1 ribosomal frameshift signal X XXY YYZ in each of the three reading frames. This raises the possibility that the F protein may also be synthesized by double −1 frameshifting events. Recently, however, it has also been suggested that transcriptional slippage at the 10-A sequence may also lead to deletion and insertion of A residues to cause the production of the F protein.[19]

Although HCV-1 contains a continuous 10-A sequence at codons 8–11, this sequence is absent in most of the HCV sequences, which are the same as the HCV-RH sequence and have an A-to-G substitution in codon 9 and an A-to-C substitution in codon 11. We have also examined whether this more common HCV sequence (i.e., the wild-type sequence) can also mediate translational ribosomal frameshift by conducting a luciferase reporter assay. In our study, we fused the luciferase reporter sequence to the core protein coding sequence in three frames (−1, 0, and +1). Our studies indicated that the wild-type HCV sequence, similar to the 10-A sequence, could also mediate both −1 and +1 ribosomal frameshifts, albeit with a reduced efficiency.[20] Triple-decoding of the HCV RNA by +1 and −1 ribosomal frameshifts at codons 8–11 will lead to the synthesis of, in addition to the core protein (0 frame) and the F protein (+1 frame), a short 13-amino acid peptide (−1 frame).[20] Antibodies reactive to this peptide were also detected in the sera of HCV patients.[20] However, since this peptide shares the first eight amino acids with the core protein, the possibility that this cross-reactivity was due to anti-core antibodies cannot be ruled out.

In addition to the slippery sequence, RNA secondary structures following the slippery sequence may also facilitate programmed ribosomal frameshift.[17] Indeed, the HCV frameshift signal at codons 8–11 is followed by a double stem-loop structure (SLV and SLVI) formed by codons 15–53. The loops of SLV and SLVI can potentially basepair to form a pseudoknot structure.[20,21] This double stem-loop structure enhanced HCV ribosomal frameshift in the presence of puromycin, a translation elongation inhibitor.[20] The increase of HCV ribosomal frameshift by puromycin may be due to the stalling of ribosomes by puromycin at the double stem-loop structure, which then promotes tRNA slippage at the frameshift site located immediately upstream. This raises the possibility that viral or cellular factors that can stall ribosomes at this RNA secondary structure may also be able to enhance the expression of the F protein.

In addition to +1/−2 ribosomal frameshift at codons 8–11, Boulant *et al.* reported that the expression of the HCV genotype 1b F protein could also be mediated by two ribosomal frameshift events: a +1 frameshift at codon 42 and a −1 frameshift at codon 144. Based on this expression mechanism, the F protein would contain the core protein sequence at the N-terminus and the C-terminus, with amino acids 42–143 derived from the +1 reading frame.[22]

The +1/−2 frameshift efficiency at codons 8–11 of the HCV-1 RNA is about 1–2% in Huh7 cells.[20] This efficiency is reduced by approximately half with the wild-type HCV sequence.[20] The ribosomal frameshift efficiency of HCV for the production of the F protein is similar to that used by the yeast L-A double-stranded RNA virus, the barley yellow dwarf virus, and HIV-1 for the expression of their polymerases, which ranges from 1% to 3%.[23–25] As mentioned above, a 16 kDa protein as well as the core protein was detected in mammalian cells that expressed the core protein sequence of either the HCV-1 or the HCV-RH strain. Although this 16 kDa protein is likely the F protein, its nature was never fully characterized. Fiorucci *et al.* fused a histidine tag to the N-terminus of the core protein of the HCV-J strain, a genotype 1b virus, and expressed the fusion protein in Huh7 hepatoma cells. By affinity-purification of the fusion protein with a nickel column, Fiorucci *et al.* detected the his-tagged core protein as well as a 17 kDa protein that cross-reacted with a monoclonal antibody directed against the F protein.[26] Since this F protein contained the his-tag, they concluded that this F protein was most likely synthesized by translational ribosomal frameshift. Whether this F protein was synthesized by +1 ribosomal frameshift at codons 8–11 or at codon 42, however, remains to be determined.

Expression of the F Protein by Translation Initiation from Internal AUG Codons

Expression studies indicate that, in addition to ribosomal frameshift, the F protein may also be produced by translation initiation from internal methionine codons.[27,28] To measure the efficiency of the F protein production, Vassilaki *et al.* fused the HCV IRES and part of the core protein coding sequence (i.e., nucleotides 9–630) of HCV-1 or HCV-H, another genotype 1a virus, to the luciferase reporter gene in all three reading frames to determine the expression efficiency of the luciferase reporter from different reading frames. They found that mutating the initiation codon of the core protein to a stop codon or introducing stop codons to the sequence upstream of methionine codon 87 (Met-87) did not affect the expression of the reporter from the +1 reading frame. In contrast, the conversion of Met-87 to a stop codon completely abolished the expression of the reporter from the +1 reading frame in BHK-21 and Huh7 cells, suggesting that the expression of the F/ARF protein might be initiated from internal AUG codons located at codon 85 and 87. They named the full-length F protein core+1/F and this shorter form of the F protein core+1/S.[27] Wolf *et al.* also conducted experiments using the luciferase reporter and obtained similar results.[28]

The expression of the core+1/S protein was found to be negatively regulated by the expression of the HCV core protein in a dose-dependent manner. When the core protein expression was abolished by a mutation, the core+1/S expression level was increased.[28,29] The biological significance of this observation is unclear, but it suggests an inverse relationship between the expression levels of the core protein and the core+1/S protein.

Expression of the F Protein from a Non-AUG Codon

The F protein may also be synthesized by translation initiation at a non-AUG codon.[30] In an attempt to understand the mechanism of F protein synthesis, Baril *et al.* fused the first 35 codons of the core protein sequence to the luciferase gene at the +1 reading frame and measured the efficiency of luciferase synthesis.[30] Introducing a stop codon at codon 3, 9, 10, 11, or 12 of the polyprotein or replacing the start codon of the polyprotein with a UAC codon did not decrease the synthesis of the reporter from the +1 reading frame. In contrast, the expression of the luciferase reporter was abolished when a stop codon was introduced at codon 30 of the +1 reading frame, but not at codon 20. Their results suggested an alternative mechanism for the synthesis of the F protein, which likely involved the sequence between codons 20 and 30. Further mutational analysis identified the translation initiation site at codon 26, which is GUG. This codon is conserved in 18% of the HCV sequences compiled in the HCV database and is GCG in 81% of the HCV sequences. Both GCG and GUG codons supported the translation initiation in an HCV IRES-dependent manner.[30]

The results of Baril *et al.*[30] contradicted with those of ours.[20] The reason for this discrepancy is unclear. However, even if the translation of the F protein can indeed initiate from codon 26, this mechanism cannot be the only one responsible for the expression of the F protein during natural HCV infection. This is because a synthetic peptide derived from the F protein sequence located upstream of codon 26 has been shown to cross-react with the antisera obtained from HCV patients.[2]

Biological Functions of the F Protein

The function of the F protein in the HCV life cycle and/or in pathogenesis is currently unknown. It is a basic protein with a pI of nearly 12 and does not show sequence similarity to any protein with a known function.[2,3] It is not required for HCV RNA replication, since the HCV subgenomic replicons without the structural protein genes including the F protein gene can still replicate efficiently in Huh7 cells.[31,32] Interestingly, the F protein, including core+1/S, has a subcellular localization similar to other HCV proteins, which are localized primarily to the endoplasmic reticulum.[29,33,34]

In an attempt to explore the biological functions of the F protein, McMullan *et al.* introduced four premature termination codons into the F protein coding sequence.[5] These mutations were created to prevent the expression of the F protein by translational ribosomal frameshift at codons 8–11 (stop codon 1), translation initiation at codon 26 (stop codon 2), and translational ribosomal frameshift at codon 42 (stop codons 3 and 4). When the HCV genomic RNA harboring these mutations were injected into an HCV-naive chimpanzee, an attenuated replication phenotype was observed. When circulating HCV was isolated from the animal for RNA sequencing, stop codons 3 and 4, which are located in the SLVI stem-loop structure, were found to have reverted back to the wild-type sequence. These four stop codon mutations also impaired the replication of HCV genomic RNA of genotype 1a and 2a in human hepatoma cells. Similarly, reversion of the two stop codons in SLVI was observed after prolonged passages of the mutant virus in cell cultures. The mutations of stop codons 1 and 2, which prevented the F protein from being expressed by ribosomal frameshift at codons 8–11 and by translation initiation at codon 26, respectively, had no effect on HCV RNA replication or its propagation in cell cultures.[5] McMullan *et al.* thus concluded that various forms of the F protein, with the exception of core+1/S which was not analyzed in their studies, were not essential for HCV replication both *in vivo* and in cell cultures. They postulated that, instead, the SLVI stem-loop structure was important for HCV replication.

Subsequently, Vassilaki *et al.* repeated the experiments of McMullan *et al.* by using a slightly different approach.[35] In their studies, besides the four stop codon mutations created by McMullan *et al.*, an additional stop codon (stop codon 5) was also introduced to abolish the possible expression of the core+1/S protein. They found that the HCV RNA of the JFH-1 strain carrying these five mutations had a delayed replication phenotype in cell cultures. The HCV RNA mutant had a lower replication rate initially. However, its replication rate was comparable to that of the wild-type viral RNA at later time points. The same results were obtained when they conducted infection studies using uPA-SCID mice carrying the human liver xenograft. They found that, in two of three animals, the HCV JFH-1 mutant virus had a low level of viremia, which then reached the wild-type level after a two- to four-week delay. The third animal did not become infected after inoculation of the mutant virus. Since providing the F protein *in trans* in cell cultures did not reverse the phenotype of the HCV RNA mutant, and an HCV mutant with only the stop codon 5 mutation, which would truncate all forms of the F protein, replicated with a wild-type efficiency, Vassilaki *et al.* concluded that the F protein was not essential for HCV replication in cell cultures. By conducting mutagenesis experiments of the stem-loop structures of SLV and SLVI, they demonstrated that the integrity of these two stem-loop structures was important for efficient HCV protein translation. That might be the reason why the HCV mutant that carried five stop-codon mutations had a delayed replication phenotype.[35]

Although the studies of McMullan *et al.* and Vassilaki *et al.* indicated that the F protein was not essential for the replication of HCV in these cell culture and animal model systems, the conservation of the F protein coding sequence in more than 90% of the HCV sequences that have been reported and in the distantly related GBV-B virus indicates that this gene must play an important role in the HCV life cycle.[3,36] In the following sections, we will review the possible functions of the F protein that have been reported.

Interaction between the F Protein and the Proteasome

The HCV F protein of the genotype 1a virus is a very unstable protein with a half-life of 10–15 minutes. It is degraded by the proteasome and can be stabilized by the proteasome inhibitor MG132.[33,34] By using a battery of genetic, biochemical, and cell biological approaches, we recently demonstrated that the HCV F protein could interact with the proteasome subunit protein α3, which reduced the level of the F protein in cells in a dose-dependent manner.[37] Mutations that removed the lysine residues from the F protein as well as the expression of the F protein in a cell line that does not have a functional ubiquitination pathway at the non-permissive temperature did not stabilize the F protein. These observations, together with the observation that the F protein could be degraded by the purified 20S proteasome *in vitro*, indicate that the F protein is degraded directly by the 20S proteasome in an ubiquitin-independent pathway, likely through its direct interaction with the proteasome subunit protein α3. The α3-interactive domain in the F protein has been mapped to amino acids 40–60, and, indeed, the fusion of this sequence as well as its preceding amino acids 20–40 to the green fluorescence protein (GFP), a stable protein, could reduce the half-life of this protein to about 15 minutes, similar to that of the F protein. In contrast, since the core+1/S protein lacks this α3-interactive domain, it is a relatively stable protein.[37] The interaction between the F protein and α3 may be a mechanism for HCV to control the expression level of this protein. Alternatively, it may also serve other functions. For example, the F protein may control the expression level of its other interactive proteins by bringing them to the proximity of the 20S proteasome for their degradation. In addition, the interaction of the 20S proteasome with the 11S regulatory complex is important for the degradation of short peptides for antigen presentation by MHC class I.[38] The HIV-1 tat protein has been shown to bind to the 20S proteasome to suppress its interaction with the 11S regulatory complex.[39] This was thought to be important for preventing HIV-1 infected cells from being removed by cytotoxic T lymphocytes.[40] It is conceivable that the binding of the F protein to α3 may also inhibit the association of the 11S regulator to the 20S proteasome and contribute to HCV immune evasion by decreasing viral antigen presentation by host cells. The ability of the F protein to bind α3 is reminiscent of the activity of the hepatitis B virus (HBV) X protein, which can bind to the proteasome subunit α7 to suppress cellular protein degradation to exert its regulatory functions.[41,42] It remains to be determined whether the HCV F protein also possesses such regulatory functions.

Regulation of Expression of Cytokines by the F Protein

Since the F protein coding sequence resides entirely in the core protein gene, many of the activities that have previously been shown to be associated with the HCV core protein may be attributed to the F protein. To address this issue, Fiorucci *et al.* introduced multiple stop codons in the +1 reading frame of the core protein coding sequence to abolish the expression of the various forms of the F protein and examined how these stop codons might affect the immuno-modulatory activities of the HCV core protein gene by performing the human cytokine array analysis.[26] They found that the expression of the intact core protein gene induced the production of pro-inflammatory cytokines, such as interleukin (IL) 6, IL-8, MCP-1, and MIP-1β, whereas the core protein gene with multiple mutations that abolished the expression of the F protein had reduced or no expression of these cytokines/chemokines in THP-1 monocytic cell line and monocyte-derived dendritic cells, which have been shown to support HCV replication.[43] Similar effects on the expression of IL-8 were observed in Huh7 cells and on IL-6 and MIP-1β in macrophages.[26] Their results indicate that the core protein expression by itself is not sufficient for the production of these cytokines and that the expression of the F protein is also required. Furthermore, they found that the F protein increased the IL-8 production by transactivating the IL-8 promoter, whereas this effect was not observed from the core protein construct that did not express the F protein. Their results indicate an immuno-modulatory role of the F protein during natural HCV infection,[26] and the possible role of this protein in HCV pathogenesis.[44,45]

Possible Role of the F Protein in Hepatocellular Carcinogenesis

Since the HCV core protein gene has been known to regulate the promoters of *p53*, *p21*, *c-myc*, and *hTERT* genes.[46,47] Basu *et al.* examined the possible role of the F protein in these regulations by comparing the activities of the wild-type core protein, the core protein mutant with a lysine-to-arginine mutation at codon 9 (core-9mt), and the F protein that had a 10-amino acid leader sequence from the core protein (i.e., the F protein synthesized by ribosomal frameshift at codons 8–11).[48] Their studies, however, did not test the functions of the F protein that may be expressed through other mechanisms. Based on the level of the reporter expressed, they found that, unlike the core protein, the F protein did not have any significant effect on *p53*, *c-myc*, and *hTERT* promoters, although it down-regulated *p21* expression.[48] The core protein has also been shown to protect the HepG2 cells against TNF-α-induced apoptosis. They also analyzed the effect of the core protein and the F protein on TNF-α-induced apoptosis and found that the TNF-α-mediated apoptosis of HepG2 cells was not inhibited by the F protein.[48] In contrast to their studies, a more recent study demonstrated that the expression of the F protein resulted in a nearly fourfold increase of the *c-myc* mRNA level and twofold decrease of the *p53* mRNA.[49] This same group also reported that the F protein could suppress TNF-α-induced apoptosis by activating NF-κB.[50] The reason for these discrepancies is unclear, but it may be related to the experimental system used. For example, in the latter studies, the F protein used for the experiments started at codon 13 of the +1 reading frame and hence was truncated at the N-terminus.

Besides the regulation of the expression levels of oncogenes and tumor suppressor genes, in a separate study it was demonstrated that the *myc*-modulator protein MM-1 could bind to the HCV F protein.[51] MM-1, initially identified as *c-myc* interacting protein, is a nuclear protein that binds to the N-terminal domain of *c-myc*, leading to the repression of the transcriptional activity of *c-myc*.[52] The interaction of the F protein with MM-1 increases the transactivation activity of *c-myc* by antagonizing the inhibitory effect of MM-1 on *c-myc*.[51]

The authors of the above reports suggested that the F protein might play a role in the development of hepatocellular carcinoma by regulating the expression levels and the activities of cellular oncogenes and tumor suppressor genes.[48,49,51] In an attempt to understand the possible role of the F protein in hepatocarcinogenesis, Sobesky *et al.* used laser capture microdissection to isolate cancerous and non-cancerous liver tissues from HCV patients with hepatocellular carcinoma.[53] They found that there was a higher selective pressure on the F protein coding sequence than on the core protein sequence in cancerous cells. They postulated that this selective pressure for the F protein might be immune-driven or caused by the adaptation of the F protein to the transformed cellular environment.[53] Alam *et al.* also analyzed the HCV sequences in cancerous and non-cancerous liver tissues from HCV patients.[54] Interestingly, they found that, in one patient, 15% of the HCV sequences isolated from non-cancerous tissues contained more than eight consecutive adenosine residues at codons 8–11. These HCV-1-like sequences were not detected in the cancerous tissues of the same patient, suggesting a selection against this sequence in cancer cells.[54] The significance of these findings, however, requires further study.

HCV F Protein and Microtubules

MM-1 is also known as PFD5, which is a subunit of prefoldin (PFD), a cytosolic hexameric molecular chaperone complex that is also known as Gim (Genes involved in microtubule biogenesis) complex. PFD is composed of six different subunits (PFD1-PFD6), binds to the nascent polypeptide chain, and functions in the folding of actin and tubulin.[55] Interestingly, in addition to MM-1/PFD5, the HCV F protein was also found to bind to PFD2 in a yeast two-hybrid screening experiment.[56] The binding of the F protein to PFD2 had been shown to interfere with the interaction of PFD1 and PFD2 and to suppress the polymerization of tubulin monomers. This interference affected the organization of microtubules, which may be important for cellular transformation.[56] It will be interesting to determine whether the interaction between the F protein and MM-1/PFD5 also leads to a change in the formation of microtubules.

As mentioned above, the HCV F protein is associated primarily with the endoplasmic reticulum during the interphase of the cell cycle.[29,33,34] Curiously, it had also been reported that the F protein was associated with the microtubules of the mitotic spindles during cell division.[29] Whether this association of the F protein with microtubules is related to its interaction with PDF2 and/or PDF5 remains to be determined. Note that since the HCV replication requires intact microtubules and microfilaments,[57,58] the possibility that the HCV F protein may regulate HCV replication via its interaction with PDF and microtubules during natural infection may require careful examination.

Conclusion

The coding sequence of the HCV F protein is conserved in the great majority of the HCV isolates and the distantly related GBV-B virus, indicating that this protein plays an important role in the HCV life cycle. The HCV F protein is expressed during natural infection, since its reactive antibodies and T cells are present in HCV patients. However, the molecular mechanisms that regulate the expression of the HCV F protein remain controversial, and the biological function of this protein is still unclear. The HCV F protein coding sequence is embedded entirely within the core protein coding sequence. Thus, many of the biological functions previously attributed to the core protein may also be caused by the F protein. For example, it has been demonstrated that the HCV core protein can transform primary rat embryo fibroblasts in cooperation with the ras oncogene,[9] induce hepatocellular carcinoma in transgenic mice,[59] and bind to tumor necrosis factor (TNF) receptor and sensitize cells to TNF-induced apoptosis.[60] Since the core protein gene used in these expression studies also has the potential to express the F protein, the possible contribution of the F protein to these observed activities of the core protein gene needs to be carefully re-evaluated. Many of the studies conducted to understand the biological functions of the F protein relied on the over-expression of this protein using heterologous expression systems. The caveat of such an experimental approach is that it does not reflect natural HCV infection, which likely produces only a low level of the F protein. Thus, to understand the biological functions of the F protein, it is necessary that the experiments be carefully designed. This, however, will require the full understanding of the mechanisms that regulate the expression of the F protein.

References

1. Simmonds P, Bukh J, Combet C, *et al.* (2005) Consensus proposals for a unified system of nomenclature of hepatitis C virus genotypes. *Hepatology* **42**: 962–973.
2. Walewski JL, Keller TR, Stump DD, Branch AD. (2001) Evidence for a new hepatitis C virus antigen encoded in an overlapping reading frame. *Rna* **7**: 710–721.
3. Xu Z, Choi J, Yen TS, *et al.* (2001) Synthesis of a novel hepatitis C virus protein by ribosomal frameshift. *Embo J* **20**: 3840–3848.
4. Varaklioti A, Vassilaki N, Georgopoulou U, Mavromara P. (2002) Alternate translation occurs within the core coding region of the hepatitis C viral genome. *J Biol Chem* **277**: 17713–17721.
5. McMullan LK, Grakoui A, Evans MJ, *et al.* (2007) Evidence for a functional RNA element in the hepatitis C virus core gene. *Proc Natl Acad Sci USA* **104**: 2879–2884.
6. Smith DB, Simmonds P. (1997) Characteristics of nucleotide substitution in the hepatitis C virus genome: constraints on sequence change in coding regions at both ends of the genome. *J Mol Evol* **45**: 238–246.
7. Lo SY, Selby M, Tong M, Ou JH. (1994) Comparative studies of the core gene products of two different hepatitis C virus isolates: two alternative forms determined by a single amino acid substitution. *Virology* **199**: 124–131.
8. Lo SY, Masiarz F, Hwang SB, *et al.* (1995) Differential subcellular localization of hepatitis C virus core gene products. *Virology* **213**: 455–461.
9. Ray RB, Lagging LM, Meyer K, Ray R. (1996) Hepatitis C virus core protein cooperates with ras and transforms primary rat embryo fibroblasts to tumorigenic phenotype. *J Virol* **70**: 4438–4443.
10. Komurian-Pradel F, Rajoharison A, Berland JL, *et al.* (2004) Antigenic relevance of F protein in chronic hepatitis C virus infection. *Hepatology* **40**: 900–909.
11. Troesch M, Jalbert E, Canobio S, *et al.* (2005) Characterization of humoral and cell-mediated immune responses directed against hepatitis C virus F protein in subjects co-infected with hepatitis C virus and HIV-1. *Aids* **19**: 775–784.
12. Cohen M, Bachmatov L, Ben-Ari Z, *et al.* (2007) Development of specific antibodies to an ARF protein in treated patients with chronic HCV infection. *Dig Dis Sci* **52**: 2427–2432.

13. Bain C, Parroche P, Lavergne JP, *et al.* (2004) Memory T-cell-mediated immune responses specific to an alternative core protein in hepatitis C virus infection. *J Virol* **78**: 10460–10469.
14. Gao DY, Zhang XX, Hou G, *et al.* (2008) Assessment of specific antibodies to F protein in serum samples from Chinese hepatitis C patients treated with interferon plus ribavarin. *J Clin Microbiol* **46**: 3746–3751.
15. Chuang WC, Allain JP. (2008) Differential reactivity of putative genotype 2 hepatitis C virus F protein between chronic and recovered infections. *J Gen Virol* **89**: 1890–1900.
16. Gesteland RF, Atkins JF. (1996) Recoding: dynamic reprogramming of translation. *Annu Rev Biochem* **65**: 741–768.
17. Dinman JD. (2006) Programmed ribosomal frameshifting goes beyond viruses: Organisms from all three kingdoms use frameshifting to regulate gene expression, perhaps signaling a paradigm shift. *Microbe Wash DC* **1**: 521–527.
18. Brierley I. (1995) Ribosomal frameshifting viral RNAs. *J Gen Virol* **76**(Pt 8):1885–1892.
19. Ratinier M, Boulant S, Combet C, *et al.* (2008) Transcriptional slippage prompts recoding in alternate reading frames in the hepatitis C virus (HCV) core sequence from strain HCV-1. *J Gen Virol* **89**: 1569–1578.
20. Choi J, Xu Z, Ou JH. (2003) Triple decoding of hepatitis C virus RNA by programmed translational frameshifting. *Mol Cell Biol* **23**: 1489–1497.
21. Vassilaki N, Kalliampakou KI, Mavromara P. (2008) Differences in the expression of the hepatitis C virus core+1 open reading frame between a nuclear and a cytoplasmic expression system. *J Gen Virol* **89**: 222–231.
22. Boulant S, Becchi M, Penin F, Lavergne JP. (2003) Unusual multiple recoding events leading to alternative forms of hepatitis C virus core protein from genotype 1b. *J Biol Chem* **278**: 45785–45792.
23. Paul CP, Barry JK, Dinesh-Kumar SP, *et al.* (2001) A sequence required for -1 ribosomal frameshifting located four kilobases downstream of the frameshift site. *J Mol Biol* **310**: 987–999.
24. Dinman JD, Icho T, Wickner RB. (1991) A -1 ribosomal frameshift in a double-stranded RNA virus of yeast forms a gag-pol fusion protein. *Proc Natl Acad Sci USA* **88**: 174–178.
25. Grentzmann G, Ingram JA, Kelly PJ, *et al.* (1998) A dual-luciferase reporter system for studying recoding signals. *Rna* **4**: 479–486.
26. Fiorucci M, Boulant S, Fournillier A, *et al.* (2007) Expression of the alternative reading frame protein of Hepatitis C virus induces cytokines involved in hepatic injuries. *J Gen Virol* **88**: 1149–1162.
27. Vassilaki N, Mavromara P. (2003) Two alternative translation mechanisms are responsible for the expression of the HCV ARFP/F/core+1 coding open reading frame. *J Biol Chem* **278**: 40503–40513.

28. Wolf M, Dimitrova M, Baumert TF, Schuster C. (2008) The major form of hepatitis C virus alternate reading frame protein is suppressed by core protein expression. *Nucleic Acids Res* **36**: 3054–3064.
29. Vassilaki N, Boleti H, Mavromara P. (2007) Expression studies of the core+1 protein of the hepatitis C virus 1a in mammalian cells. The influence of the core protein and proteasomes on the intracellular levels of core+1. *Febs J* **274**: 4057–4074.
30. Baril M, Brakier-Gingras L. (2005) Translation of the F protein of hepatitis C virus is initiated at a non-AUG codon in a +1 reading frame relative to the polyprotein. *Nucleic Acids Res* **33**: 1474–1486.
31. Blight KJ, Kolykhalov AA, Rice CM. (2000) Efficient initiation of HCV RNA replication in cell culture. *Science* **290**: 1972–1974.
32. Lohmann V, Korner F, Koch J, *et al.* (1999) Replication of subgenomic hepatitis C virus RNAs in a hepatoma cell line. *Science* **285**: 110–113.
33. Xu Z, Choi J, Lu W, Ou JH. (2003) Hepatitis C virus f protein is a short-lived protein associated with the endoplasmic reticulum. *J Virol* **77**: 1578–1583.
34. Roussel J, Pillez A, Montpellier C, *et al.* (2003) Characterization of the expression of the hepatitis C virus F protein. *J Gen Virol* **84**: 1751–1759.
35. Vassilaki N, Friebe P, Meuleman P, *et al.* (2008) Role of the hepatitis C virus core+1 open reading frame and core cis-acting RNA elements in viral RNA translation and replication. *J Virol* **82**: 11503–11515.
36. Bukh J, Apgar CL, Yanagi M. (1999) Toward a surrogate model for hepatitis C virus: an infectious molecular clone of the GB virus-B hepatitis agent. *Virology* **262**: 470–478.
37. Yuksek K, Chen WL, Chien D, Ou JH. (2009) Ubiquitin-independent degradation of Hepatitis C Virus F protein. *J Virol* **83**: 612–621.
38. Johnston SC, Whitby FG, Realini C, *et al.* (1997) The proteasome 11S regulator subunit REG alpha (PA28 alpha) is a heptamer. *Protein Sci* **6**: 2469–2473.
39. Viscidi RP, Mayur K, Lederman HM, Frankel AD. (1989) Inhibition of antigen-induced lymphocyte proliferation by Tat protein from HIV-1. *Science* **246**: 1606–1608.
40. Seeger M, Ferrell K, Frank R, Dubiel W. (1997) HIV-1 tat inhibits the 20 S proteasome and its 11 S regulator-mediated activation. *J Biol Chem* **272**: 8145–8148.
41. Zhang Z, Protzer U, Hu Z, *et al.* (2004) Inhibition of cellular proteasome activities enhances hepadnavirus replication in an HBX-dependent manner. *J Virol* **78**: 4566–4572.
42. Hu Z, Zhang Z, Doo E, *et al.* (1999) Hepatitis B virus X protein is both a substrate and a potential inhibitor of the proteasome complex. *J Virol* **73**: 7231–7240.

43. Ponzetto A, Gennero L, Cutufia M, *et al.* (2005) Effect of HCV infection on THP-1 monocytoid cells. *Cell Biochem Funct* **23**: 347–352.
44. Feldmann G, Nischalke HD, Nattermann J, *et al.* (2006) Induction of interleukin-6 by hepatitis C virus core protein in hepatitis C-associated mixed cryoglobulinemia and B-cell non-Hodgkin's lymphoma. *Clin Cancer Res* **12**: 4491–4498.
45. Miyoshi H, Fujie H, Shintani Y, *et al.* (2005) Hepatitis C virus core protein exerts an inhibitory effect on suppressor of cytokine signaling (SOCS)-1 gene expression. *J Hepatol* **43**: 757–763.
46. Lee MN, Jung EY, Kwun HJ, *et al.* (2002) Hepatitis C virus core protein represses the p21 promoter through inhibition of a TGF-beta pathway. *J Gen Virol* **83**: 2145–2151.
47. Lai MM, Ware CF. (2000) Hepatitis C virus core protein: possible roles in viral pathogenesis. *Curr Top Microbiol Immunol* **242**: 117–134.
48. Basu A, Steele R, Ray R, Ray RB. (2004) Functional properties of a 16 kDa protein translated from an alternative open reading frame of the core-encoding genomic region of hepatitis C virus. *J Gen Virol* **85**: 2299–2306.
49. Wu WB, Shao SW, Zhao LJ, *et al.* (2007) Hepatitis C virus F protein up-regulates c-myc and down-regulates p53 in human hepatoma HepG2 cells. *Intervirology* **50**: 341–346.
50. Shao SW, Wu WB, Bian ZQ, *et al.* (2008) Hepatitis C virus F protein inhibits cell apoptosis by activation of intracellular NF-kappaB pathway. *Hepatol Res.*
51. Ma HC, Lin TW, Li H, *et al.* (2008) Hepatitis C virus ARFP/F protein interacts with cellular MM-1 protein and enhances the gene trans-activation activity of c-Myc. *J Biomed Sci* **15**: 417–425.
52. Mori K, Maeda Y, Kitaura H, *et al.* (1998) MM-1, a novel c-Myc-associating protein that represses transcriptional activity of c-Myc. *J Biol Chem* **273**: 29794–29800.
53. Sobesky R, Feray C, Rimlinger F, *et al.* (2007) Distinct hepatitis C virus core and F protein quasispecies in tumoral and nontumoral hepatocytes isolated via microdissection. *Hepatology* **46**: 1704–1712.
54. Alam SS, Nakamura T, Naganuma A, *et al.* (2002) Hepatitis C virus quasispecies in cancerous and noncancerous hepatic lesions: the core protein-encoding region. *Acta Med Okayama* **56**: 141–147.
55. Hartl FU, Hayer-Hartl M. (2002) Molecular chaperones in the cytosol: from nascent chain to folded protein. *Science* **295**: 1852–1858.
56. Tsao ML, Chao CH, Yeh CT. (2006) Interaction of hepatitis C virus F protein with prefoldin 2 perturbs tubulin cytoskeleton organization. *Biochem Biophys Res Commun* **348**: 271–277.
57. Bost AG, Venable D, Liu L, Heinz BA. (2003) Cytoskeletal requirements for hepatitis C virus (HCV) RNA synthesis in the HCV replicon cell culture system. *J Virol* **77**: 4401–4408.

58. Lai CK, Jeng KS, Machida K, Lai MM. (2008) Association of hepatitis C virus replication complexes with microtubules and actin filaments is dependent on the interaction of NS3 and NS5A. *J Virol* **82**: 8838–8848.
59. Moriya K, Fujie H, Shintani Y, *et al.* (1998) The core protein of hepatitis C virus induces hepatocellular carcinoma in transgenic mice. *Nat Med* **4**: 1065–1067.
60. Zhu N, Khoshnan A, Schneider R, *et al.* (1998) Hepatitis C virus core protein binds to the cytoplasmic domain of tumor necrosis factor (TNF) receptor 1 and enhances TNF-induced apoptosis. *J Virol* **72**: 3691–3697.

Chapter 11

Lymphotropism of Hepatitis C Virus

Keigo Machida* and Michael M.C. Lai*,†

1. Introduction

HCV is primarily a hepatotropic virus, causing mostly persistent infection of the liver, leading to chronic hepatitis, liver cirrhosis, and hepatocellular carcinoma.[54,95] Besides the liver-associated symptoms, another serious complication of HCV infection is the symptoms associated with B cell anomalies,[60] particularly those of B lymphocyte proliferative disorders, including mixed cryoglobulinemia (representing oligoclonal B-cell proliferation),[2,47,56] and non-Hodgkin's B cell lymphoma (NHL).[34,111,133] In addition, many reports have suggested that HCV can somehow alter immune response (for review, see Ref. 34), resulting in ineffective immune responses in HCV patients. These immune abnormalities raised a possibility that HCV may infect immune cells, including B and T cells, impairing their functions. Other extra-hepatic manifestations of HCV infection,[42,78,102,132] including dermatological (porphyria cutanea tarda, lichen planus),[92] renal (membranoproliferative glomerulonephritis), endocrine (Type II diabetes mellitus), and autoimmune disease (peripheral neuropathy,

* Dept of Molecular Microbiology and Immunology, University of Southern California, Keck School of Medicine, 2011 Zonal Ave., Los Angeles, CA 90033 USA.
† Institute of Molecular Biology, Academia Sinica, Nankang, Taipei 11529, Taiwan, Tel: 886-2-2789-2365, Fax: 886-2-2782-6085, e-mail: michlai@usc.edu

polyarteritis nodosa, uveitis, corneal ulceration, saladenitis) have also been reported.[29] Such extra-hepatic symptoms could be the direct or indirect results of HCV infection of these tissues. In the literature, there are many reports of detection of positive- and negative-strand HCV RNA in B and T cells and various other tissues, including lymph nodes, pancreas, adrenal gland, thyroid, bone marrow, and spleen.[60] These data suggest the possible presence of HCV replication sites outside the liver, particularly in AIDS patients. Whether these findings are linked to various extra-hepatic manifestations of HCV infection remains to be determined. In most of these studies, typically, the viral RNA was detected by RT-PCR of positive- and negative-strand RNA of the virus. There have been questions raised concerning the specificity of these PCR assays and also the possible contamination from circulating viral RNA in the serum. Nevertheless, in at least some of the publications, stringent controls have been applied; thus, these reports should not be dismissed lightly. In this chapter, we will focus our discussions on HCV infection of lymphoid cells.

2. Review of Evidence for Extrahepatic Infection of HCV

2.1. *Epidemiological Evidence for the HCV Infection of B and T Cells in Hepatitis C Patients*

There have been numerous reports showing that peripheral blood mononuclear cells (PBMCs) from hepatitis C patients are positive for HCV RNA.[117,120] Both positive- and negative-strand RNAs were detected, indicating that viral replication had taken place and that the RNA did not come from serum viral RNA contamination. Also, the viral RNA quasi-species in the PBMC is different from that of the serum viral RNA.[120] The evolution pattern of viral RNA quasi-species in the serum of clinical course is also different from that in the liver, indicating that the viral RNA detected in the serum cannot be accounted for by the virus from the liver.[37] When different patients were compared, usually only a subset of patients had detectable HCV RNA in the PBMC, suggesting the variations in the viral tropism or different genetic factors of the host. In one particular series of studies,[76,77,129] the HCV RNA from the patients with mixed cryoglobulinemia (MC) had 3.3 times lower non-synonymous substitution rates than those without MC, showing positive selection. Significant HCV quasi-species was compartmentarized in the PBMC (4/9) and cryoprecipitates (7/9).[129] Reduced HCV quasi-species heterogeneity in MC patients is related to a low selection pressure.[129] The HCV quasi-species compartmentalization is frequently observed in patients' PBMC,[129] suggesting that HCV RNA replicates in PBMC and plays a role in HCV pathogenesis.

Compelling epidemiological evidence shows strong association between HCV infection and lymphoproliferative diseases. Especially, reports from the Mediterranean countries tend to show higher frequencies of HCV-associated lymphoproliferative diseases, whereas those from the North America tend to be lower. There are two confounding factors for the variability of the conclusions, including geographical region (ethnic group or geographical factors) and viral genotypes.

The following studies indicated a stronger association of lymphoproliferative diseases with HCV infection (Table 1).

Table 1. Representative Literature Reporting HCV Infection of B Cells

Paper	Conclusions	Ref.
Agnello *et al.*	Type II cryoglobulinemia is linked to HCV infection.	2
Idilman *et al.*	Mixed cryoglobulinemia is associated with malignant disorder.	47
Silvestri *et al.*	HCV infection is associated with essential mixed cryoglobulinemia (EMC)	111
Zukerman *et al.*	HCV infection is associated with B-cell non-Hodgkin lymphoma than in controls.	133

In conclusion, there is epidemiological evidence to indicate that HCV does infect and replicate in the PBMCs of hepatitis C patients and that this infection may relate to the clinical picture or immune status of the patients.

2.2. *Epidemiological Association Between HCV Infection and Lymphoma*

Because a high percentage of hepatitis C patients have mixed cryoglobulinemia, which represents a condition of oligoclonal B cell proliferation, it is logical to suspect the association of HCV infection with frank B cell lymphoma. Epidemiological evidence has strongly implicated HCV infection with the occurrence of non-Hodgkin's lymphoma. Many clinical reports, in particular those studying the populations of Mediterranean countries, showed that HCV patients have a very high frequency of non-Hodgkin's lymphoma.[27,121] In some reports, as many as 20–40% of hepatitis C patients have B-cell lymphoma.[97] For example, in a study by Mele *et al.* HCV prevalence was 17.5% among 400 lymphoma patients but only 5.6% among the 396 controls.[13] The study in North America shows 22% HCV-positive rate among the B cell lymphoma patients.[134] In contrast, other regions showed a very low prevalence rate of hepatitis C among lymphoma patients.[87,90] The basis for such a difference among these reports is not clear. It may reflect geographical variations in the genetic sequences of the virus or the ethnic composition of the host populations. A recent study on the meta-analysis of HCV patients[87] showed no clear association between HCV infection and B cell lymphoma. Another study showed that HCV infection increased the risk of lymphoma in 150,000 HCV patients (20–30%); the adjusted hazard ratio was 1.28 (95% CI, 1.12–1.45). Meta-analysis of large populations of HCV patients may have obliterated the geographical, racial, or viral genetic differences. In addition, the percentage of lymphocytes supporting active HCV replication in patients may be very minor, and not all the patients have HCV replication in their PBMC.[87] These confounding factors make it very difficult to show clear association between HCV infection and B cell lymphoma. It would be interesting to study whether immuno-deficient patients due to AIDS or immuno-suppressive agent-treatment have more active HCV replication (using negative strand detection techniques). Probably because of small patient population and the small number of lymphocytes supporting active HCV replication, it is very challenging to obtain clear evidence of HCV replication in lymphocytes. Further detailed study will be required to demonstrate clear association.

2.3. *Evidence Negating the Association between HCV Infection and Lymphoma*

There are a few reports demonstrating that HCV RNA is not correlated with lymphomas, especially in Asian countries and the United States. Using a specific RT-PCR protocol, Lanford *et al.* showed that HCV RNA is not detected in PBMC.[57] However, in studies from Mediterranean countries, there is a higher percentage of lymphoproliferative diseases with HCV RNA detection.[121,125] It is noted that HCV in Asia and North America are predominantly genotype 1a and 1b with other minor genotypes, whereas those in the Mediterranean are predominantly genotypes 2 and 3. However, no specific genotype is clearly linked to the lymphoproliferative diseases.[125] Instead, viral genome evolution in patients' bodies may generate lymphotropic viral strains or at least lymphotropic viral envelope sequences with high affinity to lymphocytes. The latter possibility has been experimentally demonstrated in a study that showed that in vitro long-term passage of H77 strain in T lymphocyte culture generated viruses with lymphotropic envelope sequences.[109] Since HCV RNA-dependent RNA polymerase does not possess proof-reading function, this highly mutagenic RNA replicase may be prone to generate and evolve high-affinity envelope sequences for lymphocytes during infections in patients. This may explain the controversial reports, namely, the predominant HCV strain may retain hepatotropism, but the minor population of emerging HCV may acquire lymphotropic properties. Therefore, analysis of the predominant strains in patients cannot rule out the possibilities that some minor HCV strains can replicate in lymphocytes. HIV–HCV coinfected patients have less antiviral immunity due to T cell-based acquired immunity, allowing lymphotropic viral infection. In contrast, AIDS patients or transplanted patients receiving immuno-suppressive agents may have a more permissive environment allowing for HCV replication in B lymphocytes.

The HCV infection is associated with lymphoma in a large study which found no statistical association between HCV infection and lymphoma (OR = 1.6, 95% CI, 0.3–8.5).[107] HCV infection is not associated with lymphoproliferative disorder in liver transplant patients,[88] since lymphoproliferative disorders associated with HCV infection are caused by chronic antigenic stimulation (not by the HCV proteins in infected cells).

2.4. *Evidence for the Causal Association Between HCV Infection and B Cell Lymphoma: Interferon Treatment Induces the Regression of HCV-Positive Lymphoma*

Despite the controversial nature of the reports on the association between HCV infection and the occurrence of B cell lymphoma, there are several reports of the efficacy of antiviral therapy in inducing the regression of B-cell lymphoma in patients who are HCV-positive. Several reports[44,65] have shown that in HCV-positive patients with B cell lymphoma, the interferon treatment, which was intended as an antiviral therapy for HCV, induced regression of the lymphoma. In contrast, such a treatment did not benefit the patients who were HCV-negative. Furthermore, the interferon treatment also induced the remission of the chromosomal aberration typically associated with lymphoma.[134] Such reports indicate that the presence of HCV in the B cell lymphoma is causal, but not merely pedestrian.

2.5. *Physical Evidence of HCV in B Cell Lymphoma*

The indisputable evidence for the replication of HCV in B cell lymphoma came from the isolation of an HCV virus from a B cell lymphoma in our laboratory.[117] This patient was HCV-positive and had mixed cryoglobulinemia but very few liver-related symptoms. Splenectomy was performed as a therapeutic measure. We established a permanent cell line from the spleen tissues, and this cell line continues to produce infectious virus in culture. We have subsequently successfully isolated a unique strain of HCV, named SB virus, from the cell line. Furthermore, the viral RNA sequence was cloned from the spleen tissue and made into an infectious virus, which preferentially infects lymphoid cells.[120] The presence of infectious virus in the spleen tissues gives indisputable evidence that at least some virus strains can infect and replicate in B cells. Furthermore, these viruses may be preferentially lymphotropic. The characterization of SB virus is described in Section 6. The difficulty of demonstrating HCV replication in B cell lymphomas, or even HCC, in most studies may have been due to the hit-and-run mechanisms employed by the virus.

Transgenic mice with expression of HCV genome or viral protein expression in transgenic mice is associated with lymphoproliferative diseases and B cell lymphomas, demonstrating that viral gene products directly cause lymphoproliferative diseases and B cell lymphomas.[50,75] A possible explanation for HCV-associated B lymphomagenesis is through Bcl-2, since enforced bcl-2 expression in B lymphoid cells causes autoimmune diseases,[115] and bcl-2-immunoglobulin transgenic mice develop follicular lymphoproliferation.[80] Indeed, HCV-infected patients express elevated levels of Bcl-2[18] and frequently have a t[14,18] chromosomal translocation.[51,134] In these publications, viral genotype, geographical factors, subsets of lymphocytes, and positivity of negative strand RNA of HCV should be carefully considered to evaluate the relationship between lymphomagenesis and lymphotropic HCV infection. Clearly, this issue requires additional study.

2.6. *Limitations for the Detection of HCV RNA in Human PBMCs and Lymphocytes*

Lymphocytes may not be able to support robust HCV replication in culture since non-transformed normal lymphocytes have poor proliferative ability and eventually die during in vitro culture conditions except for long-lived cell types in immune systems. By comparison, immortalized lymphocytes may support HCV replication. However, too-frequent proliferation of cells may dilute HCV RNA levels in the case that HCV replication rate cannot catch up with the high proliferation rate of cells. Thus, slower-growing lymphocytes may have an advantage in retaining a low level of HCV RNA.

Another possibility is that HCV evolves to replicate slowly in the host to escape from the innate and acquired immune response of the host, since robust replication of the virus may kill host cells and trigger antiviral immune responses through B and T lymphocytes and other immunological components. Slowly replicating HCV can hide in lymphocytes without being attacked from host immune systems. Therefore, a low level of HCV replication may be advantageous for the survival of HCV inside lymphocytes. This non-cytopathic nature of HCV may be one reason for the extremely high ratio of chronic HCV infection. Another theory is that HCV may have a latent phase like other chronic viral infections. Expression of host factors or extraneous stimuli may activate viral replication in these lymphocytes.

3. Review of HCV Receptors: A Lymphocyte-Specific Receptor?

Lymphocytes express certain surface co-receptor for HCV, CD81 tetraspanin, scavenger receptor class B type I (SR-BI), mannose-binding lectins DC-SIGN and L-SIGN, low-density lipoprotein receptor, heparan sulphate proteoglycans and the asialoglycoprotein receptor,[105] Claudin-1,[21,30] and Occludin[101] (Table 2). HCV is physically associated with lipoproteins and captured by L-SIGN of liver sinusoidal endothelial cells (LSEC), and transferred to hepatocytes by glycosaminoglycans, including highly sulfated heparan sulfate (HS), SR-BI, or claudin-1.[7] The data is confirmed cell-culture system (HCVcc).[126] Co-expression of HCV E1 and E2 chimera envelope glycoproteins has separable ligand sensitivity.[84] However, co-expression of CD81 and SR-BI does not allow HCV entry in non-hepatic cell lines, indicating that other molecule(s) are necessary for HCV entry. Another molecule, Claudin-1 was identified as a co-receptor.[30]

Table 2. HCV Receptor Candidates

Receptor Candidate	Putative Function	Tissue Expression and Character	Ref.
CD81	HCV early stage entry, Tetraspanin protein	Ubiquitous (except erythrocytes and platelets)	10, 24, 100, 126
Claudin-1/-7 (CLDN1)	Hepatotropic HCV late stage entry, Tight junction component	**Highly expressed in liver** and epithelial tissues, **rare expression in lymphocytes**	30, 135
Occludin-1	Hepatotropic HCV late stage entry, Tight junction component	**Highly expressed in liver** and epithelial tissues, **rare expression in lymphocytes**	101
SR-BI	HCV binding/entry, Ligand: high-density lipoprotein (HDL)	Hepatocyte, steroldogenic tissue, dendritic cell	9, 10, 62, 100
LDL-R	HCV binding/entry	Ubiquitous	1, 83
L-SIGN and DC-SIGN	HCV capture/binding, C-type lectins	LSEC, Kupffer cell	23, 68
Highly sulfated HS	HCV capture and binding	Hepatocyte	8, 11, 12
Asialoglycoprotein receptor	HCV binding, c-type lectin	Liver (HBV receptor)	105

Abbreviations: Scavenger receptor B type I (SR-BI), Low-density lipoprotein receptor, Highly sulfate heparan sulfate (HS), liver sinusoidal endothelial cells (LSEC).

Claudin-1, CD81, and SR-BI are involved in HCV entry.[10,21,30,100] Sulfated homologues of heparin inhibit HCV entry into mammalian cells.[12] Low-density lipoprotein receptor inhibits E1 pseudotype virus infectivity, while CD81 interfere with E2 pseudotype virus infectivity.[83] Exomes is associated with HCV envelope proteins.[79] The hypervariable region 1 of the E2 glycoprotein binds to glycosaminoglycans.[11] Furthermore, complement enhances antibody function for neutralization of pseudotype virus containing HCV E2 chimeric glycoprotein.[82]

Virus entry requires sequential interactions between viral proteins and multiple cellular factors, clathrin a low pH compartment.[46] Glycosaminoglycans, scavenger receptor class B member I (SR-BI) and the tetraspanin CD81, are associated with HCV entry, although these molecules are insufficient for viral entry because some cell lines do not support HCV entry.[9,10,24,62,100,126] These molecules are not expressed, or expressed at very low amounts, on the surface of lymphocytes. Thus, clearly, there must be other molecules that can serve as lymphotropic receptors. The nature of this molecule has not yet been identified.

Other viruses, including EBV, HIV, LCMV and LFV, also target lymphocytes.[106] EBV uses CD21 (CCP family) to infect B cells, leading to infectious mononucleosis and Burkitt's lymphoma. HIV uses CD4 (Ig superfamily) and chemokine receptors (TM7) GalC (Glycolipid) to infect macrophage and T cells, leading to lymphadenopathy and AIDS. Both LCMV and LFV use alpha-dystoglycan for lymphoid tissue and brain infections.

4. Previous Molecular Studies on *in vitro* HCV Infection of B and T Cells

4.1. *Replication of HCV RNA in Lymphocytes in vitro*

There have been several reports of HCV infection of lymphoid cells in tissue culture: Dr. Shimizu's group first demonstrated that the H77 strain of HCV could be serially passaged in several T cell lines, including MOLT4 cells, and that viral sequences changed during the passage.[64,109] HCV negative-strand RNA could be detected, indicating that HCV replicates in these T cells. Several other groups validated the detection of negative strand RNA of HCV in lymphoid cells. Our laboratory further demonstrated that HCV infection of T cells altered the immune properties of T cells by affecting interferon-gamma signaling in T cell lines.[53] By using a lymphotropic HCV strain (the SB virus, see Section 6), we showed that HCV could infect T cell lines (Molt-4 and Jurkat cells) *in vitro*. Both positive- and negative-strand HCV RNA were detected for several weeks after infection. Viral proteins could also be detected by immunofluorescence studies. Moreover, infectious HCV particles were produced from Molt-4 cell cultures, and could be used to infect naive T cell lines. HCV could also infect human primary CD4+ T cells, particularly naive (CD45RA+CD45RO–) CD4+ cells, in culture. HCV replication in T cells suppressed the IFN-γ/STAT-1/T-bet signaling due to the reduction of STAT-1 and inhibition of its activation (phosphorylation).[53] HCV infection of T cells also inhibits T cell proliferation and enhances fas-mediated apoptosis by down-regulating the expression of CD44 splicing variant 6.[52] These phenotypic changes after HCV infection clearly indicate that the SB strain of HCV infects and replicates in T cells. The SB virus can also infect B cells, including Raji cell lines and primary B cells, with detection of viral negative-strand RNA and viral proteins. Other groups also showed that HCV infection or expression of HCV protein is associated with apoptosis of T and B cells.[43,123] HCV infection also causes a variety of biological responses from cells, such as induction of oxidative stress, chromosomal aberration, hypermutation of immunoglobulin and cellular oncogenes, and inhibition of DNA damage repair in B cells.[55,72,74] These data clearly indicate that certain strains of HCV can infect and replicate in B and T cells.

4.2. *Controversies on the Detection of HCV RNA in Lymphocytes in vivo*

As discussed above, many reports have documented the detection of HCV positive- and negative-strand RNA in PBMC of HCV patients.

However, a recent study from Dr. Jane McKeating's group disputed these findings and showed further that there is no replication of HCV in lymphocytes.[114] They argued that there is only a casual but not real association between HCV RNA replication and lymphocytes. This casual association was explained by the binding of virus particles to the surface of lymphocytes, without HCV replication. Clearly, this issue will require further studies to resolve. Isolation and characterization of more lymphotropic HCV strains may help to resolve this issue.

5. Previous Molecular Characterization of the Lymphotropic HCV

Previous studies of HCV passages in T and B cell lines indicated that viral sequences undergo molecular evolution and acquire characteristic genetic changes reflecting the acquired lymphotropism of the new virus species.[108] These genetic changes were used to infer the genetic regions responsible for lymphotropism of HCV. These presumptive lymphotropism determinants include the following:

5.1. *Structural vs. Non-Structural Viral Proteins*

Structural proteins, in particular the envelope proteins, of HCV may be responsible for the lymphotropism, since viral envelope proteins are responsible for the virus binding to the receptors on target cells. The receptors on hepatocytes and lymphocytes are likely different. However, whether the envelope-receptor binding indeed led to productive infection of lymphoid cells is not certain. The non-structural proteins, which are responsible for viral replication, may contribute to the viral lymphotropism as well since some of the hepatotropic strains (e.g., H77 strain) have weak replication ability in lymphocytes in comparison to those in hepatocytes. Thus, cellular factors responsible for HCV replication in lymphocytes may be different from those in hepatocytes, resulting in differential abilities to infect hepatocytes or lymphoid cells. Different non-structural proteins may cooperate with cellular factors to different degrees in different cell types.

5.2. *The Role of 5′- and 3′-UTR in Lymphotropism*

The HCV strains isolated after serial passages in lymphoid cell lines have been previously identified with some sequence variations in the 5′-untranslated region (5′-UTR) of the viral genome.[64] These sequence variations in the 5′-UTR region may be linked to B and T lymphocyte replication.[64,93,108] The HCV 3′-untranslated region (3′-UTR) or a poly(A) tract promote efficient translation subsequent to the initiation phase.[16] Recent studies suggested that 3′-UTR also contributes to efficient RNA replication of HCV.[48,116] The 3′-end of NS5B-coding region has been shown to form part of the secondary structure involving the 3′-UTR.[36] The hepatotropic and lymphotropic HCV strains may differ in both 5′- and 3′-ends of viral RNA.

5.3. *Other Considerations*

Recent study shows that amino acid K33 and R35 of p7 are important for virus release. The K33 residue is conserved, but R35 is K in some variant strains.[49] This difference may contribute to the low level of virus production of the parental SB strain, which has K35 in p7 (unpublished data). In addition, NS2 protease activity is critical for virus release.[49] SB virus also differs from other HCV strains in this region.

6. Characteristics of SB Virus

One of the best characterized lymphotropic HCV strains is SB virus, which was isolated from an HCV-positive B cell lymphoma cell line.[117] This cell line was derived from splenic B cells of a patient with B-cell lymphoma and mixed cryoglobulinemia without liver symptoms. The SB cell line is an EBV-negative, HCV-producing cell line (yielding approximately 20,000 copies/mL of viral RNA in the culture media) and shows significant apoptosis. The SB virus (genotype 2b) infects primary B cells and the established B cell lines, but not Huh7.5 or HepG2 cells,[117] suggesting a preferential lymphotropism. In contrast, the much-studied JFH strain (cloned from a fulminant hepatitis patient)[126] replicates in hepatocytes, but not in B cells. JFH-1 strain and SB strain from SB cells (spleen B cells) have been used for *in vitro* infection studies[117,126] and showed distinctly different cellular tropism.

We have determined the genetic basis of the viral lymphotropism by using molecularly cloned infectious RNA of SB and JFH-1 viruses and the chimera constructs between them. The preliminary data showed that viral envelope is the primary determinant for the lymphotropism, but the non-structural proteins and 5′-UTR also contribute (unpublished data). The ability of SB-HCV RNA to replicate in B cells raised a possibility that some HCV strains may also infect and replicate in T cells. Indeed, the H77 variant strain of HCV can infect and replicate in several established T cell lines, including Molt-4 and Jurkat cells, in long-term cell culture.[108] SB-HCV RNA can replicate in Molt-4 cells at least transiently,[53] whereas transfection of the JFH-HCV full-length RNA into Molt-4 cells showed that JFH-HCV RNA cannot replicate in Molt-4 cells.[53] These data together suggested that certain HCV strains or variants can infect and replicate in B and T cells (Table 3).

Table 3. Characteristics of HCV Strains Currently Being used in the Laboratories

Strain	Genotype	Tropism	Replication	Source	Reference
H77	1a	Hepatotropic	+	Isolated from HCV-infected patient serum	127
Con1	1b	Hepatotropic	+	Isolated from liver of patients	6
JFH-1	2a	Hepatotropic	+++	Isolated from fulminant hepatitis serum	126
SB	2b	Lymphotropic	+	Isolated from SB cells (Spleen tumor-derived B cell line)	117

7. Possible Biological Implication of HCV Lymphotropism

Viral infections of lymphocytes can result in the alteration of T and B cell functions and their cytokine production, thus affecting both the innate and acquired immunity.

7.1. *Immune Escape from T Cell Function by HCV Infection*

HCV infection of T cells *in vitro* inhibits T cell function,[53] allowing HCV to escape from the host's immune defense. Mutational escape from cellular immunity in hepatitis C patients has been reported.[94] The relationship between cell-mediated immunity and the outcome of HCV infection has been established by numerous studies. Memory CD8+ cytotoxic T cells (CTL) are required for protection against persistent HCV infection; at the same time, durable intra-hepatic T cell memory is likely established during acute HCV infection, since T cells recognizing HCV antigens have been recovered from the livers of chimpanzees several years after spontaneous clearance of infection.[110] Immune evasion leading to persistent infection, in contrast to recovery from viral infection, after acute HCV infection from a shared source has been reported,[119] reinforcing the general relevance of this immune evasion mechanism to the persistence of RNA viruses in humans.[14] Amino-acid changes of HCV proteins also can alter CTL recognition of variant peptide-MHC complexes.[15]

Successful immune responses in HCV infection generally target multiple major histocompatibility complex (MHC) class I-restricted epitopes in structural and non-structural HCV proteins.[22,110] Activated CTL populations failed to secrete IFN-γ, in patients recovered as viremia declined.[63] In long-term HCV infected patients, CTL responses were more common in patients with cleared viremia than patients with persistent viremia, although the HCV-specific CTL was infrequently observed in patients during and after resolution of acute HCV infection.[63] Infection of T cells, with accompanying loss of T cell functions[54] will alter these immune responses.

7.2. *HCV Infection of B Cells Induces Immunoglobulin (Ig) Hypermutation, Altering B Cell Immunity*

The HIV-associated B cell diseases are associated with interactions with HCV envelope proteins. CD81 is a co-receptor for HCV,[46,81,100,131] is a component of the multimeric B cell antigen receptor complex.[66] CD81 is co-receptor for HCV[46,81,100,131] which is a component of the multimeric B cell antigen receptor complex.[66] CD81 is associated with the signaling molecule CD19, complement receptor 2 (CD21), and interferon-inducible Leu-13 (CD225) protein.[66,118] Interaction between CD81 and E2 or certain MAbs against CD81 induces B cell aggregation, inhibits proliferation in Daudi cell line,[35] stimulates T cells[112] and inhibits natural killer cell functions.[25,124] The CD81 signaling in B cells induces tumor necrosis factor-α (TNF-α).[4] HCV infection of primary macrophages induces TNF-α production.[104] Stimulation of the CD19-CD21-CD81 complex and the B-cell antigen receptor lowers the B cell activation threshold by lipopolysaccharide antigen-presenting cells.[17] Lymphocytes develop normally but have defective proliferative responses and are deficient in antibody production, suggesting that CD81 is required for the production of antibodies.[86] These observations suggest that HCV may modify the B cell receptor–associated signaling pathway by binding to CD81 in B cells.

Infection of HCV alters B cell functions[69,70,117] and induces double-strand DNA breaks in many cellular genes.[72] The engagement of B cells by purified E2 alone induces double-strand DNA breaks specifically in the variable region of Ig gene locus and hypermutation in Ig of B cells.[69] Incubation with the anti-CD81 MAb blocks this effect. Interaction between E2-CD81 triggers the enhanced expression of activation-induced cytidine deaminase (AID) and also stimulates the production of tumor necrosis factor-α in B cells.[69] The sRNA-mediated knockdown of AID blocked the E2-induced double-strand DNA breaks (DSBs) and hypermutation of Ig gene.[69] Therefore, HCV infection modulates the adaptive immune response by activation of AID and enhanced hypermutation of Ig gene in B cells. These effects will generate antibodies with altered immunological properties, thus contributing to HCV persistence and B cell lymphoproliferative diseases.

In addition to the increased mutation frequency in Ig gene, HCV infection also induces hypermutation of proto-oncogenes and p53.[72] The HCV infection caused an increase in mutation frequency (a 5- to 10 fold) in Ig heavy chain, p53, BCL-6, and β-catenin genes of *in vitro* HCV-infected B cell lines and HCV-associated peripheral blood mononuclear cells (PBMC), lymphomas, and HCCs. The nucleotide-substitution pattern of p53 and β-catenin was different from that of Ig heavy chain in HCV-infected cells, suggesting two different mechanisms of mutation. HCV induces a mutator phenotype and may transform cells by a hit-and-run mechanism.[72] The HCV-induced mutations of somatic genes, such as p53, are mediated by nitric oxide (NO) and reactive oxygen species (ROS) in subsequent studies.[70] In contrast, the Ig mutations are mediated by AID activation through the binding of HCV E2 protein to CD81. AID is involved in both the somatic hypermutation and class-switching recombination of Ig gene in normal B cell development; it causes deamination of deoxycytidine to deoxyuracil (dU) in the template DNA strand, with preference for certain hot-spot motifs.[99] The resulting dU/dG pairs are repaired by the mismatch repair system,[96] uracil glycosylase endonuclease[28] and error-prone DNA polymerases, including[104] Pol ι, Pol η, and Pol ζ.[32,128,130] Interestingly, AID, Pol ζ, and Pol ι are induced in HCV-infected B cells,[72] caused by the binding of HCV E2 protein to CD81, which in turn triggers DSBs and subsequent Ig hypermutations.

The E2-CD81 interaction enhances mutation frequencies in Ig gene, but not in p53 gene, since E2-CD81 interaction triggers a signaling response similar to that triggered by IL4, antiCD40, and other cytokines in B cells.[19,89] The normal somatic hypermutation mechanism of Ig gene in B cells typically affects the genomic sequences within ~2 kbp downstream from the transcription initiation site of Ig gene.[103,40] The HCV E2-induced Ig hypermutation mirrors precisely this pattern. E2 likely will bind to most of cell types since CD81 is expressed ubiquitously. However, E2 does not induce enhancement of expression of AID or DSBs in hepatocytes or other non-B cell lines.[69] The CD21 and CD19 (other components of the CD81 complex), are important for the signal transduction involved in the induction of AID. Several protein kinases associated with CD19 and CD21 but not with CD81.[33]

Ig hypermutation enhances the specificity and avidity of antibodies produced from B cells HCV infection. The neutralization activity may decline after HCV infection, which enables the virus to escape from humoral immunity. The HCV can infect human hybridoma cells producing E2-specific MAbs and reduces the avidity and specificity of the HCV-specific antibodies.[73] HCV infects antibody-producing B cells lowers the affinity and specificity of the HCV-specific antibodies through the resultant hypermutation of V(H). The V(H) sequences in the HCV-infected hybridomas displayed a significantly higher mutation frequency in complementarity determining region 3, which lowered the antibody affinity against the targeting protein and the virus-neutralizing activity of anti-E2 antibodies. Furthermore, complement-dependent cytotoxicity with the antibodies secreted from the HCV-infected hybridomas was impaired, suggesting that HCV infection could cause some anti-HCV-antibody-producing hybridoma B cells to make less-protective antibodies,[73] enabling HCV to escape from immune surveillance. Similarly, HCV infection affects T cell functions, including IFN-γ signaling,[53] which may contribute to T cell anergy and immune escape of HCV.

Furthermore, HCV infection induces toll-like receptor 4 expression, which enhances production of beta interferon and interleukin-6.[71] HCV infection significantly increased beta interferon (IFN-~) and interleukin-6 (IL-6) secretion from B cells, particularly after lipopolysaccharide stimulation. TLR4 is induced in (a three- to sevenfold up-regulation) HCV-infected cells. Peripheral blood mononuclear cells from HCV-infected individuals have a higher expression level of TLR4.[71]

In sum, HCV infection of B and T cells can directly and indirectly alter immune responses, which may contribute to the alterations of the host's innate and acquired immune responses.[71]

8. Coinfection of HCV and HIV in Lymphocytes

Another group of patients worthy of discussion are those with HCV–HIV coinfection, because the impairment of T cell functions by HIV makes it more likely that T cells can be infected with HCV. In the USA, HIV patients are mostly intravenous drug users (IVDU) and people who received blood products prior to 1985, while only a small portion (<5%) is caused by sexual transmission. Sustained response rates for the standard HCV therapy (pegylated interferon plus ribavirin) in this population are in the range of 26–40%.[20,98,122] Relapses are more frequent in this population.[113] Coinfected patients (HIV + HCV) display progressive deficits in CD4 lymphocyte numbers and function, while their liver disease is reported to be more progressive than HCV-infected patients without HIV.[31] In HIV + HCV coinfected patients, the cellular immune response is impaired and consequently higher levels of HCV RNA have been documented.[26,58,59] Coinfected patients with coinfection develop clinical liver failure more frequently than patients with HCV infection alone.[31] Fas antigen expression on the surface of infected hepatocytes may lead to apoptosis and/or cytolysis,[45] possibly via a non-HLA-restricted pathway. This situation is unlike coinfection with HIV and hepatitis B virus (HBV). Liver disease is milder than in patients with hepatitis B alone due to the lack of immune attack on infected hepatocytes.[39,85]

HIV+HCV coinfected patients display "CD4 knockdown phenotype." Because these patients develop more severe liver injury and higher HCV RNA titers with accentuated immunosuppression, these patients will either display exclusively Th1 lymphocytes or increased NK cell activity in liver tissues. Coinfection with HCV and HIV needs more machanistic insights regarding immunologic and virologic factors associated with HCV infection and liver injury at the cellular level.

HCV-specific T-cell responses are not frequently detected in peripheral blood in the presence of HIV coinfection. HCV-specific T cell responses are very weak in groups with HCV and HCV–HIV infections in the level of gamma interferon (IFN-γ).[3] Blocking antibodies against transforming growth factor beta-1 (TGF-β1), -2, and -3 and IL-10 specifically increased the HCV-specific T cell response.[3] HIV-1–HCV coinfected persons have an impairment in cellular immune responses to HCV as compared to HIV-1 infection only.[61] The cell type infected and targeted for this immune aberrations by HCV and HIV–HCV coinfected individuals is not clearly indentified.

HIV-infected individuals are very susceptible to HCV infection since HCV in HIV-positive patients is more prevalent than HIV in HCV-infected individuals. Immunosuppression induced by HIV infection could accelerate HCV infection in lymphocytes. In healthy individuals, cellular innate immunity protects the host from HCV infection. However, once the normal immune surveillance system collapses, lymphocytes become more vulnerable to HCV infection, suggesting that lymphocytes have the ability to counteract HCV infection, but once the immune surveillance system collapses, HCV coinfection will be significantly increased. Virus-host interactions will play a significant role in shielding HCV infection from immune systems.

Poor response and side effects are the main limitations of current anti-HCV therapy in HIV–HCV coinfected patients. With HIV, the use of Anti-Retroviral Therapy, or ART promotes more rapidly progressive liver disease. HIV can accelerate HCV liver disease especially when CD4-positive T cells decline. For example, within 10–15 years of infection, patients with HIV–HCV (15–25%) developed cirrhosis, in comparison with of those without HIV (2.6–6.5%). Furthermore, HCC appeared to occur at a younger age after a shorter duration of HCV infection.[38] (38). A Swiss cohort study showed that HCV accelerated the progression of HIV-associated diseases and less CD4 cell increases by HAART.[41]

Although data are limited, IFN and ribavirin combination therapy led to a high level of sustained virological response (SVR) with IFN and ribavirin combination therapy.[5,67] Many (non-responders and relapsers) need additional treatment options. The combination of pegylated IFN-α and ribavirin is the most effective therapy available, with high level of SVR treatment-naive patients and a small fraction

(10–20%) of non-responders in the HIV-negative patients.[91] The treatment-naive patients are sparse in HIV–HCV coinfected patients.[91] Coinfection of hepatitis C leads to morbidity and mortality in patients coinfected with HIV and HCV.

9. Concluding Remarks

HCV infection of lymphoid cells represents a new facet of HCV infection. The availability of lymphotropic HCV strains will offer new perspectives on HCV biology and pathogenesis. Many aspects of the host–virus interaction can now be explored, particularly the pro-viral factors that are required for viral production and the anti-viral factors that control HCV infection in lymphocytes as well as hepatocytes. In addition to the deeper understanding of basic HCV biology that is forthcoming from this system, it creates a new opportunity that could greatly accelerate the discovery of antiviral drugs and contribute to the development of HCV vaccines. The study of lymphotropic HCV will enlighten the molecular mechanism of lymphoproliferative diseases in HCV patients and thus open up an opportunity for rigorous study of lymphocyte infection with HCV. Studies on the genetic basis of lymphotropism of HCV will open up an avenue for studying extra-hepatic infection of HCV *in vivo* and *in vitro*. However, the genetic basis for the lymphotropism of these viruses has not been established. Understanding the biology of HCV-lymphoid cell interactions will lead toward defining those interactions that cause disease.

The yet-to-be identified lymphotropic receptor will become a new target for lymphotropic HCV therapeutic intervention that may complement the ongoing efforts to block intracellular replication with inhibitors of HCV proteases and polymerase. Future study needs to focus on understanding the virus–host interactions of HCV with B cells in order to clear all the potential reservoirs of HCV in the body.

In summary, there is still no universal agreement on the association between HCV replication and lymphoproliferation and B cell lymphoma. Nevertheless, dismissing this association is not justified in light of the increasing supportive evidence for this association. HCV binding to cell surface may be one reason for overestimating the HCV and lymphoma linkage, but the HCV–B cell connection does occur in at least a subset of hepatitis C patients. Further investigation, particularly molecular-based studies, will be required to show clear association.

References

1. Agnello V, Abel G, Elfahal M, *et al.* (1999) Hepatitis C virus and other flaviviridae viruses enter cells via low density lipoprotein receptor. *Proc Natl Acad Sci USA* **96**: 12766–12771.
2. Agnello V, Chung RT, and Kaplan LM. (1992) A role for hepatitis C virus infection in type II cryoglobulinemia. *N Engl J Med* **327**: 1490–1495.
3. Alatrakchi N, Graham CS, van der Vliet HJ, *et al.* (2007) Hepatitis C virus (HCV)-specific CD8+ cells produce transforming growth factor beta that can suppress HCV-specific T-cell responses. *J Virol* **81**: 5882–5892.
4. Altomonte M, Montagner R, Pucillo C, and Maio M. (1996) Triggering of target of an antiproliferative antibody-1 (TAPA-1/CD81) up-regulates the release of tumour necrosis factor-alpha by the EBV-B lymphoblastoid cell line JY. *Scand J Immunol* **43**: 367–373.
5. Ballesteros AL, Franco S, Fuster D, *et al.* (2004) Early HCV dynamics on Peg-interferon and ribavirin in HIV/HCV co-infection: indications for the investigation of new treatment approaches. *Aids* **18**: 59–66.
6. Bartenschlager R and Lohmann V. (2000) Replication of hepatitis C virus. *J Gen Virol* **81**: 1631–1648.
7. Barth H, Liang TJ, and Baumert TF. (2006) Hepatitis C virus entry: molecular biology and clinical implications. *Hepatology* **44**: 527–535.
8. Barth H, Schafer C, Adah MI, *et al.* (2003) Cellular binding of hepatitis C virus envelope glycoprotein E2 requires cell surface heparan sulfate. *J Biol Chem* **278**: 41003–41012.
9. Bartosch B, Verney G, Dreux M, *et al.* (2005) An interplay between hypervariable region 1 of the hepatitis C virus E2 glycoprotein, the scavenger receptor BI, and high-density lipoprotein promotes both enhancement of infection and protection against neutralizing antibodies. *J Virol* **79**: 8217–8229.
10. Bartosch B, Vitelli A, Granier C, *et al.* (2003) Cell entry of hepatitis C virus requires a set of co-receptors that include the CD81 tetraspanin and the SR-B1 scavenger receptor. *J Biol Chem* **278**: 41624–41630.
11. Basu A, Beyene A, Meyer K, Ray R. (2004) The hypervariable region 1 of the E2 glycoprotein of hepatitis C virus binds to glycosaminoglycans, but this

binding does not lead to infection in a pseudotype system. *J Virol* **78**: 4478–44786.

12. Basu A, Kanda T, Beyene A, *et al.* (2007) Sulfated homologues of heparin inhibit hepatitis C virus entry into mammalian cells. *J Virol* **81**: 3933–3941.
13. Bianco E, Marcucci F, Mele A, *et al.* (2004) Prevalence of hepatitis C virus infection in lymphoproliferative diseases other than B-cell non-Hodgkin's lymphoma, and in myeloproliferative diseases: an Italian multi-center case-control study. *Haematologica* **89**: 70–76.
14. Bowen DG and Walker CM. (2005) Adaptive immune responses in acute and chronic hepatitis C virus infection. *Nature* **436**: 946–952.
15. Bowen DG and Walker CM. (2005) Mutational escape from CD8+ T cell immunity: HCV evolution, from chimpanzees to man. *J Exp Med* **201**: 1709–1714.
16. Bradrick SS, Walters RW, and Gromeier M. (2006) The hepatitis C virus 3′-untranslated region or a poly(A) tract promote efficient translation subsequent to the initiation phase. *Nucleic Acids Res* **34**: 1293–1303.
17. Carter RH and Fearon DT. (1992) CD19: lowering the threshold for antigen receptor stimulation of B lymphocytes. *Science* **256**: 105–107.
18. Casato M, Mecucci C, Agnello V, *et al.* (2002) Regression of lymphoproliferative disorder after treatment for hepatitis C virus infection in a patient with partial trisomy 3, Bcl-2 overexpression, and type II cryoglobulinemia. *Blood* **99**: 2259–2261.
19. Chaudhuri J, Tian M, Khuong C, *et al.* (2003) Transcription-targeted DNA deamination by the AID antibody diversification enzyme. *Nature* **422**: 726–730.
20. Chung RT, Andersen J, Volberding P, *et al.* (2004) Peginterferon Alfa-2a plus ribavirin versus interferon alfa-2a plus ribavirin for chronic hepatitis C in HIV-coinfected persons. *N Engl J Med* **351**: 451–459.
21. Cocquerel L, Voisset C, and Dubuisson J. (2006) Hepatitis C virus entry: potential receptors and their biological functions. *J Gen Virol* **87**: 1075–1084.
22. Cooper S, Erickson AL, Adams EJ, *et al.* (1999) Analysis of a successful immune response against hepatitis C virus. *Immunity* **10**: 439–449.
23. Cormier EG, Durso RJ, Tsamis F, *et al.* (2004) L-SIGN (CD209L) and DC-SIGN (CD209) mediate transinfection of liver cells by hepatitis C virus. *Proc Natl Acad Sci U S A* **101**: 14067–14072.
24. Cormier EG, Tsamis F, Kajumo F, *et al.* (2004) CD81 is an entry coreceptor for hepatitis C virus. *Proc Natl Acad Sci USA* **101**: 7270–7274.
25. Crotta S, Stilla A, Wack A, *et al.* (2002) Inhibition of natural killer cells through engagement of CD81 by the major hepatitis C virus envelope protein. *J Exp Med* **195**: 35–41.

26. Dammacco F, Gatti P, and Sansonno D. (1998) Hepatitis C virus infection, mixed cryoglobulinemia, and non-Hodgkin's lymphoma: an emerging picture. *Leuk Lymphoma* **31**: 463–476.
27. Dammacco F, Sansonno D, Piccoli C, *et al.* (2001) The cryoglobulins: an overview. *Eur J Clin Invest* **31**: 628–638.
28. Di Noia J and Neuberger MS. (2002) Altering the pathway of immunoglobulin hypermutation by inhibiting uracil-DNA glycosylase. *Nature* **419**: 43–48.
29. El-Serag HB, Hampel H, Yeh C, Rabeneck L. (2002) Extrahepatic manifestations of hepatitis C among United States male veterans. *Hepatology* **36**: 1439–1445.
30. Evans MJ, von Hahn T, Tscherne DM, *et al.* (2007) Claudin-1 is a hepatitis C virus co-receptor required for a late step in entry. *Nature* **446**: 801–805.
31. Eyster ME, Diamondstone LS, Lien JM, *et al.* (1993) Natural history of hepatitis C virus infection in multitransfused hemophiliacs: effect of coinfection with human immunodeficiency virus. The Multicenter Hemophilia Cohort Study. *J Acquir Immune Defic Syndr* **6**: 602–610.
32. Faili A, Aoufouchi S, Flatter E, *et al.* (2002) Induction of somatic hypermutation in immunoglobulin genes is dependent on DNA polymerase iota. *Nature* **419**: 944–947.
33. Fearon DT, Carter RH. (1995) The CD19/CR2/TAPA-1 complex of B lymphocytes: linking natural to acquired immunity. *Annu Rev Immunol* **13**: 127–149.
34. Ferri C, La Civita L, Zignego AL, Pasero G. (1997) Viruses and cancers: possible role of hepatitis C virus. *Eur J Clin Invest* **27**: 711–718.
35. Flint M, Maidens C, Loomis-Price LD, *et al.* (1999) Characterization of hepatitis C virus E2 glycoprotein interaction with a putative cellular receptor, CD81. *J Virol* **73**: 6235–6244.
36. Friebe P, Boudet J, Simorre JP, and Bartenschlager R. (2005) Kissing-loop interaction in the 3′ end of the hepatitis C virus genome essential for RNA replication. *J Virol* **79**: 380–392.
37. Gao G, Z. Buskell Z, Seeff L, and Tabor E. (2002) Drift in the hypervariable region of the hepatitis C virus during 27 years in two patients. *J Med Virol* **68**: 60–67.
38. Garcia-Samaniego J, Rodriguez M, Berenguer J, *et al.* (2001) Hepatocellular carcinoma in HIV-infected patients with chronic hepatitis C. *Am J Gastroenterol* **96**: 179–183.
39. Goldin RD, Fish DE, Hay A, *et al.* (1990) Histological and immunohistochemical study of hepatitis B virus in human immunodeficiency virus infection. *J Clin Pathol* **43**: 203–205.
40. Goyenechea B, Klix N, Yelamos J, *et al.* (1997) Cells strongly expressing Ig(kappa) transgenes show clonal recruitment of hypermutation: a role for both MAR and the enhancers. *Embo J* **16**: 3987–3994.

41. Greub G, Ledergerber B, Battegay M, *et al.* (2000) Clinical progression, survival, and immune recovery during antiretroviral therapy in patients with HIV-1 and hepatitis C virus coinfection: The Swiss HIV Cohort Study. *Lancet* **356**: 1800–1805.
42. Hadziyannis SJ. (1996) Nonhepatic manifestations and combined diseases in HCV infection. *Dig Dis Sci* **41**: 63S–74S.
43. Hahn CS, Cho YG, Kang BS, *et al.* (2000) The HCV core protein acts as a positive regulator of fas-mediated apoptosis in a human lymphoblastoid T cell line. *Virology* **276**: 127–137.
44. Hermine O, Lefrere F, Bronowicki JP, *et al.* (2002) Regression of splenic lymphoma with villous lymphocytes after treatment of hepatitis C virus infection. *N Engl J Med* **347**: 89–94.
45. Hiramatsu N, Hayashi N, Katayama K, *et al.* (1994) Immunohistochemical detection of Fas antigen in liver tissue of patients with chronic hepatitis C. *Hepatology* **19**: 1354–1359.
46. Hsu M, Zhang J, Flint M, *et al.* (2003) Hepatitis C virus glycoproteins mediate pH-dependent cell entry of pseudotyped retroviral particles. *Proc Natl Acad Sci USA* **100**: 7271–7276.
47. Idilman R, Colantoni A, De Maria N, *et al.* (2004) Lymphoproliferative disorders in chronic hepatitis C. *J Viral Hepat* **11**: 302–309.
48. Inoue Y, Miyazaki M, Ohashi R, *et al.* (1998) Ubiquitous presence of cellular proteins that specifically bind to the 3′ terminal region of hepatitis C virus. *Biochem Biophys Res Commun* **245**: 198–203.
49. Jones CT, Murray CL, Eastman DK, *et al.* (2007) Hepatitis C virus p7 and NS2 proteins are essential for production of infectious virus. *J Virol* **81**: 8374–8383.
50. Kasama Y, Sekiguchi S, Saito M, *et al.* (2010) Persistent expression of the full genome of hepatitis C virus in B cells induces spontaneous development of B-cell lymphomas *in vivo*. *Blood* **116**: 4926–4933.
51. Kitay-Cohen Y, Amiel A, Hilzenrat N, *et al.* (2000) Bcl-2 rearrangement in patients with chronic hepatitis C associated with essential mixed cryoglobulinemia type II. *Blood* **96**: 2910–2912.
52. Kondo Y, Machida K, Liu HM, *et al.* (2009) Hepatitis C virus infection of T cells inhibits proliferation and enhances fas-mediated apoptosis by down-regulating the expression of CD44 splicing variant 6. *J Infect Dis* **199**: 726–736.
53. Kondo Y, Sung VM, Machida K, *et al.* (2007) Hepatitis C virus infects T cells and affects interferon-gamma signaling in T cell lines. *Virology* **361**: 161–173.
54. Kulik LM. (2006) Can therapy of hepatitis C affect the development of hepatocellular carcinoma? *J Natl Compr Canc Netw* **4**: 751–757.
55. Lai CK, Jeng KS, Machida K, *et al.* (2008) Hepatitis C virus NS3/4A protein interacts with ATM, impairs DNA repair and enhances sensitivity to ionizing radiation. *Virology* **370**: 295–309.

56. Landau DA, Saadoun D, Calabrese LH, Cacoub P. (2007) The pathophysiology of HCV induced B-cell clonal disorders. *Autoimmun Rev* **6**: 581–587.
57. Lanford RE, Chavez D, Chisari FV, Sureau C. (1995) Lack of detection of negative-strand hepatitis C virus RNA in peripheral blood mononuclear cells and other extrahepatic tissues by the highly strand-specific rTth reverse transcriptase PCR. *J Virol* **69**: 8079–8083.
58. Laskus T, Radkowski M, Wang LF, *et al.* (1998) Hepatitis C virus quasispecies in patients infected with HIV-1: correlation with extrahepatic viral replication. *Virology* **248**: 164–171.
59. Laskus T, Radkowski M, Wang LF, *et al.* (1998) The presence of active hepatitis C virus replication in lymphoid tissue in patients coinfected with human immunodeficiency virus type 1. *J Infect Dis* **178**: 1189–1192.
60. Laskus T, Radkowski M, Wang LF, *et al.* (1998) Search for hepatitis C virus extrahepatic replication sites in patients with acquired immunodeficiency syndrome: specific detection of negative-strand viral RNA in various tissues. *Hepatology* **28**: 1398–1401.
61. Lauer GM, Nguyen TN, Day CL, *et al.* (2002) Human immunodeficiency virus type 1-hepatitis C virus coinfection: intraindividual comparison of cellular immune responses against two persistent viruses. *J Virol* **76**: 2817–2826.
62. Lavillette D, Tarr AW, Voisset C, *et al.* (2005) Characterization of host-range and cell entry properties of the major genotypes and subtypes of hepatitis C virus. *Hepatology* **41**: 265–274.
63. Lechner F, Wong DK, Dunbar PR, *et al.* (2000) Analysis of successful immune responses in persons infected with hepatitis C virus. *J Exp Med* **191**: 1499–1512.
64. Lerat H, Shimizu YK, and Lemon SM. (2000) Cell type-specific enhancement of hepatitis C virus internal ribosome entry site-directed translation due to 5′ nontranslated region substitutions selected during passage of virus in lymphoblastoid cells. *J Virol* **74**: 7024–7031.
65. Levine AM, Shimodaira S, Lai MM. (2003) Treatment of HCV-related mantle-cell lymphoma with ribavirin and pegylated interferon Alfa. *N Engl J Med* **349**: 2078–2079.
66. Levy S, Todd SC, Maecker HT. (1998) CD81 (TAPA-1): a molecule involved in signal transduction and cell adhesion in the immune system. *Annu Rev Immunol* **16**: 89–109.
67. Liu CH, Liu CJ, Lin CL, *et al.* (2008) Pegylated interferon-alpha-2a plus ribavirin for treatment-naive Asian patients with hepatitis C virus genotype 1 infection: a multicenter, randomized controlled trial. *Clin Infect Dis* **47**: 1260–1269.

68. Lozach PY, Amara A, Bartosch B, *et al.* (2004) C-type lectins L-SIGN and DC-SIGN capture and transmit infectious hepatitis C virus pseudotype particles. *J Biol Chem* **279**: 32035–32045.
69. Machida K, Cheng KT, Pavio N, *et al.* (2005) Hepatitis C virus E2-CD81 interaction induces hypermutation of the immunoglobulin gene in B cells. *J Virol* **79**: 8079–8089.
70. Machida K, Cheng KT, Sung VM, *et al.* (2004) Hepatitis C virus infection activates the immunologic (type II) isoform of nitric oxide synthase and thereby enhances DNA damage and mutations of cellular genes. *J Virol* **78**: 8835–8843.
71. Machida K, Cheng KT, Sung VM, *et al.* (2006) Hepatitis C virus induces toll-like receptor 4 expression, leading to enhanced production of beta interferon and interleukin-6. *J Virol* **80**: 866–874.
72. Machida K, Cheng KT, Sung VM, *et al.* (2004) Hepatitis C virus induces a mutator phenotype: enhanced mutations of immunoglobulin and protooncogenes. *Proc Natl Acad Sci USA* **101**: 4262–4267.
73. Machida K, Kondo Y, Huang JY, *et al.* (2008) Hepatitis C virus (HCV)-induced immunoglobulin hypermutation reduces the affinity and neutralizing activities of antibodies against HCV envelope protein. *J Virol* **82**: 6711–6720.
74. Machida K, McNamara G, Cheng KT, *et al.* (2010) Hepatitis C virus inhibits DNA damage repair through reactive oxygen and nitrogen species and by interfering with the ATM-NBS1/Mre11/Rad50 DNA repair pathway in monocytes and hepatocytes. *J Immunol* **185**: 6985–6998.
75. Machida K, Tsukiyama-Kohara K, Sekiguch S, *et al.* (2009) Hepatitis C virus and disrupted interferon signaling promote lymphoproliferation via type II CD95 and interleukins. *Gastroenterology* **137**: 285–296, 296 e1–11.
76. Maggi F, Fornai C, Morrica A, *et al.* (1999) Divergent evolution of hepatitis C virus in liver and peripheral blood mononuclear cells of infected patients. *J Med Virol* **57**: 57–63.
77. Maggi F, Fornai C, Vatteroni ML, *et al.* (1997) Differences in hepatitis C virus quasispecies composition between liver, peripheral blood mononuclear cells and plasma. *J Gen Virol* **78**: 1521–1525.
78. Manns MP, Rambusch EG. (1999) Autoimmunity and extrahepatic manifestations in hepatitis C virus infection. *J Hepatol* **31**(Suppl 1): 39–42.
79. Masciopinto F, Giovani C, Campagnoli S, *et al.* (2004) Association of hepatitis C virus envelope proteins with exosomes. *Eur J Immunol* **34**: 2834–2842.
80. McDonnell TJ, Deane N, Platt FM, *et al.* (1989) bcl-2-immunoglobulin transgenic mice demonstrate extended B cell survival and follicular lymphoproliferation. *Cell* **57**: 79–88.

81. McKeating JA, Zhang LQ, Logvinoff C, *et al.* (2004) Diverse hepatitis C virus glycoproteins mediate viral infection in a CD81-dependent manner. *J Virol* **78**: 8496–8505.
82. Meyer K, Basu A, Przysiecki CT, *et al.* (2002) Complement-mediated enhancement of antibody function for neutralization of pseudotype virus containing hepatitis C virus E2 chimeric glycoprotein. *J Virol* **76**: 2150–2158.
83. Meyer K, Basu A, and Ray R. (2000) Functional features of hepatitis C virus glycoproteins for pseudotype virus entry into mammalian cells. *Virology* **276**: 214–226.
84. Meyer K, Beyene A, Bowlin TL, *et al.* (2004) Coexpression of hepatitis C virus E1 and E2 chimeric envelope glycoproteins displays separable ligand sensitivity and increases pseudotype infectious titer. *J Virol* **78**: 12838–12847.
85. Mills CT, Lee E, Perrillo R. (1990) Relationship between histology, aminotransferase levels, and viral replication in chronic hepatitis B. *Gastroenterology* **99**: 519–524.
86. Miyazaki T, Muller U, Campbell KS. (1997) Normal development but differentially altered proliferative responses of lymphocytes in mice lacking CD81. *Embo J* **16**: 4217–4225.
87. Morgensztern D, Rosado M, Silva O, *et al.* (2004) Prevalence of hepatitis C infection in patients with non-Hodgkin's lymphoma in South Florida and review of the literature. *Leuk Lymphoma* **45**: 2459–2464.
88. Morton LM, Landgren O, Chatterjee N, *et al.* (2007) Hepatitis C virus infection and risk of posttransplantation lymphoproliferative disorder among solid organ transplant recipients. *Blood* **110**: 4599–4605.
89. Muramatsu M, Sankaranand VS, Anant S, *et al.* (1999) Specific expression of activation-induced cytidine deaminase (AID), a novel member of the RNA-editing deaminase family in germinal center B cells. *J Biol Chem* **274**: 18470–18476.
90. Musolino C, Campo S, Pollicino T, *et al.* (1996) Evaluation of hepatitis B and C virus infections in patients with non-Hodgkin's lymphoma and without liver disease. *Haematologica* **81**: 162–164.
91. Myers RP, Benhamou Y, Bochet M, *et al.* (2004) Pegylated interferon alpha 2b and ribavirin in HIV/hepatitis C virus-co-infected non-responders and relapsers to IFN-based therapy. *Aids* **18**: 75–79.
92. Nagao Y, Sata M, Noguchi S, *et al.* (2000) Detection of hepatitis C virus RNA in oral lichen planus and oral cancer tissues. *J Oral Pathol Med* **29**: 259–266.
93. Nakajima N, Hijikata M, Yoshikura H, Shimizu YK. (1996) Characterization of long-term cultures of hepatitis C virus. *J Virol* **70**: 3325–3329.
94. Neumann-Haefelin C, Timm J, Spangenberg HC, *et al.* (2008) Virological and immunological determinants of intrahepatic virus-specific CD8+ T-cell failure in chronic hepatitis C virus infection. *Hepatology* **47**: 1824–1836.

95. Okuda K. (2000). Hepatocellular carcinoma. *J Hepatol* **32**: 225–237.
96. Papavasiliou FN and Schatz DG. (2002) Somatic hypermutation of immunoglobulin genes: merging mechanisms for genetic diversity. *Cell* **109**(Suppl): S35–44.
97. Park BB, Kim JS, Lee YY, *et al.* (2008) Clinical characteristics and outcome for hepatitis C virus-positive diffuse large B-cell lymphoma. *Leuk Lymphoma* **49**: 88–94.
98. Perez-Olmeda M, Nunez M, Romero M, *et al.* (2003) Pegylated IFN-alpha2b plus ribavirin as therapy for chronic hepatitis C in HIV-infected patients. *Aids* **17**: 1023–1028.
99. Petersen-Mahrt SK, Harris RS, Neuberger MS. (2002) AID mutates E. coli suggesting a DNA deamination mechanism for antibody diversification. *Nature* **418**: 99–103.
100. Pileri P, Uematsu Y, Campagnoli S, *et al.* (1998) Binding of hepatitis C virus to CD81. *Science* **282**: 938–941.
101. Ploss A, Evans MJ, Gaysinskaya VA, *et al.* (2009) Human occludin is a hepatitis C virus entry factor required for infection of mouse cells. *Nature* **457**: 882–886.
102. Pyrsopoulos NT and Reddy KR. (2001) Extrahepatic manifestations of chronic viral hepatitis. *Curr Gastroenterol Rep* **3**: 71–78.
103. Rada C, Milstein C. (2001) The intrinsic hypermutability of antibody heavy and light chain genes decays exponentially. *Embo J* **20**: 4570–4576.
104. Radkowski M, Bednarska A, Horban A, *et al.* (2004) Infection of primary human macrophages with hepatitis C virus *in vitro*: induction of tumour necrosis factor-alpha and interleukin 8. *J Gen Virol* **85**: 47–59.
105. Saunier B, Triyatni M, Ulianich L, *et al.* (2003) Role of the asialoglycoprotein receptor in binding and entry of hepatitis C virus structural proteins in cultured human hepatocytes. *J Virol* 77: 546–559.
106. Schneider-Schaulies J. (2000) Cellular receptors for viruses: links to tropism and pathogenesis. *J Gen Virol* **81**: 1413–1429.
107. Schollkopf C, Smedby KE, Hjalgrim H, *et al.* (2008) Hepatitis C infection and risk of malignant lymphoma. *Int J Cancer* **122**: 1885–1890.
108. Shimizu YK, Iwamoto A, Hijikata M, *et al.* (1992) Evidence for *in vitro* replication of hepatitis C virus genome in a human T-cell line. *Proc Natl Acad Sci USA* **89**: 5477–5481.
109. Shimizu YK, Purcell RH, Yoshikura H. (1993) Correlation between the infectivity of hepatitis C virus *in vivo* and its infectivity *in vitro*. *Proc Natl Acad Sci USA* **90**: 6037–6041.
110. Shoukry NH, Cawthon AG, Walker CM. (2004) Cell-mediated immunity and the outcome of hepatitis C virus infection. *Annu Rev Microbiol* **58**: 391–424.
111. Silvestri F, Pipan C, Barillari G, *et al.* (1996) Prevalence of hepatitis C virus infection in patients with lymphoproliferative disorders. *Blood* **87**: 4296–4301.

112. Soldaini E, Wack A, D'Oro U, *et al.* (2003) T cell costimulation by the hepatitis C virus envelope protein E2 binding to CD81 is mediated by Lck. *Eur J Immunol* **33**: 455–464.
113. Soriano V, Perez-Olmeda M, Rios P, *et al.* (2004) Hepatitis C virus (HCV) relapses after anti-HCV therapy are more frequent in HIV-infected patients. *AIDS Res Hum Retroviruses* **20**: 351–353.
114. Stamataki Z, Shannon-Lowe C, Shaw J, *et al.* (2009). Hepatitis C virus association with peripheral blood B lymphocytes potentiates viral infection of liver-derived hepatoma cells. *Blood* **113**: 585–593.
115. Strasser A, Harris AW, Bath ML, Cory S. (1990). Novel primitive lymphoid tumours induced in transgenic mice by cooperation between myc and bcl-2. *Nature* **348**: 331–333.
116. Sugiyama K, Kato N, Mizutani T, *et al.* (1997). Genetic analysis of the hepatitis C virus (HCV) genome from HCV-infected human T cells. *J Gen Virol* **78**: 329–336.
117. Sung VM, Shimodaira S, Doughty AL, *et al.* (2003). Establishment of B-cell lymphoma cell lines persistently infected with hepatitis C virus *in vivo* and *in vitro*: the apoptotic effects of virus infection. *J Virol* 77: 2134–2146.
118. Takahashi S, Doss C, Levy S, Levy R. (1990). TAPA-1, the target of an antiproliferative antibody, is associated on the cell surface with the Leu-13 antigen. *J Immunol* **145**: 2207–2213.
119. Tester I, Smyk-Pearson S, Wang P, *et al.* (2005). Immune evasion versus recovery after acute hepatitis C virus infection from a shared source. *J Exp Med* **201**: 1725–1731.
120. Thelu MA, Leroy V, Ramzan M, *et al.* (2007). IRES complexity before IFN-alpha treatment and evolution of the viral load at the early stage of treatment in peripheral blood mononuclear cells from chronic hepatitis C patients. *J Med Virol* **79**: 242–253.
121. Timuraglu A, Colak D, Ogunc D, *et al.* (1999). Hepatitis C virus association with non-Hodgkin's lymphoma. *Haematologia (Budap)* **29**: 301–304.
122. Torriani FJ, Rodriguez-Torres M, Rockstroh JK, *et al.* (2004) Peginterferon Alfa-2a plus ribavirin for chronic hepatitis C virus infection in HIV-infected patients. *N Engl J Med* **351**: 438–450.
123. Toubi E, Kessel A, Peri R, *et al.* (2004) Enhanced apoptosis of peripheral CD5-negative B lymphocytes from chronically hepatitis C virus-infected patients: reversal after antiviral treatment. *J Virol* **78**: 11379–11384.
124. Tseng CT, Klimpel GR. (2002) Binding of the hepatitis C virus envelope protein E2 to CD81 inhibits natural killer cell functions. *J Exp Med* **195**: 43–49.
125. Viguier M, Rivet J, Agbalika F, *et al.* (2002) B-cell lymphomas involving the skin associated with hepatitis C virus infection. *Int J Dermatol* **41**: 577–582.

126. Wakita T, Pietschmann T, Kato T, *et al.* (2005) Production of infectious hepatitis C virus in tissue culture from a cloned viral genome. *Nat Med* **11**: 791–796.
127. Yanagi M, Purcell RH, Emerson SU, Bukh J. (1997) Transcripts from a single full-length cDNA clone of hepatitis C virus are infectious when directly transfected into the liver of a chimpanzee. *Proc Natl Acad Sci USA* **94**: 8738–8743.
128. Zan H, Komori A, Li Z, *et al.* (2001) The translesion DNA polymerase zeta plays a major role in Ig and bcl-6 somatic hypermutation. *Immunity* **14**: 643–653.
129. Zehender G, De Maddalena C, Bernini F, *et al.* (2005) Compartmentalization of hepatitis C virus quasispecies in blood mononuclear cells of patients with mixed cryoglobulinemic syndrome. *J Virol* **79**: 9145–9156.
130. Zeng X, Winter DB, Kasmer C, *et al.* (2001) DNA polymerase eta is an A-T mutator in somatic hypermutation of immunoglobulin variable genes. *Nat Immunol* **2**: 537–541.
131. Zhang J, Randall G, Higginbottom A, *et al.* (2004) CD81 is required for hepatitis C virus glycoprotein-mediated viral infection. *J Virol* **78**: 1448–1455.
132. Zignego AL, Giannini C, and Ferri C. (2007) Hepatitis C virus-related lymphoproliferative disorders: an overview. *World J Gastroenterol* **13**: 2467–2478.
133. Zuckerman E, Zuckerman T, Levine AM, *et al.* (1997) Hepatitis C virus infection in patients with B-cell non-Hodgkin lymphoma. *Ann Intern Med* **127**: 423–428.
134. Zuckerman E, Zuckerman T, Sahar D, *et al.* (2001) The effect of antiviral therapy on t(14;18) translocation and immunoglobulin gene rearrangement in patients with chronic hepatitis C virus infection. *Blood* **97**: 1555–1559.
135. Harris HJ, Davis C, Mullins JG, *et al.* (2010) *J Biol Chem* **285**: 21092–21102. Epub 2010 Apr 7.

Chapter 12

Host Genetic Factors Control Hepatitis C Virus Infection

Yasuhito Tanaka*,† and Masashi Mizokami‡

Diversity in Outcomes of HCV Infection

Chronic infection with hepatitis C virus (HCV) is a significant health problem worldwide, with approximately 3% of the world population affected, i.e., more than 170 million people. According to different studies, 10–30% of HCV-infected individuals recover spontaneously, while 70–90% develop chronic hepatitis. Without treatment, chronic infection progresses to liver fibrosis and cirrhosis (LC) resulting in hepatic failure.[1] An estimated 8% of patients with LC develop hepatocellular carcinomas (HCC) annually.[2]

The current standard of care (SOC) for HCV-infected patients comprises pegylated interferon (peg-IFN)-α plus ribavirin (RBV). Results from clinical trials carried out in the past few years

* Corresponding Author: Yasuhito Tanaka, M.D., Ph.D., Department of Virology & Liver Unit, Nagoya, City University Graduate School of Medical Sciences, Kawasumi, Mizuho, Nagoya 467-8601, Japan, Fax: +81-52-842-0021, e-mail: ytanaka@med.nagoya-cu.ac.jp.

† Department of Virology & Liver Unit, Nagoya City University Graduate School of Medical Sciences, 1 Kawasumi, Mizuho, Mizuho, Nagoya 467-0001, Japan.

‡ The Research Center for Hepatitis and Immunology, National Center for Global Health and Medicine, 1-7-1 Kohnodai, Ichikawa 272-8516, Japan;[4].

on specifically targeted antiviral therapy for HCV (STAT-C) agents in combination with SOC provide great promise of higher rates of sustained virological response (SVR) and, potentially, shorter duration of therapy than standard therapy alone achieves.[3]

Treatment success means achieving sustained virological response, defined by undetectable HCV RNA levels six months after completing the therapy. Current HCV treatment permanently eradicates infection in 30–80% of cases. A portion of patients who fail to achieve SVR have transient viral response (TVR), defined by undetectable HCV RNA during treatment with relapse of the viral RNA level after discontinuation of treatment. Another category of patients who fail to achieve SVR is non viral responders (NVR), defined by detectable viremia during and after treatment.

Viral genotype, viral load, and stage of liver fibrosis/cirrhosis are the main factors recognized to have strong correlation with the treatment outcome.[4–6] Compared among different HCV genotypes, the lowest effectiveness of standard treatment is observed in patients infected with genotype 1 and 4; i.e., only 42–52% of patients with HCV genotype 1 achieve SVR in the US and Europe[6–8] versus 82% in genotypes 2 and 3.[9]

Other factors reported with less consistency are age and ethnicity. Effectiveness of the standard treatment is lower in elderly patients. Among different ethnicities, the SVR rate was the lowest in African Americans — 20–28% versus 40–52% in Caucasians.[5,10,11]

Research on factors associated with the probability of natural (without treatment) and/or treatment-induced clearance of HCV infection were carried intensively in the past decade. Two major groups of factors were revealed that may have independent or cumulative effect on viral clearance: viral factors and host factors.

Viral Factors

Several viral factors — such as genotype 1 (HCV-1), high baseline viral load, viral kinetics during treatment, and amino acid pattern in the IFN sensitivity-determining region — have been found to be significantly associated with outcome by a number of independent studies.[12–14]

Host Factors

Host factors associated with viral clearance include, gender, age, and advanced liver disease stage (LC).[15,16] Difference in HCV infection outcomes observed clinically in different ethnicity groups as well as studies on cohorts of patients infected from a single source[17,18] indicate that host genetic factors can be an important determinant of the infection outcome.

The human genome contains more than 3.3 billion base pairs; about 10 million of these may vary among individuals. This variability is determined as single-nucleotide polymorphism (SNP). Some of this variation may result in altered expression of the gene, altered processing of the gene product (post-transcriptional and post-translational modification), or altered functional activity (such as receptor binding). Identifying SNP that could be associated with clinical expression (phenotype) has been a difficult task. A few candidate genes were found by targeted genotyping of candidate genes in association with spontaneous or treatment-induced clearance, including type I IFN receptor-1 (IFNAR1), mitogen-activated protein kinase-activated protein kinase 3 (MAPKAPK3),[19,20] killer cell immunoglobulin-like receptor KIR2DL3 and its human leukocyte antigen C group1 (HLA-C1) ligand,[21] and HLA type.[22,23]

Genome-Wide Association Study (GWAS) of Treatment-Induced HCV Clearance

The success of the human genome project[24,25] accelerated studies on genetic factors involved in different outcomes of HCV infection.

A significant breakthrough has occurred in identifying phenotype-associated SNPs when the genome-wide association study (GWAS) approach was implemented. Using high-throughput genotyping technology usually for SNPs, ranging in number from 300,000 to 900,000 for each sample, it became possible to identify genes and genomic regions that may play an important role in phenotypic differences without any *a priori* hypotheses.[26–28]

Four independent groups simultaneously published results of the GWAS they carried out to assess the role of genetic variation in response to peg-IFN and RBV combination therapy for chronic hepatitis C patients.[26–30] Interestingly, despite different cohorts and GWAS platforms (the different platforms contain slightly different sets and numbers of SNPs included in testing throughout the whole genome), all four of them revealed strong association between genetic polymorphism near the IL-28B/Interferon-lambda3 gene and treatment-induced HCV clearance. The detailed results are summarized in Fig. 1.

Ge *et al.* and Suppiah *et al.* studied genetic variants associated with SVR to peg-IFN&RBV therapy in individuals infected with HCV genotype 1.[26,27] The former examined genetic factors associated with treatment response in patients from the IDEAL trial (individualized dosing efficacy versus flat dosing to assess optimal pegylated interferon therapy),[31] a large randomized controlled trial involving Caucasian, African-American, and Hispanic individuals in North America (n = 1137) (Table 1). The latter study group analyzed Caucasians, consisting of 293 Australian individuals (Northern European ancestry) with HCV genotype 1, and also validated an independent replication cohort consisting of 555 Europeans from the UK, Germany, Italy, and Australia. These two study groups mainly investigated GWAS in Caucasians, and analyzed host factors associated with SVR.

Tanaka *et al.* studied host factors associated with response to peg-IFN and RBV treatment in 142 Japanese patients with chronic hepatitis C of genotype 1 for GWAS, and prepared an independent replication cohort of 172 Japanese (Table 1).[28] In this study, patients were divided into three groups, SVR, TVR, or NVR. NVR versus virological responder (VR), consisting of SVR and TVR, was used for the predication of NVR factors. The data set of NVR versus VR (SVR and TVR) was constructed to discover the host factors related to NVR. In high-density association mapping and variation screening, seven SNPs (Rs. 8105790, Rs. 11881222, Rs. 8103142, Rs. 28416813, Rs. 4803219, Rs. 8099917, and Rs. 7248668) located in the IL28B region showed the significant associations ($P = 5.52 \times 10^{-28}$–2.68×10^{-32}; OR = 22.3–27.1). Of them, Rs. 8099917 showed the most significant associations ($p = 2.68 \times 10^{-32}$, OR = 27.1) in Japanese population. Compared to NVR, these SNPs were also associated with SVR (Table 1).

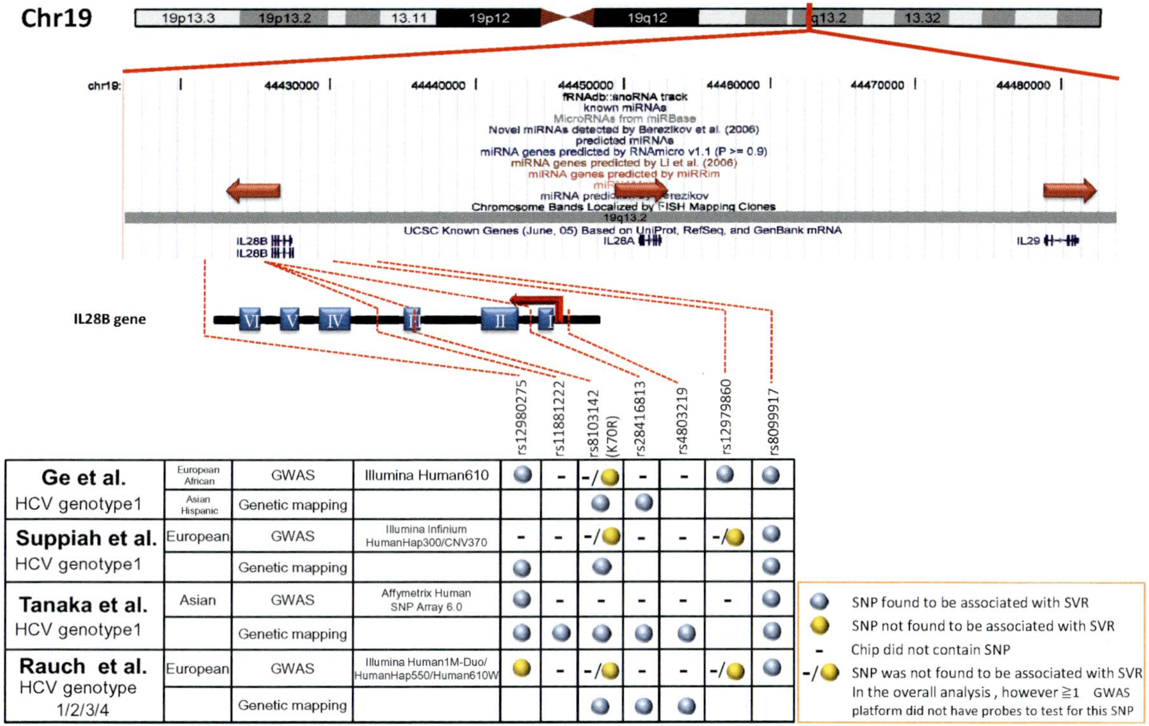

Fig. 1. SNPs in IL28B gene cluster associated with the response to treatment or natural clearance of HCV. The IFN-λ genes cluster is shown in the top panel, indicating its position on chromosome 19. IL28B is upstream and in reverse orientation compared with IL28A. The third panel depicts the genomic structure of IL28B, including its 6 exons, intervening introns, and flanking putative regulatory regions. The bottom panel shows which SNPs were associated with response to treatment in each of the indicated studies, including those that were found by genetic mapping studies. The only SNP that is in a coding region encodes K70R. As shown in the cartoon, symbols are used to indicate if a given SNP was associated with SVR in each of 4 GWAS and genetic mapping studies.

Table 1. IL28B-Related Papers Based on GWAS

Study	Ge *et al.*	Suppiah *et al.*	Tanaka *et al.*	Rauch *et al.*
Journal	*Nature* 16 Aug 2009	*Nature Genetics* online 13 Sep 2009	*Nature Genetics* online 13 Sep 2009	*Gastroenterology* online 2010
Region	Northern America	Northern Europe Australia	Japan	Switzerland
Ancestry	Caucasian/African/ Hispanic	Caucasian	Japanese	Caucasian
GWAS size	871/191/75	293	142	465
Replication	No replication	555	172	No replication
Case/control	SVR vs. non-SVR	SVR vs. non-SVR	SVR vs. non-SVR VR vs. NVR	SVR vs. non-SVR
Adherence	Over 80% adherent to PEG-IFN&RBV during the first 12 weeks of therapy	Not controlled	Over 80% adherent to PEG-IFN&RBV during the first 12 weeks of therapy	Over 80% adherent to PEG-IFN & RBV during the first 12 weeks of therapy
HCV genotype	1	1	1	1, 2, 3, 4
Significant SNPs	Rs. 12979860	Rs. 8099917	Rs. 8099917	Rs. 8099917
P value*	1.37×10^{-28}	9.25×10^{-9}	1.18×10^{-18}	3.11×10^{-8}
OR* (SVR)	3.1 (2.1 – 4.7)	1.98 (1.57–2.52)	12.1 (6.5–22.4)	5.19(2.9–9.3)
Platform	Illumina 610-quad	Illumina CNV370-quad	Affymetrix SNPs 6.0	Illumina Human 1M-dup

Rauch *et al.* investigated 465 Caucasians infected with HCV genotypes 1, 2, 3, or 4 to reveal genetic variations associated with response to the combination therapy.[29] Strong predicting value of the IL28B polymorphism was observed in genotype 1 and 4 patients, but not in patients infected with genotype 2 and 3. When we compared the prediction value of three of the GWAS (except for Suppiah *et al.*) that selected patients receiving at least 80% of the recommended treatment dose to emphasize genetic associations, the association with IL28B genotypes was the strongest in East Asians who have similar haplotype structure based on the HapMap data (Fig. 2). On the other hand, the significance of Caucasians by Suppiah *et al.* was much lower due to the dose reduction of peg-IFN and/or RBV. As shown in Fig. 2, comparison of pairwise linkage disequilibrium (LD; D′ and r^2) diagrams among three populations (JPT, CEU, and YRI) showed that these SNPs were in the strongest LD in JPT, followed by CEU

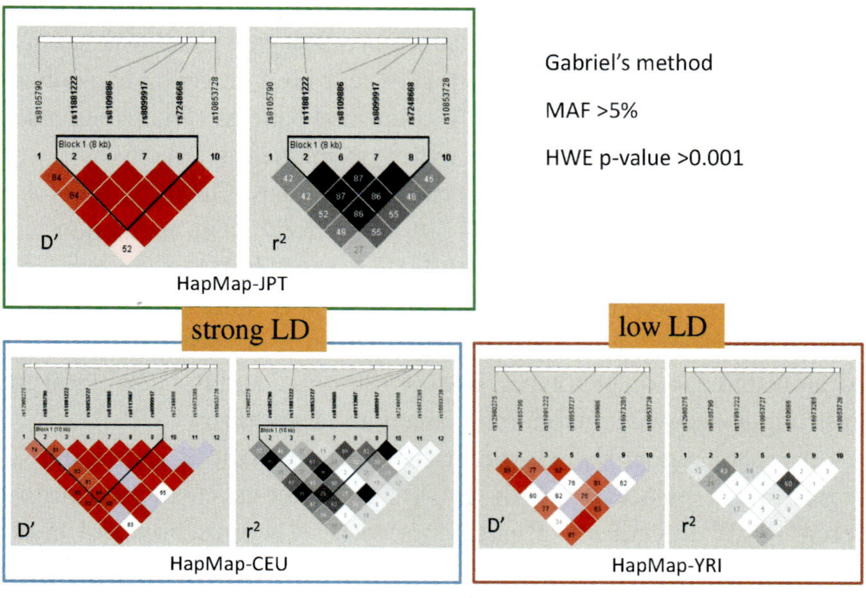

Fig. 2. LD structure on Chr19 (44423623–44436986) among 3 populations. Comparison of pairwise linkage disequilibrium (LD; D′ and r^2) diagrams between 3 populations (JPT, CEU and YRI) showed that these SNPs were in the strongest LD in JPT, followed by CEU population though low LD was predicted in YRI population. MAF: minor allele frequency. HWE: Hardy-Weinberg equilibrium.

population, though low LD was predicted in YRI population. This is one of the reasons why the OR was the highest in East Asians.

SNPs Associated with Spontaneous Clearance of HCV

GWAS on spontaneous clearance of HCV has been carried by Rauch *et al.* using HumanHap550 platform with 500K SNPs.[29] The case-control study was designed for 347 individuals with spontaneous HCV clearance, 567 individuals with chronic hepatitis C, and 448 individuals of HCV/HIV coinfection. The significant SNP found was rs8099917 (combined $P = 6.07 \times 10^{-9}$, OR = 2.31, 95%CI = 1.74–3.04). The effect of HIV coinfection was also similar to HCV mono-infection ($P = 8.25 \times 10^{-5}$, OR = 2.16, 95%CI = 1.47–3.18; $P = 1.96 \times 10^{-5}$, OR = 2.49, 95%CI = 1.64–3.79, respectively). Thomas *et al.* using targeted genotyping, investigated association of the Rs. 12979860 SNP reported by Ge *et al.* to determine whether the SNP was also associated with spontaneous clearance of HCV infection.[32] This study included 388 individuals with spontaneous HCV clearance and 620 with persistent HCV infection in a cohort consisting of HCV and HIV/HCV coinfected patients. The strong association of Rs. 12979860 with spontaneous recovery was found in European and African-American individuals (OR = 2.6, 95%CI = 1.9–3.8; OR = 3.1, 95%CI = 1.7–5.8, respectively). The association was also independent of coinfection with HIV, type of HCV transmission, and history of HBV infection. The Rs. 12979860 SNP studied by Thomas *et al.* was not present on the panel used by Rauch *et al.*; however, based on accumulated data on genotypes, both Rs. 8099917 and Rs. 12979860 are in strong linkage disequilibrium in Asians and Caucasians.[33]

A recent report also showed that genetic variations in the IL28B region were associated with HCV spontaneous clearance during recent HCV infection.[34] Participants with seroconversion and jaundice were more frequently Rs. 8099917 TT major homozygotes than other (GG/GT) genotypes (32% vs. 5%, $P = 0.047$). This suggests a stronger immune response during acute phase of HCV infection among patients with Rs. 8099917 TT genotype, resulting in a higher frequency of spontaneous clearance (Fig. 3).

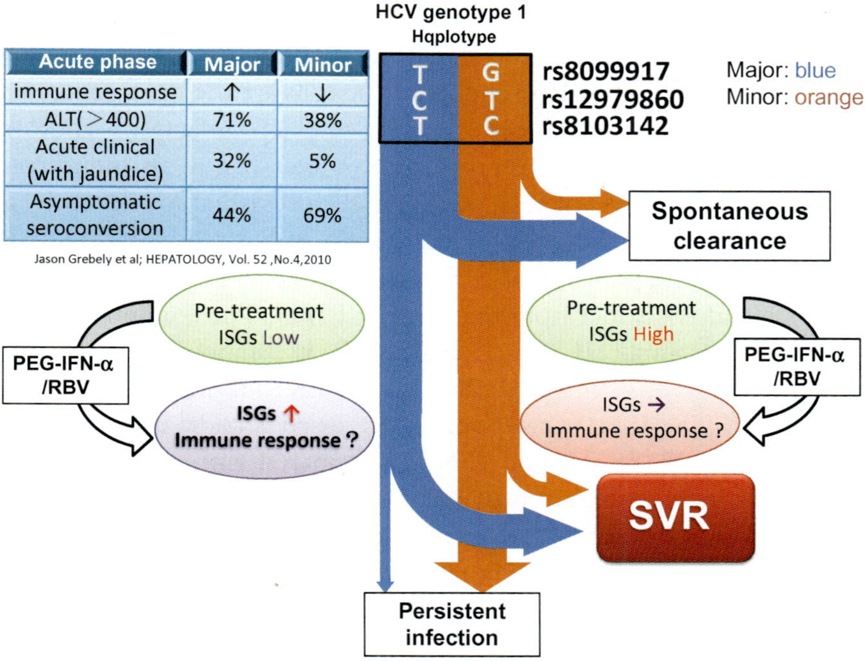

Fig. 3. Predicted natural and treatment history of HCV infection based on haplotype. Based on both GWAS and genetic mapping data, several SNPs with alleles that are in strong linkage disequilibrium have been identified in or near the *IL28B* gene locus. These SNPs collectively delineate haplotypes associated with HCV clearance or persistence. After acute HCV infection, a greater proportion of patients with major haplotypes spontaneously clear HCV due to stronger immune response, compared with patients with the other (hetero/minor) haplotypes. Of the patients who develop persistent HCV infection, intrahepatic expression of ISGs in this group posttreatment is lower relative to the HCV clearance haplotype (major). Patients with the major haplotype are more likely to achieve SVRs due to more induction of ISGs, when they receive treatment with PEG-IFN-α/RBV than patients with the other haplotypes.

Biological Characteristics of IL-28B and the IFN-λ Family

IL-28B, referred to as IFN-λ3, belongs to the IFN-λ family, which consists of IL-29/IFN-λ1 and IL-28A/IFN-λ2, and IL-28B. The IL-28B gene was recently discovered and classified into type III IFN, i.e., a member of the class II cytokine family. This class II family includes type I, II, and III IFN and the IL-10 family (IL-10, IL-19, IL-20, IL-22, IL-24, IL-26, IL-28, and IL-29). Peripheral blood mononuclear cells (PBMCs) and dendritic cells are the main source of IFN-λ,[35,36] which is induced by IFN-α, viral infection, and/or stimulations of toll-like receptors. IFN-λ behaves as an interferon stimulated gene (ISG) of IFN-α, which is expressed at low levels by a broad variety of human cells, similar to IFN-α.[37]

The signaling pathway of IFN-λ is initiated through a membrane receptor that is distinct from that of type I IFN. The common receptor for IFN-λ family (IL28A, IL28B, IL29) is composed of heterodimer molecules consisting of IL-28RA/IFN-λR1 subunit and IL-10R2 subunit.[35,36] The IL-10R2 subunit is expressed broadly and is shared by IL-10, IL-22, IL-26, and IFN-λ. Compared with the IFN-α/-β receptor, which is ubiquitously expressed, the IL-28RA receptor displays a more restricted tissue distribution. The signal transduction of IFN-λ receptor is mediated via Jak1 and Tyk2, which can induce the phosphorylation of STAT1 and STAT2 molecules and with subsequent expression of ISGs (Fig. 4).[38]

The Antiviral Effect of IFN-λ Against HCV in Research and Clinical Trials

Antiviral effects of IFN-λs against HCV were reported before the discovery of their association with the response to HCV therapy. Treatment with IFN-α or IFN-λ1 inhibited HCV replication at similar levels at low concentrations.[39] The combination treatment of IFN-α and IL-29/28A enhanced the antiviral effect against HCV replicon synergistically.[40] Using HG-U133 (Affymetrix, Santa Clara, CA microarray chip representing 39,000 transcripts derived from approximately 33,000 known human genes), Marcello *et al.* studied ISG induction by IFN-α/β or IFN-λ1 in Huh-7 cells. IFN-λ1 showed a unique pattern on ISG expression compared to that of IFN-α/β.[39] For example, a total of 19 genes, which were not detected in the IFN-α treated cells, were specifically up-regulated by IFN-λ1 at the late phase of treatment. Therefore, the signal pathway downstream of IL-28R1 might partially differ from that of IFN-α

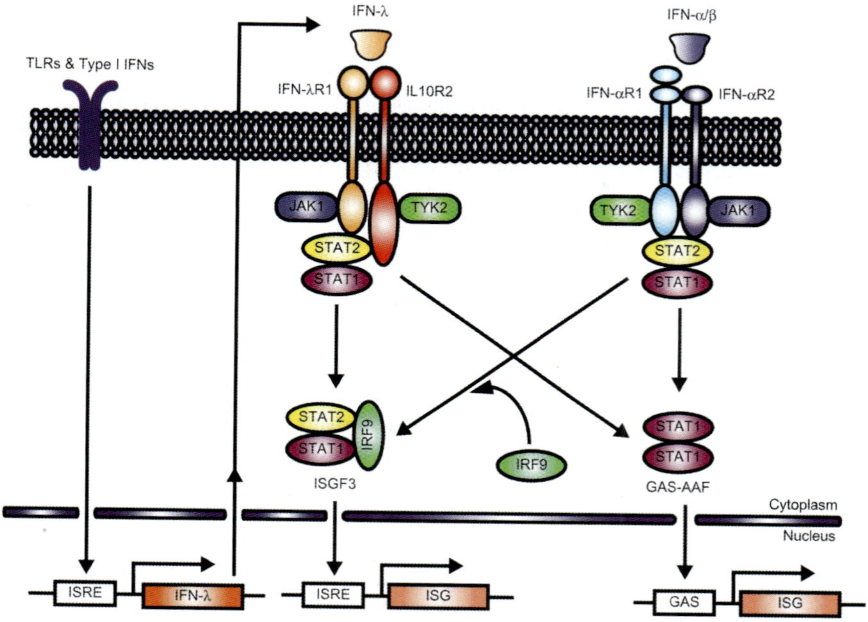

Fig. 4. Schematics of expression pathway of IFN-λ and IFN-α. IFN-λ and IFN-λ activate similar antiviral response pathway. ISGs induced by IFN-λ and IFN-α are similar, but some of them might be different (39). IL, interleukin; IRF, interferon regulatory factor; STAT, signal transducer and activator of transcription; TLR, Toll-like receptor.

and possess an important biological function, although the pattern of signal transduction is supposed to be similar to that of IFN-αR1/2 in current knowledge.[41]

The expression pattern of IFN-λ receptor is restricted to specific organs (Fig. 4). High expression of the receptor was observed in the pancreas, liver, prostate, or thyroid, whereas the central nerve system (the bone marrow or the brain) showed low expression.[35,36] This feature of interferon lambda system has been shown clinically beneficial in comparison to IFN-α/β in terms of toxicity.[42]

IL28B Genotype and Hepatic ISGs

There has been recent interest in the relationship between intrahepatic ISG expression in patients chronically infected with HCV and response to IFN-based treatment regimens (Fig. 3). A growing body of evidence, in both the human and the chimpanzee models, supports a paradigm in which IFN non-response is associated with pre-activated, perhaps maximally or inappropriately stimulated endogenous intrahepatic ISG expression.[43–47] In contrast, pre-treatment ISG from quiescent liver has been associated with sensitivity to exogenous IFN therapy and viral eradication. The mechanistic basis for this association is unknown, and it remains unclear whether it is driven by the host, the virus, or both. In this context, the IL28B genotype is a very attractive candidate for such a genetic factor, all the more because the IL28B gene product, IFNλ3, activates similar signal transduction pathways as type I IFNs and induces a similar set of target genes.

Two recent studies indeed revealed an association between the IL28B genotype and the expression level of hepatic ISGs (Fig. 3).[48,49] However, it is difficult to establish whether intrahepatic ISG expression level contributes to viral clearance independently of IL28B genotypes or not. Another recent study addressed that question and the result suggested that there is no absolute correlation between them.[50] Furthermore, IFN-lambda transcript levels measured in PBMCs or liver revealed inconsistent results in the context of its association with IL28B genotype.[27,28,49,50]

Conclusions

The GWAS approach has revealed overwhelming evidence of an association between genetic polymorphism around the IL28B gene and natural/treatment-induced clearance of HCV infection. This was confirmed in different populations using targeted genotyping.

The IL28B genotype appears to be a promising clinical marker allowing for estimates of the chance of treatment-induced clearance in HCV-infected patients before initiation of the treatment. It is important to note, however, that approximately 20–30% of homozygotes for the minor allele in Caucasians and 10–20% of heterozygotes/homozygotes of the minor allele in the Japanese population still achieve SVR. Furthermore, not all patients with major homozygotes achieve SVR,[26–29] indicating that the response to a combination therapy is not absolutely restricted to the IL28B genotype. To improve the prediction rates, other host and viral factors should be considered.[51]

Considering its biological properties and genetic association, IFN-lambda may play an important role in host immune response to HCV infection. However, whether any of the highlighted SNPs modulates the effect of the cytokine is still an open question. Accumulated data presently do not allow for identification of the place of the IFN-lambda and intrahepatic ISGs in the pathway of interaction between virus and host.

References

1. Hoofnagle JH. (2002) Course and outcome of hepatitis C. *Hepatology* **36** (Suppl 1): S21–29.
2. Shiratori Y, Shiina S, Imamura M, *et al.* (1995) Characteristic difference of hepatocellular carcinoma between hepatitis B- and C-viral infection in Japan. *Hepatology* **22**: 1027–1033.
3. Pereira AA, Jacobson IM. (2009) New and experimental therapies for HCV. *Nat Rev Gastroenterol Hepatol* **6**: 403–411.
4. Kau A, Vermehren J, Sarrazin C. (2008) Treatment predictors of a sustained virologic response in hepatitis B and C. *J Hepatol* **49**: 634–651.
5. Muir AJ, Bornstein JD, Killenberg PG. (2004) Peginterferon alfa-2b and ribavirin for the treatment of chronic hepatitis C in blacks and non-Hispanic whites. *N Engl J Med* **350**: 2265–2271.
6. Manns MP, McHutchison JG, Gordon SC, *et al.* (2001) Peginterferon alfa-2b plus ribavirin compared with interferon alfa-2b plus ribavirin for initial treatment of chronic hepatitis C: A randomised trial. *Lancet* **358**: 958–965.
7. Fried MW, Shiffman ML, Reddy KR, *et al.* (2002) Peginterferon alfa-2a plus ribavirin for chronic hepatitis C virus infection. *N Engl J Med* **347**: 975–982.
8. Hadziyannis SJ, Sette H, Jr., Morgan TR, *et al.* (2004) Peginterferon-alpha2a and ribavirin combination therapy in chronic hepatitis C: a randomized study of treatment duration and ribavirin dose. *Ann Intern Med* **140**: 346–355.
9. Shiffman ML, Suter F, Bacon BR, *et al.* (2007) Peginterferon alfa-2a and ribavirin for 16 or 24 weeks in HCV genotype 2 or 3. *N Engl J Med* **357**: 124–134.
10. Conjeevaram HS, Fried MW, Jeffers LJ, *et al.* (2006) Peginterferon and ribavirin treatment in African American and Caucasian American patients with hepatitis C genotype 1. *Gastroenterology* **131**: 470–477.
11. Jeffers LJ, Cassidy W, Howell CD, *et al.* (2004) Peginterferon alfa-2a (40 kd) and ribavirin for black American patients with chronic HCV genotype 1. *Hepatology* **39**: 1702–1708.

12. Akuta N, Suzuki F, Sezaki H, *et al.* (2005) Association of amino acid substitution pattern in core protein of hepatitis C virus genotype 1b high viral load and non-virological response to interferon-ribavirin combination therapy. *Intervirology* **48**: 372–380.
13. Pascu M, Martus P, Hohne M, *et al.* (2004) Sustained virological response in hepatitis C virus type 1b infected patients is predicted by the number of mutations within the NS5A-ISDR: a meta-analysis focused on geographical differences. *Gut* **53**: 1345–1351.
14. Shirakawa H, Matsumoto A, Joshita S, *et al.* (2008) Pretreatment prediction of virological response to peginterferon plus ribavirin therapy in chronic hepatitis C patients using viral and host factors. *Hepatology* **48**: 1753–1760.
15. Walsh MJ, Jonsson JR, Richardson MM, *et al.* (2006) Non-response to antiviral therapy is associated with obesity and increased hepatic expression of suppressor of cytokine signalling 3 (SOCS-3) in patients with chronic hepatitis C, viral genotype 1. *Gut* **55**: 529–535.
16. Gao B, Hong F, Radaeva S. (2004) Host factors and failure of interferon-alpha treatment in hepatitis C virus. *Hepatology* **39**: 880–890.
17. Kenny-Walsh E. (1999) Clinical outcomes after hepatitis C infection from contaminated anti-D immune globulin. Irish Hepatology Research Group. *N Engl J Med* **340**: 1228–1233.
18. Wiese M, Grungreiff K, Guthoff W, *et al.* (2005) Outcome in a hepatitis C (genotype 1b) single source outbreak in Germany — a 25-year multicenter study. *J Hepatol* **43**: 590–598.
19. Matsuyama N, Mishiro S, Sugimoto M, *et al.* (2003) The dinucleotide microsatellite polymorphism of the IFNAR1 gene promoter correlates with responsiveness of hepatitis C patients to interferon. *Hepatol Res* **25**: 221–225.
20. Tsukada H, Ochi H, Maekawa T, *et al.* (2009) A polymorphism in MAPKAPK3 affects response to interferon therapy for chronic hepatitis C. *Gastroenterology* **136**: 1796–1805.
21. Khakoo SI, Thio CL, Martin MP, *et al.* (2004) HLA and NK cell inhibitory receptor genes in resolving hepatitis C virus infection. *Science* **305**: 872–874.
22. Wang JH, Zheng X, Ke X, *et al.* (2009) Ethnic and geographical differences in HLA associations with the outcome of hepatitis C virus infection. *Virol J* **6**: 46.
23. Montes-Cano MA, Caro-Oleas JL, Romero-Gomez M, *et al.* (2005) HLA-C and KIR genes in hepatitis C virus infection. *Hum Immunol* **66**: 1106–1109.
24. Venter JC, Adams MD, Myers EW, *et al.* (2001) The sequence of the human genome. *Science* **291**: 1304–1351.
25. Lander ES, Linton LM, Birren B, *et al.* (2001) Initial sequencing and analysis of the human genome. *Nature* **409**: 860–921.
26. Ge D, Fellay J, Thompson AJ, *et al.* (2009) Genetic variation in IL28B predicts hepatitis C treatment-induced viral clearance. *Nature* **461**: 399–401.

27. Suppiah V, Moldovan M, Ahlenstiel G, *et al.* (2009) IL28B is associated with response to chronic hepatitis C interferon-alpha and ribavirin therapy. *Nat Genet* **41**: 1100–1104.
28. Tanaka Y, Nishida N, Sugiyama M, *et al.* (2009) Genome-wide association of IL28B with response to pegylated interferon-alpha and ribavirin therapy for chronic hepatitis C. *Nat Genet* **41**: 1105–1109.
29. Rauch A, Kutalik Z, Descombes P, *et al.* (2010) Genetic variation in IL28B is associated with chronic hepatitis C and treatment failure: a genome-wide association study. *Gastroenterology* **138**: 1338–1345, 1345 e1331–1337.
30. Rauch A, Kutalik Z, Descombes P, *et al.* (2010) Genetic variation in IL28B is associated with chronic hepatitis C and treatment failure: a genome-wide association study. *Gastroenterology* **138**: 1338–1345.
31. McHutchison JG, Lawitz EJ, Shiffman ML, *et al.* (2009) Peginterferon alfa-2b or alfa-2a with ribavirin for treatment of hepatitis C infection. *N Engl J Med* **361**: 580–593.
32. Thomas DL, Thio CL, Martin MP, *et al.* (2009) Genetic variation in IL28B and spontaneous clearance of hepatitis C virus. *Nature* **461**: 798–801.
33. Tanaka Y, Nishida N, Sugiyama M, *et al.* (2010) lambda-Interferons and the single nucleotide polymorphisms: a milestone to tailor-made therapy for chronic hepatitis C. *Hepatol Res* **40**: 449–460.
34. Grebely J, Petoumenos K, Hellard M, *et al.* (2010) Potential role for interleukin-28B genotype in treatment decision-making in recent hepatitis C virus infection. *Hepatology* **52**: 1216–1224.
35. Kotenko SV, Gallagher G, Baurin VV, *et al.* (2003) IFN-lambdas mediate antiviral protection through a distinct class II cytokine receptor complex. *Nat Immunol* **4**: 69–77.
36. Sheppard P, Kindsvogel W, Xu W, *et al.* (2003) IL-28, IL-29 and their class II cytokine receptor IL-28R. *Nat Immunol* **4**: 63–68.
37. Siren J, Pirhonen J, Julkunen I, Matikainen S. (2005) IFN-alpha regulates TLR-dependent gene expression of IFN-alpha, IFN-beta, IL-28, and IL-29. *J Immunol* **174**: 1932–1937.
38. Donnelly RP, Sheikh F, Kotenko SV, Dickensheets H. (2004) The expanded family of class II cytokines that share the IL-10 receptor-2 (IL-10R2) chain. *J Leukoc Biol* **76**: 314–321.
39. Marcello T, Grakoui A, Barba-Spaeth G, *et al.* (2006) Interferons alpha and lambda inhibit hepatitis C virus replication with distinct signal transduction and gene regulation kinetics. *Gastroenterology* **131**: 1887–1898.
40. Brand S, Zitzmann K, Dambacher J, *et al.* (2005) SOCS-1 inhibits expression of the antiviral proteins 2′,5′-OAS and MxA induced by the novel interferon-lambdas IL-28A and IL-29. *Biochem Biophys Res Commun* **331**: 543–548.

41. O'Brien TR. (2009) Interferon-alfa, interferon-lambda and hepatitis C. *Nat Genet* **41**: 1048–1050.
42. Muir AJ, Shiffman ML, Zaman A, *et al.* (2010) Phase 1b study of pegylated interferon lambda 1 with or without ribavirin in patients with chronic genotype 1 hepatitis C virus infection. *Hepatology* **52**: 822–832.
43. Asselah T, Bieche I, Narguet S, *et al.* (2008) Liver gene expression signature to predict response to pegylated interferon plus ribavirin combination therapy in patients with chronic hepatitis C. *Gut* **57**: 516–524.
44. Chen L, Borozan I, Feld J, *et al.* (2005) Hepatic gene expression discriminates responders and nonresponders in treatment of chronic hepatitis C viral infection. *Gastroenterology* **128**: 1437–1444.
45. Feld JJ, Nanda S, Huang Y, *et al.* (2007) Hepatic gene expression during treatment with peginterferon and ribavirin: identifying molecular pathways for treatment response. *Hepatology* **46**: 1548–1563.
46. Lanford RE, Guerra B, Bigger CB, *et al.* (2007) Lack of response to exogenous interferon-alpha in the liver of chimpanzees chronically infected with hepatitis C virus. *Hepatology* **46**: 999–1008.
47. Sarasin-Filipowicz M, Oakeley EJ, Duong FH, *et al.* (2008) Interferon signaling and treatment outcome in chronic hepatitis C. *Proc Natl Acad Sci USA* **105**: 7034–7039.
48. Honda M, Sakai A, Yamashita T, *et al.* (2010) Hepatic ISG expression is associated with genetic variation in interleukin 28B and the outcome of IFN therapy for chronic hepatitis C. *Gastroenterology* **139**: 499–509.
49. Urban TJ, Thompson AJ, Bradrick SS, *et al.* (2010) IL28B genotype is associated with differential expression of intrahepatic interferon-stimulated genes in patients with chronic hepatitis C. *Hepatology* **52**: 1888–1896.
50. Dill MT, Duong FH, Vogt JE, *et al.* (2011) Interferon-Induced Gene Expression Is a Stronger Predictor of Treatment Response than IL28B Genotype in Patients with Hepatitis C. *Gastroenterology* **140**: 1021–1031.
51. Kurosaki M, Tanaka Y, Nishida N, *et al.* (2011) Pre-treatment prediction of response to pegylated-interferon plus ribavirin for chronic hepatitis C using genetic polymorphism in IL28B and viral factors. *J Hepatol* **54**: 439–448.

Chapter 13

Treatment of Hepatitis C Virus Infection: Past, Present and Future

Chun-Jen Liu* and Ding-Shinn Chen*,†

Introduction

Health Burden Worldwide

Hepatitis C virus (HCV) infection is a worldwide health burden.[1] World Health Organization estimates that as many as 170 million persons are infected with HCV. For example, in the United States, the Centers for Disease Control and Prevention estimates that there are more than 2.7 million people with chronic HCV infection.[2] In Asia, the prevalence ranges from 0.3% of the general population in New Zealand to 4% in Cambodia. The African and Eastern Mediterranean countries and Mongolia reported the highest prevalence rates of HCV infection (>10%). Data available on the incidence of new cases of hepatitis C are scanty because of the difficulty in differentiating between new cases

* Graduate Institute of Clinical Medicine; Department of Internal Medicine, and Hepatitis Research Center, National Taiwan University College of Medicine and National Taiwan University Hospital, Taipei, Taiwan.

† Corresponding Author: Distinguished Chair Professor, Department of Internal Medicine, National Taiwan University College of Medicine, 1 Chang-Te Street, Taipei 10002, TAIWAN, Tel: 886-2-23123456 ext 67120, Fax: 886-2-23317624, e-mail: chends@ntu.edu.tw

and the initial diagnosis of chronically infected subjects. Since the 1990s, with the implementation of universal screening of blood products and the abolition of paid blood donation, a main route of HCV transmission has already been controlled. The introduction of effective antiviral therapy will further eradicate HCV infection in existing carriers.

Natural History

Natural history studies revealed that 55–85% of patients who develop acute hepatitis C will become chronic HCV carriers. Among these individuals, 5–20% were reported to develop cirrhosis over a period of approximately 20 to 25 years.[3,4] Patients with HCV-related cirrhosis then are at risk of developing end-stage liver diseases (~3%/year) as well as hepatocellular carcinoma (HCC) (1–2%/year).[5] The 15–45% of patients with acute hepatitis C who do recover are not subject to long-term complications and do not need antiviral treatment. In general practice, the majority of patients already have chronic hepatitis C. For these patients, evolution to cirrhosis is the primary concern. Significant fibrosis on liver biopsy is an important predictor of future progression of liver disease and the need for HCV treatment.

Treatment of HCV Infection

Benefits of Anti-HCV Therapy

The goal of antiviral treatment is to prevent progression and complications of HCV infection; this is principally achieved by eradication of the virus. Accordingly, treatment responses are frequently characterized by the results of the serum HCV RNA tests. The virus is considered eradicated when there is a sustained virologic response (SVR), defined as the absence of HCV RNA in serum by a sensitive qualitative test at the end of treatment and at least six months later.[3,6,7]

An SVR is the best correlate of beneficial changes in hepatic fibrosis, prevention of HCC and improvement in other clinical outcomes.

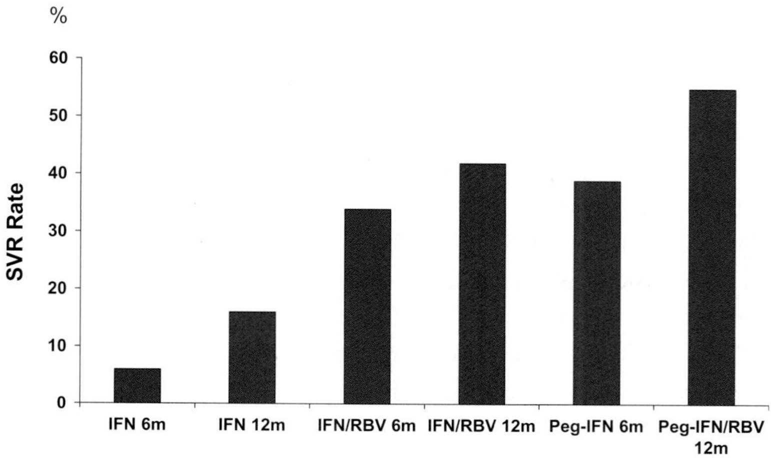

Fig. 1. Progress in the treatment of chronic hepatitis C virus infection (m, months; IFN, interferon; RBV, ribavirin; SVR, sustained virologic response)

An SVR has been shown to have the following beneficial effects: (i) fibrotic regression; (ii) reduced rate of HCC; (iii) decreased rate of other complications including liver failure and liver-related death; and (iv) improved quality of life.

Progress in the Past Two Decades

There have been substantial improvements in the success of HCV treatment in the past two decades (Fig. 1).[3] Reviewing the development history of treatment for chronic HCV infection, the rates of SVR with interferon (IFN) monotherapy are only approximately 6–12%, increasing to 38–42% with the combination therapy of conventional IFN and ribavirin (RBV), and increasing to as high as 55% in major clinical trials of pegylated IFN (peg-IFN) plus RBV combination therapy,[8–11] which represents the current standard of care.

IFN-α Monotherapy

When treatment with IFN was first introduced, a six-month course of thrice-weekly injections of three million units (MU) IFN was approved. In early studies, the primary end point was a biochemical response, defined as normalization of serum alanine aminotransferase (ALT) levels.[12,13] Two meta-analyses on randomized controlled trials (RCTs) of treatment using IFN monotherapy for 3–6 months[14,15] showed normalization of serum ALT levels at the end of treatment in 51% and 45% of subjects, respectively. However, the ALT normalization rate decreased to only 22% and 21%, respectively, 3–6 months after discontinuation of therapy (or post treatment).

When virologic assays became available to detect serum HCV RNA, response rates were found to be lower than those reported with less stringent biochemical parameters. In a meta-analysis of 32 RCTs between 1986 and 1996 among patients with chronic hepatitis C receiving IFN-α-2b (at least 2 MU thrice weekly for 24 weeks), IFN compared with placebo or no treatment was evaluated in 20 trials, and different regimens of treatment were compared in 12 trials.[16] Normalization of serum ALT levels at the end of treatment was seen in 47% of the treated patients compared with 4% of controls ($P < 0.0001$); and at six months after stopping treatment in 23% of the treated patients compared with 2% of controls ($P < 0.0001$). End-of-treatment virologic responses, however, were observed in only 29% of the treated patients compared with 5% of controls ($P < 0.001$); and at six months post treatment, SVRs were documented in 8% of the treated patients compared with 1% of controls ($P < 0.001$). In addition, several studies already demonstrated lower response rates among patients infected with genotype 1, cirrhotic patients, and previous nonresponders to IFN.

Addition of Ribavirin to Conventional IFNs

Addition of the synthetic guanosine analogue RBV to IFN was a major breakthrough in the treatment of chronic HCV infection. Although RBV monotherapy was shown to be ineffective, pilot studies had suggested that combination therapy with IFN and RBV was more effective than IFN alone. For example, we conducted a small pilot RCT study and showed an SVR in 9 (43%) of the 21 patients in the IFN-α-2b and RBV group compared with 1 (6%) of the 19 patients in the IFN monotherapy group ($P = 0.017$).[17] Several landmark studies consistently demonstrated the dramatically improved responses to combination therapy, especially for patients with genotypes 2 and 3.[10,11,17,18] The studies by McHutchison *et al.* and Poynard *et al.* reinforced the need for therapy to be extended to 48 weeks in patients with genotype 1 infection.

Replacing Conventional IFNs by peg-IFNs

Peg-IFNs are produced by binding of the inert polyethylene glycol moiety to IFN molecules, thus decreasing renal clearance, altering metabolism, and increasing the half-life of the peg-IFN molecule. There are two licensed products worldwide, the 12-kd peg-IFN-α-2b (Peg-Intron®) and the 40-kd peg-IFN-α-2a (Pegasys®). Because of their prolonged half-lives, they can be administered by subcutaneous injection once weekly. In large RCTs, higher SVR rates have been achieved with the combination of weekly injections of peg-IFN-α plus RBV given twice daily than with conventional IFN-α given by injection thrice weekly together with RBV or with peg-IFN-α alone.[8,9,19] In two of these trials, peg-IFN-α-2b was dosed by weight (1.5 mcg/kg) and coupled with 800 mg of RBV; peg-IFN-α-2a was given as a fixed dose of 180 mcg along with a weight-adjusted, higher dose of RBV (1,000 mg if < 75 kg and 1,200 mg if > 75 kg).[9,10] In the third trial, 180 mcg peg-IFN-α-2a was used with either 800 mg or the higher, weight-adjusted doses of RBV.[19] Since the two peg-IFN-α compounds have not been compared head to head using similar RBV doses in the above studies, their relative efficacies cannot be assessed reliably. However, there were similar indicators of treatment response and adverse events.

Baseline Predictors of Response

The likelihood of achieving an SVR can be predicted by the pre-treatment patient's characteristics as well as viral factors including genotype and viral load, and on-treatment viral dynamics.

Hepatitis C Viral Factors

Genotype. There are six major HCV genotypes with multiple subtypes. Although genotype does not predict the natural outcome of infection, it does predict the likelihood of treatment response. Genotyping can be performed by direct sequence analysis, by reverse hybridization to genotype-specific oligonucleotide probes, or by the use of restriction fragment length polymorphism.

Chronic infection with HCV genotype 1 is associated with lower SVR rates than other genotypes.[8,9,19] For example, in patients who were treated with peg-IFN-α-2a together with RBV, the independent variables associated with an SVR included genotype non-1,[8] age less than 40 years, and body weight less than 75 kg. Using peg-IFN-α-2b and RBV, SVR rates were higher in patients who had genotype 2 or 3 HCV infection, lower pre-treatment HCV RNA level, younger age, lower body weight, and absence of bridging fibrosis and cirrhosis.[8,9,19] The majority of patients in the first 2 peg-IFN combination trials who had genotype non-1 infection were infected with genotype 2 or 3, but a small number were infected with genotypes 4, 5, and 6. In these two registration trials, in patients with genotype 1 infection, SVRs were 42% to 46%, while the response rates in those with genotype 2 or 3 were higher — 76% to 82%.

Viral load. Although HCV RNA quantification was not shown to be predictive for the degree of HCV-related liver injury or the progression of disease, a low baseline viral load (< 600,000–800,000 IU/mL) was shown to be an independent predictor of SVR.[9,20] In the study that evaluated peg-IFN-α-2a, the data were re-analyzed by combining genotype and viral load. Patients with genotype 1 infection and a high viral load ($\geq 2 \times 10^6$ copies/mL, equivalent to ≥ 800,000 IU/mL) who received the combination therapy had an SVR of 41%, whereas the rate among those with genotype 1 and a low viral load ($< 2 \times 10^6$ copies/mL) who were treated with the same regimen was 56%. In contrast, in patients with genotype 2 or 3 infection and a high viral load who were given peg-IFN-α-2a and RBV, the SVR rate was 74%, while those with genotype 2 or 3 and a low viral load who were treated similarly had an SVR of 81%.

Overall response rates to peg-IFN plus RBV according to baseline genotype and pre-treatment HCV RNA levels are shown in Table 1.

Host Factors

Fibrosis and cirrhosis. Although patients with advanced fibrosis or well-compensated cirrhosis generally have lower response rates, they can be successfully treated and may achieve SVRs.[21] In a trial of 1,311 patients with advanced liver disease, 63% of all patients and 52% of patients with genotype 1 treated with peg-IFN plus RBV achieved SVRs.[19] However, the success of antiviral therapy diminishes in the face of decompensated cirrhotic disease owing to the severity of adverse effects in these severely ill patients.[23] In addition, in a subgroup analysis of the HALT-C (Hepatitis C Antiviral Long-term Treatment Against Cirrhosis) study conducted by Everson *et al.* the presence of advanced fibrosis and cirrhosis was a major independent predictor of non-response to antiviral therapy. This study compared four groups of patients with increasingly severe liver disease, as determined by Ishak scores and platelet counts. SVR rates decreased from 23% to 9% as Ishak scores increased and platelet counts decreased ($P < 0.0001$), confirming the effect of advanced liver disease in the responses to antiviral therapy.[24]

Ethnicity. HCV kinetics and drug pharmacokinetics are influenced by ethnic characteristics.[25] For example, HCV-infected African American patients are less likely to be responsive to antiviral therapy than the non-Hispanic white population.[26,27] In 401 patients infected with HCV genotype 1, peg-IFN-α-2a and RBV produced SVR rates of 28% for African American patients and 52% for white patients ($P < 0.0001$).[28] Importantly, rates of serious adverse events, dose reductions and discontinuations were similar between two populations, suggesting unidentified mechanisms for the observed efficacy difference. One possible explanation is a high prevalence of HCV genotype 1 among African American patients (88–96%),[27,29,30] but this cannot convincingly explain the observation. Layden–Almer *et al.*[29] noted that compared with white patients, African Americans infected with genotype 1 exhibited significantly lower decreases in first-phase viral RNA, slower elimination of infected cells, and smaller declines in mean viral RNA over one month. This suggests that African Americans may have a genetically impaired ability to block viral production.

Similarly, the rate of SVR achieved with peg-IFN-α plus RBV therapy in Taiwanese patients infected with HCV genotype 1 also appears to be higher than that reported in Caucasian patients.[31–33] The better SVR rates in Taiwanese patients may be due to a lower pre-treatment HCV RNA level and a lower mean body weight in Taiwanese patients as compared with the Caucasian patients. Thus the Taiwanese patients received relatively higher average doses of peg-IFN and RBV on body weight basis, which may contribute to the better virologic response. Besides, Taiwanese patients have been shown to have higher rates of rapid and early virologic response, which may also account for the high HCV SVR rates.[34]

Regarding this unresolved issue, a relatively new methodology in genetics, the genome-wide association study (GWAS), has just been applied successfully for the studies of the pharmacogenomics of chronic hepatitis C.[35–37] Ge *et al.* analyzed 1,137 patients with HCV infection who were enrolled in a clinical trial to test the relative effectiveness of two commercial interferon preparations.[35] They identified several single nucleotide polymorphisms (SNPs) near the IL28B gene on chromosome 19 that were significantly more common in patients that responded to interferon therapy than in non-responders. Two independent studies, one by Suppiah *et al.*[36] and the other by Tanaka *et al.*[37] report similar findings. These studies partly explained the mechanisms behind a lack of response in African Americans and a beneficial response in Asian populations. Most importantly, these findings raise the prospect of a personalized approach to treating HCV by tailoring treatment to patients who are most likely to benefit.

Age. In all large prospective studies of peg-IFN and RBV combination therapy, younger age correlated significantly with an SVR when assessed by univariate and multivariate analyses. Patients younger than 40–45 years showed the best response rates.

In the elderly population (age > 65 years), immunological suppression, chronic co-morbidity, and concurrent medications adversely affect response and heighten the probability of adverse reactions to antiviral therapy. However, research focusing on the efficacy and safety of antiviral therapy in the older population is limited to a few small studies.[38] In a retrospective cohort study of 84 elderly patients age ≥ 65 years without genotype 1b or high viral load, 30 patients (36%) receiving IFN monotherapy achieved SVRs.[38] The likelihood of SVR was significantly lower among those with high baseline viremia (HCV RNA >100 KIU/mL; $P < 0.0001$) and advanced liver fibrosis ($P = 0.04$). Eleven patients (13%) withdrew because of adverse events, and univariate analysis showed a higher likelihood of withdrawal owing to adverse events among those age > 70 years than those age ≤ 70 ($P < 0.009$).

Obesity, Hepatic Steatosis, and Insulin Resistance

About 20–37% of HCV-infected patients are obese, a potential barrier to treatment success.[39] The body mass index inversely correlates with SVRs.[39] One explanation for the interaction between obesity and antiviral therapy response[40] focuses on hepatic steatosis, because obesity is an independent risk factor for non-alcoholic fatty liver disease.[39,41,42] Other explanations include an obesity-triggered inflammatory reaction that decreases response and impairs IFN absorption owing to high levels of subcutaneous fat.[40,43] Regardless of the cause of lower response rates in obese HCV-infected patients,[43,44] weight loss is an important component of treatment strategy, as it may lower elevated liver enzymes and improve liver fibrosis.[45] Weight-based dosing of antiviral medications also becomes an important consideration in patients with high body mass indices.[46]

Steatosis is common in patients with HCV infection, but the pathogenic mechanisms differ. The frequency of significant steatosis in chronic hepatitis C ranges between 25% and 80% depending on additional risk factors of fatty liver disease. Steatosis in HCV-related disease relates to both viral factors (HCV genotype 3) and host factors (alcohol consumption, overweight, hyperlipidemia, diabetes mellitus). In HCV-related disease, steatosis impacts both fibrosis progression and response to treatment.

Insulin resistance (IR) is extremely common and frequently associated with co-morbid conditions such as cardiovascular disease, hypertension, and obesity. In addition, IR is the driving force for type 2 diabetes mellitus, and is crucial for the development of non-alcoholic fatty liver disease. Interestingly, co-existence of IR and chronic hepatitis C occurs more often than predicted by chance, with recent estimates indicating that 30–70% of patients with chronic hepatitis C display some evidence of IR. Recent research has revealed that several molecules — including tumor necrosis factor alfa, suppressor of cytokine signaling 1 and 3 proteins, insulin-receptor substrates 1 and 2, and other adipocytokines — are potentially involved in the development of IR in patients with chronic hepatitis C.[44,47] Clinically, baseline IR has a negative impact on treatment outcomes in patients with chronic hepatitis C.[48,49] Management of insulin resistance or diabetes mellitus in these patients thus might improve likelihood of successful outcomes with antiviral therapy. Likewise, eradication of HCV in patients with IR or diabetes mellitus might improve glucose metabolism.[50] Although adjunctive therapies such as insulin sensitizers and weight loss are often recommended, their ability to improve antiviral treatment response in patients with chronic hepatitis C is still under active investigation.

On-Treatment Predictors

The virologic response has become an important parameter to monitor during treatment of chronic hepatitis C. Change in serum HCV load after 4 and 12 weeks of therapy predicts the likelihood of SVR. Patients who test negative for HCV RNA (< 50 IU/mL) at four weeks are defined as having a rapid virologic response (RVR). Early virologic response (EVR) is defined as undetectable HCV RNA or at least a 2-log decrease in serum HCV RNA from baseline level after 12 weeks of therapy.

EVR

In the study of peg-IFN-α-2a with RBV, the predictability of an SVR based on the EVR was assessed. Overall, 65% of patients with an EVR subsequently achieved an SVR. Conversely, among those who did not have an EVR, 97% failed to develop an SVR.[8] Similar data were noted in the study that used peg-IFN-α-2b together with RBV.[51] Among treated persons who had an EVR, 72% ultimately achieved an SVR; of those who did not have an EVR, none developed an SVR. Week-12 results form the basis of "stopping rules" during hepatitis C therapy. Monitoring of patients is thus important throughout therapy as it allows identification of patients with inadequate treatment response who may avoid further treatment-related adverse effects and might benefit from early introduction of alternative therapies.

RVR

An RVR is increasingly recognized as one of the most important independent predictors of SVR. In a recent retrospective analysis of 1,383 patients, it was shown that achieving RVR correlates with a high probability (86–100%) of SVR to peg-IFN/RBV combination therapy, regardless of genotype.[52] In genotype 1-infected patients, treatment shortening (24 instead of 48 weeks) is feasible if a low (< 600,000–800,000 IU/mL) baseline viral load and an RVR are present. SVR rates were > 75% in these patients as reported in a number of studies involving both peg-IFN-α-2a- and peg-IFN-α-2b-based therapy.

In genotype 2/3-infected patients, similar rules can be applied. Several clinical trials have pointed out the possibility of shortening treatment duration to 12–16 weeks instead of 24 weeks following a RVR.[53–55] However, data from a recently published larger trial suggested that treatment shortening to 16 weeks should be considered for patients with an RVR and low baseline viral load (< 800,000 IU/mL) only.[20]

Current Recommendations

The optimal treatment duration and ribavirin dose were ever investigated in a RCT in which all patients received peg-IFN-α-2a at a dose of 180 mcg while patients in the four arms received either 24 or 48 weeks of RBV at doses of either 800 mg or the higher, weight-based dose of 1,000 or 1,200 mg daily.[8] Data were analyzed taking into account not only the HCV genotype, but also the pre-treatment viral load ($<$ or $\geq 2 \times 10^6$ copies/mL) of those with genotype 1. In patients with genotype 1 and low-level viremia, the SVR was highest in those who had received the higher RBV dose and who were treated for 48 weeks (61%). This regimen was also optimal for patients with genotype 1 and a high viral load: 46% achieved an SVR. In contrast, in patients with genotype 2 or 3, regardless of the pre-treatment viral load, no differences were detected among the four treatment arms, suggesting that peg-IFN-α-2a plus RBV at a dose of 800 mg given for 24 weeks is adequate. Based on the data from this registration trial and recent data regarding on-treatment viral kinetics, the following recommendations are made.

In chronic HCV genotype 1 infection, treatment with peg-IFN and ribavirin for 48 weeks is recommended. In patients who achieve an RVR at week 4 and have pre-treatment low-level viremia, treatment can be discontinued after 24 weeks. In patients who achieve an EVR at week 12, treatment should be continued up to 48 weeks. Treatment may be stopped in patients who do not achieve an EVR at week 12.

In chronic HCV genotype 2 or 3 infection, treatment with either conventional IFN-α or peg-IFN-α plus ribavirin for 24 weeks is recommended. There is some evidence that shortening duration of therapy to 16 weeks in HCV genotype 2-infected patients with RVR and pre-treatment low-level viremia provides equal SVR to 24 weeks treatment.

Treatment of HCV infection in special populations

Patients with Acute HCV Infection

The acute phase of HCV infection represents a key point in the evolution of hepatitis C. However, because acute hepatitis C is often asymptomatic, detection and diagnosis are usually difficult. Furthermore, there are no established treatment guidelines, leaving physicians to make several challenging decisions, such as whether to treat, when to treat, and what treatment regimen to use. Treatment of acute hepatitis C patients with conventional IFN-α has been investigated in several small clinical trials, which typically used 3–6 MU IFN-α administered thrice weekly for 4–24 weeks. Overall, 32–52% of patients treated with IFN-α attained an SVR compared with only 4–11% of untreated patients.[15,56]

Different strategies have been explored to optimize SVR rates in patients with acute hepatitis C treated with peg-IFN-α. When considering treatment initiation, several time points have been evaluated, ranging from immediately after diagnosis to after an observation period of several weeks. Data from trials using immediate or delayed treatment strategies (8–12 weeks) have demonstrated high SVR rates ranging from 71% to 94%.[57,58] With its once-weekly administration schedule, peg-IFN-α may ultimately become the standard of care in acute hepatitis C treatment. Unfortunately, available data lack the robust quality required to develop standardized treatment recommendations.

At present, there is no evidence that the addition of ribavirin to peg-IFN improves response rates in patients with acute hepatitis C. The role of ribavirin has yet to be established.

Patients with Normal Serum ALT Level

Approximately 30% of patients with chronic HCV have persistently normal serum ALT levels.[59] Several studies have shown that patients with persistently normal serum ALT levels demonstrate less evidence of histologic progression than patients with elevated ALT levels, as shown by serial liver biopsy specimens, during intervals of up to five years.[60,61] Therefore, some clinicians choose to monitor such patients without therapy.

Among patients with normal ALT levels monitored for five years, levels of ALT become elevated in approximately one-fifth to one-third, reinforcing the importance of close monitoring in this population. Besides, up to 20% of patients with normal ALT levels have bridging fibrosis or cirrhosis.[62] Therefore, laboratory evaluation alone without liver biopsy cannot reliably differentiate between those with mild disease and those with more advanced fibrosis. Patients with proven minimal or no fibrosis on liver biopsy may be reassured about their favorable prognosis, and they may choose to defer therapy; those with more than stage I fibrosis should be advised to consider treatment.

In early trials of IFN monotherapy among patients with normal ALT levels, response rates were low, and elevation of ALT levels during therapy was a concern. Later on, the rate of SVR with IFN/ribavirin or peg-IFN/ribavirin combination therapy in patients with persistently normal ALT levels was found to be similar to those with elevated serum ALT levels, and elevations of ALT levels during combination therapy are rare.[63,64]

Given the borderline indication for therapy in this group and the factors weighing for and against treatment, clinicians may rely in their decision-making on individual patients' features, including motivation and perspective, genotype, relative histologic activity and fibrosis, duration of HCV infection, age, occupation, symptoms, and so on. As therapy becomes more effective and better tolerated, the threshold for treatment will be lowered. For patients with genotype 2 or 3 infection, SVR rate with peg-IFN-α plus ribavirin therapy is high; these patients may elect to undergo therapy regardless of the stage of disease and thus obviate the need for liver biopsy.

Patients with no or Minimal Fibrosis

Patients with grade 1 inflammation and minimal/no fibrosis are at low risk for developing advanced liver disease.[65] Liver biopsy may be repeated in three to five years if results would change management. Treatment should be reconsidered if the liver disease has progressed. Treatment should also be provided to patients, despite minimal fibrosis, who desire treatment and/or who have significant symptoms. Patients with extra-hepatic manifestations of HCV infection should be considered for HCV therapy, regardless of the severity of their liver disease. IFN/RBV or peg-IFN/RBV combination therapy has been shown to achieve response rates comparable to those seen in patients with histologically advanced disease.[66,67] Given the borderline indication for therapy in this group and the factors weighing for and against treatment, clinicians may rely in their decision-making on individual patients' features. Particularly, in patients with genotype 2 or 3 infection, the high SVR may shift the risk-benefit ratio in favor of treatment even for those with minimal histologic disease.

Patients with Hematologic Disorders

Patients with thalassemia or other hemoglobinopathies and patients with hemophilia or other inherited coagulation disorders were at greater risk of acquiring HCV infection before the introduction of blood donor screening for HCV, as a consequence of repeated transfusions of blood components and clotting factors, respectively. Because of their hematologic disorders, however, they were usually excluded from registration trials of anti-HCV therapy. Thus their response to contemporary peg-IFN/ribavirin combination therapy has not been reported.

Thalassemia. For thalassemic patients with chronic hepatitis C, anemia represents a potential contraindication for antiviral regimens that include ribavirin. A concern in this population is the exacerbation of anemia by the dose-dependent ribavirin-induced hemolytic anemia. Moreover, many thalassemic patients have concomitant severe hemosiderosis and iron storage liver disease. Treatment for chronic hepatitis C in this population has been studied in only few small trials. Rates of SVR after IFN monotherapy were comparable to those achieved in non-thalassemic patients.[68,69] After IFN/ribavirin combination therapy, an SVR rate as high as 72% was obtained in 18 thalassemic patients. Therapy was complicated by a substantial increase in transfusion requirements during therapy. Therefore, although this chronically anemic, iron-overloaded subpopulation can be treated effectively, the risk of adverse effects is higher than in those without thalassemia. For thalassemic patients with substantial hemosiderosis, initial therapy, before considerations of antiviral therapy, may be focused on reduction of iron load.

Hemophilia. Rates of SVR among hemophiliac patients treated with IFN monotherapy or IFN/ribavirin combination therapy have been reported in a limited number of small clinical trials. Response rates tend to be similar to or appreciably lower than those in the non-hemophiliac population;[70,71] approximately 30% have been reported to respond to one year of combination therapy. For example, in the largest trial, 113 patients with inherited coagulation disorders were treated with IFN-α-2b (3 MU thrice weekly) plus ribavirin (1,000 mg/day) or IFN monotherapy. The SVR rate in the combination therapy arm was 29% compared with 7% in the monotherapy arm.[70]

In general, these patients should receive care similar to that recommended for other HCV-infected patients. Although data were not available on the use of peg-IFN and ribavirin in hemophiliac patients, the safety and efficacy profiles are likely to be similar to those of the general patients with hepatitis C. The duration of therapy could be guided by genotype, and liver biopsies can be performed safely by experienced hepatologists working in conjunction with hematologists.

Patients Infected with HIV

An estimated one million individuals are human immunodeficiency virus (HIV) positive in the United States alone, and as many as 300,000 are possibly coinfected with HCV. Coinfection is associated with substantial morbidity and mortality, including end-stage liver disease.[72] Furthermore, HIV coinfection complicates anti-HCV therapy. Common adverse effects of IFN and RBV therapy, such as depression and anemia, are often amplified in patients also receiving antiretroviral therapy,[73] and the risk of drug–drug interactions is substantial. HCV clearance may be slower and relapse is common in patients coinfected with HIV.[74–76] In a study of 89 HIV/HCV-infected patients by Soriano *et al.*[76] 29 (32.6%) achieved SVRs with peg-IFN-α-2b and RBV. In another study,[72] 133 patients infected with HCV and HIV were randomly assigned to receive either peg-IFN-α-2a or standard IFN-α-2a with RBV. Although the group receiving the peg-IFN regimen showed a higher SVR rate (27% vs. 12%, $P = 0.03$), the rate was still lower than that reported in patients infected with HCV alone.[8,9,75]

Patients Dually Infected with HBV

Dual infection with hepatitis B virus (HBV) and HCV is not uncommon, especially in areas with a high endemic level of either viral infection, with a prevalence of around 7–11% in patients with chronic HBV infection and 2–10% in chronic hepatitis C patients.[77] Patients with dual HBV and HCV infection are shown to have more severe liver disease and an increased risk of hepatocellular carcinoma (HCC) in the majority of studies.[77]

No standard of care has been established for patients who are dually infected with HBV and HCV. A previous study revealed that active hepatitis C can be found in around 70% of these patients. Small studies showed that IFN alone was not effective in clearing HCV or HBV in dually infected patients. Combining IFN with RBV improved HCV response rates in patients mono-infected with HCV as well as in dually infected patients.[78] Nevertheless, the HCV SVR rates were significantly lower in those dually infected with HCV genotype 1 compared with those with genotype 2/3 infection (44% vs. 85%, $P < 0.05$). The efficacy of treatment of dually infected patients may be enhanced through peg-IFN plus RBV. A recent large multi-center clinical trial in Taiwan found that for genotype 1 infection, HCV SVR was 72.2% in dually infected patients vs. 77.3% in mono-infected patients by using peg-IFN-α-2a plus ribavirin combination therapy.[31] For genotype 2/3 infections, SVR was 82.8% and 84.0%, respectively. Combination therapy of peg-IFN-α-2a and ribavirin appears to be just as effective for the treatment of HBsAg-positive patients chronically infected with active hepatitis C as it is in patients with HCV mono-infection. The treatment recommendations regarding therapy duration and ribavirin dosage according to genotype in HCV mono-infection appear to be applicable also for this patient group.

Virologic response of HBV to peg-IFN and the potential for reappearance of HBV DNA during treatment of co-existing chronic hepatitis C are two major clinical issues that needed to be addressed in dually infected patients receiving anti-hepatitis C therapy. In our study, HBV virologic response was obtained in 56% of the dually infected patients with pre-treatment HBV viremia. Post-treatment HBsAg clearance was found in 18 (11.2%) of the 161 dually infected patients. In contrast, spontaneous HBsAg seroclearance in chronic HBV infection occurs at an annual incidence of 1–2%.[79] On the other hand, 28 (36%) of the 77 dually infected patients whose pre-treatment serum HBV DNA was undetectable experienced a reappearance of HBV DNA post treatment. Thus, long-term follow-up is needed to clarify the clinical significance of HBV DNA reappearance in patients with dual chronic HCV/HBV infection.

While this study shows that peg-IFN-α-2a plus ribavirin can be used for the treatment of active hepatitis C in patients chronically infected with HCV/HBV, it remains to be determined whether this approach is also suitable for dually infected patients with a dominant HBV infection who are HBeAg positive and are likely to have a higher hepatitis B viral load.

Patients with End-Stage Renal Disease

Chronic HCV infection is common in dialysis patients, with the prevalence ranging from 3% to 80% worldwide.[80] However, the natural course of chronic HCV infection was only partly clarified in this special clinical setting. Consistently, previous studies revealed that HCV infection is an independent risk factor for death among dialysis patients and transplant recipients. Furthermore, recurrence of HCV-associated kidney disease can adversely affect graft survival.

Available data support treating patients who have chronic hepatitis C and are awaiting kidney transplantation, with resultant improvement of both renal function and renal histology. The long-term virologic response pre-transplant can be maintained after kidney transplantation if successful eradication of HCV by IFN-based therapy is achieved in the dialysis stage.[81] On the other hand, routine antiviral therapy for patients after kidney transplantation is not recommended because of the risk of graft rejection, and the low response rate owing to concomitant use of immunosuppressive agents.

Briefly, IFN-based monotherapy is the treatment of choice in HCV-positive dialysis patients awaiting kidney transplantation. Two meta-analyses evaluating the efficacy and safety of standard IFN monotherapy showed that the SVR rates were 37% and 33%, respectively; and the corresponding dropout rates were 17% and 29.6%, respectively.[81,82] The SVR and dropout rates were higher than those with normal renal function, possibly due to a lower IFN clearance rate. With regard to the efficacy and safety of peg-IFN-α-2a and -2b in treating dialysis patients, a more favorable outcome of patients treated with peg-IFN-α-2a (135–180 mcg/week: SVR 50–75%, good tolerance) than those treated with peg-IFN-α-2b (0.5–1.0 mcg/week: SVR 12.5%, poor tolerance) was found,[83,84] which may result from different pharmacokinetic profiles between these two peg-IFNs.

Recently, we conducted a head-to-head comparative study on the efficacy and safety of using peg-IFN vs. standard IFN in this clinical setting.[85] Overall, 50 patients were randomly assigned to receive either peg-IFN-α-2a 135 mcg weekly or standard IFN-α-2a at 3 MU thrice weekly for 24 weeks. We found that patients receiving peg-IFN-α-2a tended to have a higher SVR than those receiving standard IFN-α-2a (48% vs. 20%, $P = 0.07$). A pre-treatment serum HCV RNA level of < 800,000 IU/mL ($P = 0.007$) was independently predictive of an SVR. All patients failing to achieve a rapid RVR could not achieve an SVR. Interestingly, patients receiving peg-IFN-α-2a had a significantly lower treatment-related withdrawal rate than those receiving standard IFN-α-2a (0% vs. 20%, $P = 0.04$). These data suggested that peg-IFN-α-2a once weekly at a lower dose may provide a more effective and safer therapy than standard IFN for the treatment of naive dialysis patients with chronic hepatitis C.

RBV, which has been used in combination with IFN to treat chronic hepatitis C in general patients, is considered contraindicated in dialysis patients with chronic hepatitis C due to the risk of severe hemolytic anemia. However, the notion has been challenged recently. Some pilot studies evaluating combined standard IFN-α plus low-dose RBV (170–300 mg/day) showed SVR rates of 17–66% after 24–48 weeks of treatment. In addition, a recent study including six patients with a combination of peg-IFN-α plus low-dose RBV also showed an SVR rate of 50%.[86] Nevertheless, more prospective studies are needed to confirm the safety and the optimal dosage of RBV in patients with end-stage renal disease.

Patients Receiving Liver Transplantation

Graft reinfection in liver transplantation resulting from chronic HCV infection is common, as the virus seeded into the bloodstream can come back to the new graft.[87] Orthotopic liver transplantation affects response to HCV therapies, with SVR rates of 20–30%,[88–90] which are lower than those achieved in non-transplantation patients. A meta-analysis of 48 studies examined the safety and efficacy of both standard IFN and peg-IFN in liver transplant recipients.[89] The overall SVR rates were 24% with IFN plus RBV and 27% with peg-IFN plus RBV; discontinuation rates were 24% and 26%, respectively, and pooled rates of graft rejection were 2% and 5%, respectively. The efficacy and the rates of discontinuation and graft rejection were comparable between the two combination regimens.

Adverse effects significantly compromise outcomes. Adverse effects can prompt premature treatment discontinuation in up to 50% of patients.[87,92] Severe neutropenia and the corresponding increased risk of infection and hemolytic anemia are primary treatment-limiting toxicities associated with IFN plus RBV.[87]

Relapsers or Non-Responders to Previous Anti-HCV Therapy

Retreatment with the same agents but different strategies. The patients who do not respond to or relapse after prior therapy should be carefully evaluated before starting subsequent IFN-based retreatment. Although some factors that influence SVR rates, such as inappropriate management of adverse effects and early dose reductions, may be corrected, others, such as HCV genotype and ethnicity, cannot be specifically modified to improve treatment outcomes. Besides, ongoing alcohol and substance abuse has also been shown to contribute to the failure of IFN-based treatment.[93,94] Identifying the factors for poor response can aid clinicians in predicting response and making the retreatment decision. The goals of retreatment can be eradication of HCV or slowdown of fibrosis progression.

Recently, a large trial showed that retreatment of non-responders or relapsers with peg-IFN-α-2b/RBV is associated with an overall SVR rate of 22%. It is to be noted that 56% of the patients with an EVR achieved an SVR.[95]

The efficacy of high-dose peg-IFN was evaluated in a large trial, in which 704 non-responders to IFN plus RBV therapy were randomized to receive peg-IFN-α-2b at 3.0 or 1.5 mcg/kg/week plus RBV at 12–15 mg/kg/day.[96] SVRs were achieved in 17% of the 3.0 mcg/kg group vs. 12% of the 1.5 mcg/kg group ($P = 0.03$).

Intensified treatment with higher fixed-dose induction of peg-IFN and/or longer treatment duration may increase SVR rates in patients with prior non-response to peg-IFN/RBV treatment. This trial compared both strategies in prior non-responders to peg-IFN-α-2a/RBV.[97] Patients (n = 942) were randomized into four arms: Patients in arms A and B received peg-IFN-α-2a induction with 360 mcg/week for 12 weeks followed by peg-IFN at 180 mcg/week for further 60 or 36 weeks, patients in arms C and D received peg-IFN-α-2a at 180 mcg/week for 72 and 48 weeks, respectively. All patients were treated with RBV 1,000–1,200 mg/day. Overall, the SVR rate was 16%, 7%, 14%, and 9% in arms A, B, C, and D, respectively. The SVR rate was higher in patients receiving 72-week treatment compared with 48-week treatment. There was no difference between the induction and the non-induction arms.

Maintenance therapy with a lower dosage of IFN/peg-IFN. Preventing complications by long-term antiviral therapy is also a consideration in the management of refractory HCV infection. Small studies showed that long-term treatment strategy can maintain biochemical and virologic responses and prevent liver histologic progression in patients with an initial partial response to IFN therapy.[98,99] In one maintenance study, 53 patients who had normalized ALT values but positive serum HCV RNA after six months of IFN-α-2b at five MU thrice weekly were randomly assigned to continue a reduced dose of IFN at three MU thrice weekly or stop therapy.[99] During the initial antiviral treatment, significant reductions in serum ALT levels, viral load, and hepatic inflammation were noted ($P < 0.05$ vs. baseline). These improvements were sustained in patients receiving maintenance therapy. In contrast, serum ALT, HCV RNA levels, and hepatic inflammation returned to baseline values after IFN withdrawal, and significant increases in the mean hepatic inflammatory scores ($P = 0.004$ vs. maintenance group) and worsening hepatic histology were seen in patients not receiving long-term therapy ($P < 0.01$ vs. maintenance group).[99]

Two large trials assessed the efficacy of peg-IFN alone or with RBV as maintenance therapy for refractory chronic hepatitis C. The first trial evaluated the efficacy and safety of a low-dose peg-IFN-α-2a at 90 mcg/week for 3.5 years in chronic hepatitis C patients with bridging fibrosis or cirrhosis and persistent viremia despite previous IFN therapy.[100] Unfortunately, the results of such trials demonstrated that long-term IFN therapy did not improve the clinical and histologic outcomes in difficult-to-treat patients with chronic hepatitis C. In the other trial, 555 patients with prior non-response to IFN-based antiviral therapy were randomized to four-year peg-IFN-α-2b (0.5 mcg/kg/wk) (n = 286) or colchicine (0.6 mg twice daily) (n = 269). Again, long-term maintenance therapy does not improve clinical outcomes of patients with chronic hepatitis C.[101]

Promising Chemicals/Drugs Under Investigation

Since the introduction of peg-IFN and RBV combination therapy, there is an increase in rates of SVR. Nevertheless, there are several notable disadvantages, including significant adverse effects and limited efficacy against HCV genotypes 1 and 4. Progress in the development of selective therapy has been slowed because of the lack of cell culture systems for HCV. Fortunately, in a few years, advances in the development of *in vitro* cell culture systems have led to novel therapeutic approaches against HCV.

Drugs in development include small molecules such as the protease inhibitor (VX-950, SCH 503034), polymerase inhibitor

(NM-283), and toll-like receptor drug classes (CPG 10101).[102–104] For example, another promising strategy is the modification of IFN molecule (albuferon) with improved half-life leading to longer dosing intervals and increased response rates.[105]

While many of these drugs seem to hold promise as either a primary or an adjunctive treatment for patients with chronic hepatitis C, they are years from the market and their safety and efficacy are uncertain in difficult-to-treat patients.[102] In the meantime, the IFNs will continue to form the backbone of HCV therapy for initial therapy, retreatment, and maintenance therapy.

Perspectives and Conclusions

Since the introduction of IFN/peg-IFN plus RBV combination therapy, chronic HCV infection can be cured in more than half of the treated population. Nevertheless, multiple host and viral factors may negatively affect treatment outcomes and complicate management of patients with hepatitis C. Unfortunately, because most of these factors cannot be modified, a significant number of patients will not respond to antiviral therapy or will relapse following treatment withdrawal. Recognition of these viral, hepatic, and demographic factors that foster treatment resistance in HCV infection, as well as early viral kinetics (e.g., measuring viral load early in the treatment course), may help to predict poor response and determine the need for treatment modifications.

Therapeutic options for treatment-resistant patients are limited, and retreatment with peg-IFN plus RBV can produce SVR in only a proportion of complicated HCV infections (~15–20%). Additionally, preliminary findings suggest limited benefits in maintenance peg-IFN therapy in patients with persistent hepatitis C viremia after scheduled antiviral therapy. Although potentially beneficial, the cost, safety, and risk of non-compliance will limit this approach. Finally, the potential effects of adding new anti-HCV agents or using insulin sensitizers upon current standard-of-care combination therapy in difficult-to-treat populations are now under active investigation.

Acknowledgments

The study was supported by grants from the Department of Health, the National Science Council, Executive Yuan, Taiwan; National Taiwan University; and National Health Research Institutes, Taiwan.

References

1. Lauer GM, Walker BD. (2001) Hepatitis C virus infection. *N Engl J Med* **345**: 41–52.
2. Kim WR. (2002) The burden of hepatitis C in the United States. *Hepatology* **36**(Suppl 1): S30–S34
3. Strader DB, Wright T, Thomas DL, Seeff LB, American Association for the Study of Liver Diseases. (2004) Diagnosis, management, and treatment of hepatitis C. *Hepatology* **39**: 1147–1171.
4. Seeff LB, Hoofnagle JH. (2002) National Institutes of Health Consensus Development Conference: management of hepatitis C: 2002. *Hepatology* **36**(Suppl 1): S1–S2.
5. Fattovich G, Giustina G, Degos F, *et al.* (1997) Effectiveness of interferon alfa on incidence of hepatocellular carcinoma and decompensation in cirrhosis type C. European Concerted Action on Viral Hepatitis (EUROHEP). *J Hepatol* **27**: 201–205.
6. Liang TJ, Rehermann B, Seeff LB, Hoofnagle JH. (2000) Pathogenesis, natural history, treatment, and prevention of hepatitis C. *Ann Intern Med* **132**: 296–305.
7. Shiffman ML. (2006) Chronic hepatitis C: treatment of pegylated interferon/ribavirin nonresponders. *Curr Gastroenterol Rep* **8**: 46–52.
8. Fried MW, Shiffman ML, Reddy KR, *et al.* (2002) Peginterferon-α-2a plus ribavirin for chronic hepatitis C virus infection. *N Engl J Med* **347**: 975–982.
9. Manns MP, McHutchison JG, the International Hepatitis Interventional Therapy Group, *et al.* (2001) Peginterferon-α-2b plus ribavirin compared with interferon-α-2b plus ribavirin for initial treatment of chronic hepatitis C: a randomised trial. *Lancet* **358**: 958–965.
10. McHutchison JG, Gordon SC, for the Hepatitis Interventional Therapy Group, *et al.* (1998) Interferon-α-2b alone or in combination with ribavirin as initial treatment for chronic hepatitis C. *N Engl J Med* **339**: 1485–1492.
11. Poynard T, Bedossa P, Chevallier M, *et al.* (1995) A comparison of three interferon-α-2b regimens for the long-term treatment of chronic non-A, non-B hepatitis. Multicenter Study Group. *N Engl J Med* **332**: 1457–1462.
12. Davis GL, Balart LA, Schiff ER, *et al.* (1989) Treatment of chronic hepatitis C with recombinant interferon-α: a multicenter randomized, controlled trial. *N Engl J Med* **321**: 1501–1506.

13. Di Bisceglie AM, Martin P, Kassianides C, *et al.* (1989) Recombinant interferon-α therapy for chronic hepatitis C: a randomized, double-blind, placebo-controlled trial. *N Engl J Med* **321**: 1506–1510.
14. Niederau C, Heintges T, Haussinger D. (1996) Treatment of chronic hepatitis C with a-interferon: an analysis of the literature. *Hepatogastroenterology* **43**: 1544–1556.
15. Poynard T, Leroy V, Cohard M, *et al.* (1996) Meta-analysis of interferon randomized trials in the treatment of viral hepatitis C: effects of dose and duration. *Hepatology* **24**: 778–789.
16. Carithers RLJ, Emerson SS. (1997) Therapy of hepatitis C: meta-analysis of interferon-α-2b trials. *Hepatology* **26**(Suppl 1): 83S–88S.
17. Lai MY, Kao JH, Yang PM, *et al.* (1996) Long-term efficacy of ribavirin plus interferon-α in the treatment of chronic hepatitis C. *Gastroenterology* **111**: 1307–1312.
18. Poynard T, Marcellin P, Lee SS, *et al.* (1998) Randomized trial of interferon-α-2b plus ribavirin for 48 weeks or for 24 weeks versus interferon-α-2b plus placebo for 48 weeks for treatment of chronic infection with hepatitis C virus. *Lancet* **352**:1426–1432.
19. Hadziyannis SJ, Sette H Jr, Morgan TR, *et al.* (2004) Peginterferon-α-2a and ribavirin combination therapy in chronic hepatitis C: a randomized study of treatment duration and ribavirin dose. *Ann Intern Med* **140**: 346–355.
20. Shiffman ML, Suter F, Bacon BR, *et al.* (2007) Peginterferon-α-2a and ribavirin for 16 or 24 weeks in HCV genotype 2 or 3. *N Engl J Med* **357**: 124–134.
21. Tsubota A, Chayama K, Ikeda K, *et al.* (1994) Factors predictive of response to interferon-alpha therapy in hepatitis C virus infection. *Hepatology* **19**: 1088–1094.
23. Crippin JS, McCashland T, Terrault N, Sheiner P, Charlton MR. (2002) A pilot study of the tolerability and efficacy of antiviral therapy in hepatitis C virus-infected patients awaiting liver transplantation. *Liver Transpl* **8**: 350–355.
24. Everson GT, Hoefs JC, Seeff LB, *et al.* (2006) Impact of disease severity on outcome of antiviral therapy for chronic hepatitis C: lessons from the HALT-C trial. *Hepatology* **44**: 1675–1684.
26. Muir AJ, Bornstein JD, Killenberg PG. (2004) Peginterferon-α-2b and ribavirin for the treatment of chronic hepatitis C in blacks and non-Hispanic whites. *N Engl J Med* **350**: 2265–2271.
25. Layden-Almer JE, Layden TJ. (2003) Viral kinetics in hepatitis C virus: special patient populations. *Semin Liver Dis* **23**(Suppl 1): 29–33.
27. Reddy KR, Hoofnagle JH, for the Consensus Interferon Study Group, *et al.* (1999) Racial differences in responses to therapy with interferon in chronic hepatitis C. *Hepatology* **30**: 787–793.
28. Conjeevaram HS, Fried MW, Jeffers LJ, *et al.* (2006) Peginterferon and ribavirin treatment in African American and Caucasian American patients with hepatitis C genotype 1. *Gastroenterology* **131**: 470–477.

29. Layden-Almer JE, Ribeiro RM, Wiley T, Perelson AS, Layden TJ. (2003) Viral dynamics and response differences in HCV-infected African American and white patients treated with IFN and ribavirin. *Hepatology* **37**: 1343–1350.
30. Blatt LM, Mutchnick MG, Tong MJ, *et al.* (2000) Assessment of hepatitis C virus RNA and genotype from 6807 patients with chronic hepatitis C in the United States. *J Viral Hepat* 7: 196–202.
31. Liu CJ, Chuang WL, Lee CM, *et al.* (2009) Peginterferon-α-2a plus ribavirin for the treatment of dual chronic infection with hepatitis B and C viruses. *Gastroenterology* **136**: 496–504.
32. Lee SD, Yu ML, Cheng PN, *et al.* (2005) Comparison of a 6-month course peginterferon-α-2b plus ribavirin and interferon-α-2b plus ribavirin in treating Chinese patients with chronic hepatitis C in Taiwan. *J Viral Hepatit* **12**; 283–291.
33. Yu ML, Dai CY, Huang JF, *et al.* (2008) Rapid virological response and treatment duration for chronic hepatitis C genotype 1 patients: a randomized trial. *Hepatology* **47**: 1884–1893.
34. Hsu CS, Liu CJ, Lai MY, *et al.* (2007) Early viral kinetics during treatment of chronic hepatitis C virus infection with pegylated interferon alpha plus ribavirin in Taiwan. *Intervirology* **50**: 310–315.
35. Ge D, Fellay J, Thompson AJ, *et al.* (2009) Genetic variation in IL28B predicts hepatitis C treatment-induced viral clearance. *Nature* **461**: 399–401.
36. Suppiah V, Moldovan M, Ahlenstiel G, *et al.* (2009) IL28B is associated with response to chronic hepatitis C interferon-alpha and ribavirin therapy. *Nat Genet* **41**: 1100–1104.
37. Tanaka Y, Nishida N, Sugiyama M, *et al.* (2009) Genome-wide association of IL28B with response to pegylated interferon-alpha and ribavirin therapy for chronic hepatitis C. *Nat Genet* **41**: 1105–1109.
38. Koyama R, Arase Y, Ikeda K, *et al.* (2006) Efficacy of interferon therapy in elderly patients with chronic hepatitis C. *Intervirology* **49**: 121–126.
39. Hu K-Q, Kyulo NL, Esrailian E, *et al.* (2004) Overweight and obesity, hepatic steatosis, and progression of chronic hepatitis C: a retrospective study on a large cohort of patients in the United States. *J Hepatol* **40**: 147–154.
40. Charlton MR, Pockros PJ, Harrison SA. (2006) Impact of obesity on treatment of chronic hepatitis C. *Hepatology* **43**: 1177–1186.
41. Jianwu Y, Shuchen L, Gui-Qiang W. (2006) Effects of fatty liver and related factors on the efficacy of combination antiviral therapy in patients with chronic hepatitis C. *Liver Int* **26**: 166–172.
42. Ong JP, Younossi ZM, Speer C, *et al.* (2001) Chronic hepatitis C and superimposed non-alcoholic fatty liver disease. *Liver* **21**: 266–271.
43. Bressler BL, Guindi M, Tomlinson G, Heathcote J. (2003) High body mass index is an independent risk factor for nonresponse to antiviral treatment in chronic hepatitis C. *Hepatology* **38**: 639–644.

44. Walsh MJ, Jonsson JR, Richardson MM, *et al.* (2006) Non-response to antiviral therapy is associated with obesity and increased hepatic expression of suppressor of cytokine signalling 3 (SOCS-3) in patients with chronic hepatitis C, viral genotype 1. *Gut* **55**: 529–535.
45. Hickman IJ, Clouston AD, MacDonald GA, *et al.* (2002) Effect of weight reduction on liver histology and biochemistry in patients with chronic hepatitis C. *Gut* **51**: 89–94.
46. Gheorghe L, Iacob S, Sporea I, *et al.* (2007) Efficacy, tolerability and predictive factors for early and sustained virologic response in patients treated with weight-based dosing regimen of PegIFN alpha-2b ribavirin in real-life healthcare setting. *J Gastrointestin Liver Dis* **16**: 23–29.
47. Larrea E, Garcia N, Qian C, *et al.* (1996) Tumor necrosis factor alpha gene expression and the response to interferon in chronic hepatitis C. *Hepatology* **23**: 210–217.
48. Chu CJ, Lee SD, Hung TH, *et al.* (2008) Insulin resistance is a major determinant of sustained virological response in genotype 1 chronic hepatitis C patients receiving peginterferon Alpha-2b plus ribavirin. *Aliment Pharmacol Ther* Aug 1. [Epub ahead of print]
49. Romero-Gomez M, Del Mar Viloria M, Andrade RJ, *et al.* (2005) Insulin resistance impairs sustained response rate to peginterferon plus ribavirin in chronic hepatitis C patients. *Gastroenterology* **128**: 636–641.
50. Kawaguchi T, Ide T, Taniguchi E, *et al.* (2007) Clearance of HCV improves insulin resistance, beta-cell function, and hepatic expression of insulin receptor substrate 1 and 2. *Am J Gastroenterol* **102**: 570–576.
51. Davis GL, Wong JB, McHutchison JG, *et al.* (2003) Early virologic response to treatment with peginterferon-α-2b plus ribavirin in patients with chronic hepatitis C. *Hepatology* **38**: 645–652.
52. Fried MW, Hadziyannis SJ, Shiffman M, *et al.* (2008) Rapid viral response is a more important predictor of sustained virological response (SVR) than genotype in patients with chronic hepatitis C virus infection. *J Hepatol* **48 (Suppl 2)**: 5A.
53. Von Wagner M, Huber M, Berg T, *et al.* (2005) Peginterferon-alpha-2a (40KD) and ribavirin for 16 or 24 weeks in patients with genotype 2 or 3 chronic hepatitis C. *Gastroenterology* **129**: 522–527.
54. Mangia A, Santoro R, Minerva N, *et al.* (2005) Peginterferon-α-2b and ribavirin for 12 vs. 24 weeks in HCV genotype 2 or 3. *N Engl J Med* **352**: 2609–2617.
55. Dalgard O, Bjoro K, Ring-Larsen H, *et al.* (2008) Pegylated interferon-α and ribavirin for 14 versus 24 weeks in patients with hepatitis C virus genotype 2 or 3 and rapid virological response. *Hepatology* **47**: 35–42.
56. Camma C, Almasio P, Craxi A. (1996) Interferon as treatment for acute hepatitis C: A meta-analysis. *Dig Dis Sci* **41**: 1248–1255.

57. Wiegand J, Buggisch P, Boecher W, *et al.* (2006) Early monotherapy with pegylated interferon alpha-2b for acute hepatitis C infection: the HEP-NET acute-HCV-II study. *Hepatology* **43**: 250–256.
58. Santantonio T, Fasano M, Sinisi E, *et al.* (2005) Efficacy of a 24-week course of PEG-interferon-α-2b monotherapy in patients with acute hepatitis C after failure of spontaneous clearance. *J Hepatol* **42**: 329–333.
59. Liu CJ, Chen PJ, Shau WY, *et al.* (2003) Clinical aspects and outcomes of volunteer blood donors testing positive for hepatitis-C virus infection in Taiwan: a prospective study. *Liver Int* **23**: 148–155.
60. Mathurin P, Moussalli J, Cadranel JF, *et al.* (1998) Slow progression rate of fibrosis in hepatitis C virus patients with persistently normal alanine aminotransferase activity. *Hepatology* **27**: 868–872.
61. Persico M, Persico E, Suozzo R, *et al.* (2000) Natural history of hepatitis C virus carriers with persistently normal aminotransferase levels. *Gastroenterology* **118**: 760–764.
62. Pradat P, Alberti A, Poynard T, *et al.* (2002) Predictive value of ALT levels for histologic findings in chronic hepatitis C: a European Collaborative Study. *Hepatology* **36**: 973–977.
63. Zeuzem S, Diago M, Gane E, *et al.* (2004) Peginterferon-α-2a (40 kilodaltons) and ribavirin in patients with chronic hepatitis C and normal aminotransferase levels. *Gastroenterology* **127**: 1724–1732.
64. Jacobson IM, Ahmed F, Russo MW, *et al.* (2004) Interferon-α-2b and ribavirin for patients with chronic hepatitis C and normal ALT. *Am J Gastroenterol* **99**: 1700–1705.
65. Baccato S, Pistis R, Noventa F, *et al.* (2006) Fibrosis progression in initially mild hepatitis C. *J Viral Hepat* **13**: 297–302.
66. Verbaan HP, Widell HE, Bondeson TL, *et al.* (2002) High sustained response rate in patients with histologically mild (low grade and stage) chronic hepatitis C infection. A randomized, double blind, placebo controlled trial of interferon alpha-2b with and without ribavirin. *Eur J Gastroenterol Hepatol* **14**: 627–633.
67. Wright M, Forton D, Main J, *et al.* (2005) Treatment of histologically mild hepatitis C virus infection with interferon and ribavirin: a multicentre randomized controlled trial. *J Viral Hepat* **12**: 58–66.
68. Telfer PT, Garson JA, Whitby K, *et al.* (1997) Combination therapy with interferon alpha and ribavirin for chronic hepatitis C virus infection in thalassaemic patients. *Br J Haematol* **98**: 850–855.
69. Sievert W, Pianko S, Warner S, *et al.* (2002) Hepatic iron overload does not prevent a sustained virological response to interferon-alpha therapy: a long term follow-up study in hepatitis C-infected patients with beta thalassemia major. *Am J Gastroenterol* **97**: 982–987.

70. Fried MW, Peter J, Hoots K, *et al.* (2002) Hepatitis C in adults and adolescents with hemophilia: a randomized, controlled trial of interferon-α-2b and ribavirin. *Hepatology* **36**: 967–972.
71. Hanley JP, Jarvis LM, Andrew J, *et al.* (1996) Interferon treatment for chronic hepatitis C infection in hemophiliacs — influence of virus load, genotype, and liver pathology on response. *Blood* **87**: 1704–1709.
72. Bica I, McGovern B, Dhar R, *et al.* (2001) Increasing mortality due to end-stage liver disease in patients with human immunodeficiency virus infection. *Clin Infect Dis* **32**: 492–497.
73. Chung RT, Andersen J, Volberding P, *et al.* (2004) Peginterferon-α-2a plus ribavirin versus interferon-α-2a plus ribavirin for chronic hepatitis C in HIV-coinfected persons. *N Engl J Med* **351**: 451–459.
74. Torriani FJ, Ribeiro RM, Gilbert TL, *et al.* (2003) Hepatitis C virus (HCV) and human immunodeficiency virus (HIV) dynamics during HCV treatment in HCV/HIV coinfection. *J Infect Dis* **188**: 1498–1507.
75. Perez-Olmeda M, Nunez M, Romero M, *et al.* (2003) Pegylated IFN-a2b plus ribavirin as therapy for chronic hepatitis C in HIV-infected patients. *AIDS* **17**: 1023–1028.
76. Soriano V, Nunez M, Camino N, *et al.* (2004) Hepatitis C virus-RNA clearance in HIV-coinfected patients with chronic hepatitis C treated with pegylated interferon plus ribavirin. *Antivir Ther* **9**: 505–509.
77. Liu CJ, Liou JM, Chen DS, Chen PJ. (2005) Natural course and treatment of dual hepatitis B virus and hepatitis C virus infections. *J Formos Med Assoc* **104**: 783–791.
78. Liu CJ, Chen PJ, Lai MY, *et al.* (2003) Ribavirin and interferon is effective for hepatitis C virus clearance in hepatitis B and C dually infected patients. *Hepatology* **37**: 568–576.
79. Chu CM, Liaw YF. (2010) Hepatitis B surface antigen seroclearance during chronic HBV infection. *Antivir Ther* **15**: 133–143.
80. Fabrizi F, Poordad FF, Martin P. (2002) Hepatitis C infection and the patient with end-stage renal disease. *Hepatology* **36**: 3–10.
81. Fabrizi F, Dulai G, Dixit V, Bunnapradist S, Martin P. (2003) Metaanalysis: interferon for the treatment of chronic hepatitis C in dialysis patients. *Aliment Pharmacol Ther* **18**: 1071–1081.
82. Russo MW, Goldsweig CD, Jacobson IM, Brown RS Jr. (2003) Interferon monotherapy for dialysis patients with chronic hepatitis C: an analysis of the literature on efficacy and safety. *Am J Gastroenterol* **98**: 1610–1615.
83. Kokoglu OF, Ucmak H, Hosoglu S, *et al.* (2006) Efficacy and tolerability of pegylated-interferon alpha-2a in hemodialysis patients with chronic hepatitis C. *J Gastroenterol Hepatol* **21**: 575–580.

84. Russo MW, Ghalib R, Sigal S, Joshi V. (2006) Randomized trial of pegylated interferon alpha-2b monotherapy in haemodialysis patients with chronic hepatitis C. *Nephrol Dial Transplant* **21**: 437–443.
85. Liu CH, Liang CC, Lin JW, *et al.* (2008) Pegylated interferon alpha-2a versus standard interferon alpha-2a for treatment-naive dialysis patients with chronic hepatitis C: A randomised study. *Gut* **57**: 525–530.
86. Bruchfeld A, Lindahl K, Reichard O, *et al.* (2006) Pegylated interferon and ribavirin treatment for hepatitis C in haemodialysis patients. *J Viral Hepat* **13**: 316–321.
87. Roche B, Samuel D. (2006) Aspects of hepatitis C virus infection relating to liver transplantation. *Eur J Gastroenterol Hepatol* **18**: 313–320.
88. De Vera ME, Smallwood GA, Rosado K, *et al.* (2001) Interferon alpha and ribavirin for the treatment of recurrent hepatitis C after liver transplantation. *Transplantation* **71**: 678–686.
89. Alberti AB, Belli LS, Airoldi A, *et al.* (2001) Combined therapy with interferon and low-dose ribavirin in posttransplantation recurrent hepatitis C: a pragmatic study. *Liver Transpl* 7: 870–876.
90. Ahmad J, Dodson SF, Demetris AJ, *et al.* (2001) Recurrent hepatitis C after liver transplantation: a nonrandomized trial of interferon-α alone versus interferon-α and ribavirin. *Liver Transplant* 7: 863–869.
91. Wang CS, Ko HH, Yoshida EM, *et al.* (2006) Interferon-based combination anti-viral therapy for hepatitis C virus after liver transplantation: a review and quantitative analysis. *Am J Transplant* **6**: 1586–1599.
92. Lavezzo B, Franchello A, Smedile A, *et al.* (2002) Treatment of recurrent hepatitis C in liver transplants: efficacy of a six versus a twelve month course of interferon-α 2b with ribavirin. *J Hepatol* **37**: 247–252.
93. Sulkowski MS, Thomas DL. (2005) Epidemiology and natural history of hepatitis C virus infection in injection drug users: implications for treatment. *Clin Infect Dis* **40**(Suppl 5): S263–269.
94. Ohnishi K, Matsuo S, Matsutani K, *et al.* (1996) Interferon therapy for chronic hepatitis C in habitual drinkers: comparison with chronic hepatitis C in infrequent drinkers. *Am J Gastroenterol* **91**: 1374–1379.
95. Poynard T, Schiff E, Terg R, *et al.* (2008) Sustained viral response is dependent on baseline characteristics in the retreatment of previous-α interferon/ribavirin nonresponders: final results from the EPIC3 program. *J Hepatol* **48**(Suppl 2): S369.
96. Gross J, Johnson S, the RENEW investigators, *et al.* (2005) Double dose peginterferon-α-2b with weight-based ribavirin improves response for interferon/ribavirin non-responders with hepatitis C: final results of "RENEW". *Hepatology* **42**(Suppl 1): 219A.
97. Jensen DM, Freilich B, Andreone P, *et al.* (2007) Pegylated interferon-α-2a (40KD) plus ribavirin (RBV) in prior non-responders to pegylated

interferon-α-2b (12KD)/RBV: final efficacy and safety outcomes of the REPEAT study. *Hepatology* **46** (Suppl 1):291A.

98. Alric L, Duffaut M, Selves J, *et al.* (2001) Maintenance therapy with gradual reduction of the interferon dose over one year improves histological response in patients with chronic hepatitis C with biochemical response: results of a randomized trial. *J Hepatol* **35**: 272–278.
99. Shiffman ML, Hofmann CM, Contos MJ, *et al.* (1999) A randomized, controlled trial of maintenance interferon therapy for patients with chronic hepatitis C virus and persistent viremia. *Gastroenterology* **117**: 1164–1172.
100. Di Bisceglie AM, Schiffman ML, Everson GT, *et al.* (2007) Prolonged antiviral therapy with peginterferon to prevent complications of advanced liver disease associated with hepatitis C: results of the hepatitis C antiviral long-term treatment against cirrhosis (HALT-C) trial. *Hepatology* **46**(Suppl 1): 290A.
101. Afdhal NH, Levine R, Brown Jr. R, *et al.* (2008) Colchicine versus peg-interferon-α 2b long term therapy: results of the 4 year COPILOT trial. *J Hepatol* **48**(Suppl 2): S4.
102. Pawlotsky JM, Chevaliez S, McHutchison JG. (2007) The hepatitis C virus life cycle as a target for new antiviral therapies. *Gastroenterology* **132**: 1979–1998.
103. Lawitz E, Rodriguez-Torres M, Muir AJ, *et al.* (2008) Antiviral effects and safety of telaprevir, peginterferon-α-2a, and ribavirin for 28 days in hepatitis C patients. *J Hepatol* **49**: 163–169.
104. McHutchison JG, Bacon BR, Gordon SC, *et al.* (2007) Phase 1B, randomized, double-blind, dose-escalation trial of CPG 10101 in patients with chronic hepatitis C virus. *Hepatology* **46**: 1341–1349.
105. Zeuzem S, Yoshida EM, Benhamou Y, *et al.* (2008) Albinterferon-α-2b dosed every two or four weeks in interferon-naïve patients with genotype 1 chronic hepatitis C. *Hepatology* **48**: 407–417.

Chapter 14

Individualized Therapy for Chronic Hepatitis C

Ming-Lung Yu*,†,‡ and Wan-Long Chuang†,‡

Introduction

Hepatitis C virus (HCV) infection frequently causes chronic liver disease, leading to cirrhosis and hepatocellular carcinoma (HCC).[1] There are an estimated 170 million people or more with chronic HCV infection, and about four million persons are newly infected each year.[2] The risk for developing cirrhosis 20 years after initial HCV infection among those chronically infected varies between studies, but is estimated at around 5–15%.[3] Once cirrhosis is established, the annual risk of HCC, hepatic decompensation, and liver-related death is approximately 1–7%, 5% and 2%, respectively.[4] HCV-related end-stage liver disease and HCC have become the leading cause for liver transplantation worldwide.[5]

* Department of Internal Medicine, Kaohsiung Municipal Ta-Tung Hospital, Kaohsiung, Taiwan.

† Faculty of Internal Medicine, College of Medicine, Kaohsiung Medical University, Kaohsiung, Taiwan.

‡ Hepatobiliary Division, Department of Internal Medicine, Kaohsiung Medical University Hospital, Kaohsiung, Taiwan.

Corresponding Address: Wan-Long Chuang Hepatobiliary Division, Department of Internal Medicine, Kaohsiung Medical University Hospital, No. 100, Tzyou 1st Road, Kaohsiung 807, Taiwan, Tel: 886-7-312-1101 ext. 7475, Fax: 886-7-323-4553, e-mail: fishya@ms14.hinet.net; fish6069@gmail.com

Treatment for Chronic Hepatitis C

The objectives of hepatitis C treatment are eradication of the virus and prevention of long-term hepatic complications, such as cirrhosis, HCC, and liver-related death. Therefore, the primary goal of treatment for chronic hepatitis C is a sustained virological response (SVR), defined as polymerase chain reaction (PCR)-seronegative of HCV RNA (<50 IU/mL) throughout the 24-week post-treatment follow-up.[2] SVR has been shown to be durable up to 18 years, strongly suggesting that SVR may be considered to show eradication of HCV infection.[6] Achievement of an SVR has been associated with persistent regression of hepatic fibrosis, reduced incidence of cirrhosis, HCC, and liver-related mortality.[7–11] Given the lack of an effective vaccine, optimal treatment of chronic hepatitis C (CHC) is now perceived as a "must" in terms of public health strategies, as well as of the clinical setting for individual patients.

Conventional Interferon-based Therapy

Interferon-alfa (IFN-α) therapy has been associated with a sustained normalization of alanine aminotransferase (ALT) levels in patients diagnosed as non-A, non-B hepatitis even before HCV was identified as the major etiologic agent in the disease.[12] Until the 1990s, the only therapy of proven benefit for patients with CHC was IFN-α.[13] A 48-week regimen of IFN-α-2b, at a dose of three million units (MU) thrice weekly, had an SVR rate of 13–19%.[14,15]

Ribavirin (1-β-D-ribofuranosyl-1,2,4-triazole-3-carboxamide) is an oral purine nucleoside analogue with broad activity against viral pathogens.[16] Although ribavirin monotherapy was shown to be ineffective for CHC,[17] it was a major breakthrough in the treatment of CHC when used in combination with IFN-α. In 1996, Lai *et al.* observed that a significantly higher rate of durable SVR was observed in CHC patients treated with 24 weeks of IFN, at 3 MU thrice weekly, plus ribavirin at 1,200 mg/d than in those with 24 weeks of IFN alone (43% vs. 6%).[18] The subsequent two large randomized, international clinical trials with 1,744 CHC patients confirmed the findings.[14,15]

The overall SVR rates were 33% and 41% for 24 and 48 weeks of IFN-α-2b, at 3 MU thrice weekly, and ribavirin at 1,000–1,200 mg/d, respectively, compared with 6% and 16% for those with 24 and 48 weeks of IFN-α-2b monotherapy. In 1998, the FDA approved the combination of IFN-α and ribavirin for patients with chronic HCV infection.

Pegylated Interferon-based Therapy

Attachment of an inert polyethylene glycol polymer to conventional IFN-α — pegylated IFN-α (peg-IFN-α) — slows subcutaneous absorption and reduces degradation and clearance. This prolongs the half-life of IFN and permits less frequent, weekly dosing while maintaining higher sustained IFN blood levels. Two large randomized clinical trials were applied with fixed durations of 48 weeks. In these trials, peg-IFN-α-2b was dosed by weight (1.5 μg/kg/week was FDA approved) and coupled with 800 mg of ribavirin; peg-IFN-α-2a was given at a fixed dose of 180 μg/week along with a weight-adjusted, higher dose of ribavirin (1,000 mg/day or 1,200 mg/day for patients weighing < 75 kg or > 75 kg, respectively). They observed an overall SVR rate of 54–56%, significantly higher than those with conventional IFN/ribavirin or peg-IFN monotherapy.[19,20] Currently, peg-IFN-α plus ribavirin combination treatment has become the standard of care for all patients infected with HCV.[21,22]

Factors Associated with Treatment Efficacy

Although the introduction of new agents and regimens for the treatment of CHC, such as peg-IFNs and combination therapy with ribavirin, has resulted in substantial improvements in overall SVR rates, treatment remains a challenge, particularly for certain patient populations. Accurately predicting therapeutic response is a critical issue in the management of diseases. With the great progress in the management of CHC, predictors for SVR in CHC therapy have been elucidated. The treatment efficacy is mainly determined by viral factors, host factors, and treatment regimens,[21,23,24] which may be divided into two major categories: baseline and on-treatment predictors (Table 1). Understanding the determining factors associated with treatment efficacy will help us to tailor the treatment regimen for individualized therapy, maximize the treatment efficacy, and minimize side effects and costs. To date, the baseline and on-treatment virological factors are the most well established markers for tailoring the treatment regimens for individualized HCV therapy.

Baseline Predictors of Response to IFN-based Therapy

The pre-treatment variable most strongly predictive of an SVR is the presence of HCV-2 or -3 infection,[25] whether with conventional IFNs or peg-IFNs, alone or in combination with ribavirin.[14,15,19,20] Under the current recommendation,[21,26] SVR rates were 42–60% for HCV-1

Table 1. Factors Associated with Response to Interferon-Based Therapy for Hepatitis C

Baseline

- Virological factors
  - Hepatitis C virus genotype
  - Hepatitis C viral loads
  - Quasispecies
- Host factors
  - Host genetics
  - Bridging fibrosis/cirrhosis
  - Gender
  - Age
  - Ethnicity
  - Insulin resistance
  - Obesity
  - Hepatic steatosis
  - Coinfection with HIV
  - Nonresponse to previous interferon-based therapy

On-treatment

- Rapid virological response at week 4
- Early virological response at week 12
- Medical adherence

infection with a 48-week peg-IFN–ribavirin treatment, compared with 76–95% for HCV-2 or -3 infections with a 24-week regimen.[19,20,27–30]

Baseline viral loads, even less important than HCV genotype, is another important predictor of SVR in IFN-based therapy.[14,15,19] A higher HCV RNA level predicts a lower response rate. The impact of HCV RNA level on the response to combination therapy varied among patients with different HCV genotype infections. High viral load (with a cutoff value of 200,000 copies/mL, or 800,000 IU/mL) influenced the response rate in patients with HCV-1 (41% vs. 56%), but not in patients with HCV-2 or -3 (74% vs. 81%).[20]

On-Treatment Predictors of Response to IFN-based Therapy

During IFN-based therapy, serum levels of HCV RNA generally fall in a biphasic manner.[31] The first, rapid phase of viral suppression, from a few hours after the first IFN-α injection to the end of the first day, is associated with an inhibition of viral replication by a direct, non-specific action of IFN-α. This early initial decline in HCV RNA levels correlates poorly with the eventual response to IFN-based therapy.[31,32] The second, slower phase of viral suppression, beginning on day two and gradually leading to seroclearance of HCV RNA, is possibly related to the gradual clearance of infected cells by the patient's immune system. This phase exhibits a good response to IFN-based therapy and is an excellent marker of an SVR to the treatment.[31,33]

The role of virological response at treatment week 4 and week 12 on treatment efficacy is well established. Seroclearance of HCV RNA during first four weeks of treatment could predict an SVR to IFN/ribavirin with a high degree of accuracy in both HCV-1 and -2 patients, with positive predictive values of 78% and 93%, respectively,[34] and is shown to be the single best predictor of an SVR to peg-IFN/ribavirin for HCV-1[35,36] and HCV-2/3 patients.[29,30,37,38] By contrast, for patients not achieving PCR-seronegative, or at least a two logs decrease, of HCV RNA from baseline after 12 weeks of therapy, the likelihood of an SVR is only 0–2%.[39] The robust negative predictor, obvious in HCV-1 but not HCV-2 patients,[34] establishes the 12-week stopping rule of treatment for HCV-1 patients.[40]

Two major on-treatment virological responses to IFN-based therapy for CHC have been established,[40–42] including:

1. *Rapid virological response (RVR)*: PCR-seronegative (<50 IU/mL) of HCV RNA after four weeks of therapy.
2. *Early virological response (EVR)*: PCR-seronegative of HCV RNA or HCV RNA >2 log drop from baseline after 12 weeks of therapy. Recently, it has been suggested that EVR be subdivided into RVR, complete EVR (cEVR, no RVR, but HCV RNA <50 IU/mL at week 12), and partial EVR (pEVR, HCV RNA >2 log drop but still detectable, ≥50 IU/mL, at week 12) to further improve the prediction of patients who are likely to achieve an SVR and to allow for the possibility of tailoring of treatment duration.[43,44]

Individualized HCV Therapy

Individualized therapy has become a major consideration for clinicians. It is desirable to expose CHC patients to the lowest doses and shortest durations of treatment possible to reduce the likelihood of adverse events and to minimize costs, without compromising treatment efficacy. On the other hand, some difficult-to-treat patients have to receive longer and/or higher-dose therapy to ensure response.

The first concept of individualized HCV therapy came from the results of the two landmark studies of IFN plus ribavirin.[14,15] SVRs to 48 or 24 weeks of combination therapy occurred in 29% and 17% of HCV-1 patients, respectively, and in 65% and 66% of HCV-2 or -3 patients, respectively. The two studies reinforced the importance of longer duration therapy of 48 weeks in patients with HCV-1 infection. Similarly, SVRs to 48 or 24 weeks of combination therapy occurred in 38% and 27% of patients with pre-treatment HCV RNA levels of $>2 \times 10^6$ copies/mL, respectively, but the SVR rates were no different for those with lower levels (45% and 43%, respectively). Therefore, in 1999, the EASL International Consensus Conference on Hepatitis C recommended an individualized HCV therapy: for treatment-naive, CHC patients, (a) standard therapy should consist of 24 weeks of

IFN-α and ribavirin; and (b) treatment should be extended to 48 weeks in patients with both HCV genotype I (HCV-1) and viral load greater than 2×10^6 copies/mL.[45]

Genotype-guided Therapy

The optimal treatment duration and ribavirin dose were investigated in a large randomized clinical trial in which all CHC patients received peg-IFN-α-2a at a dose of 180 μg, plus ribavirin at a dose of 800 mg or standard 1,000–1,200 mg daily, for either 24 or 48 weeks.[28] The study observed that the 48-week regimen with standard dose of ribavirin had the highest SVR rate at 52% for HCV-1 patients; whereas the four treatment arms had similar SVR rates for HCV non-1 patients, (79–84%, average 81%). Therefore, a 48-week regimen of peg-IFN-α plus standard doses of ribavirin for HCV-1/4 patients and a 24-week regimen with lower dose of ribavirin (800 mg/day) for HCV-2/3 patients become the current standard of care (SOC) for CHC.[5,21,26] It is noted, however, that this study actually enrolled very few HCV-4 cases. With the recommended SOC, SVR was around 50% (42–52%) and 80% (76–84%), respectively, for HCV-1 and HCV-2/3 patients in Caucasian-predominant clinical trials,[19,20,28,30,46] and 55–77% and 80–94%, respectively, in Asian people.[47–54]

It is now important to tailor optimal treatment options for HCV-4, -6 patients with comparable treatment approaches as for HCV-1, -2, and -3 patients. Data for HCV-4 are mainly from the Middle East and Egypt. Treatment with a 48-week regimen plus standard dose of ribavirin results in substantially higher SVR rates of 55–69% than those treated with a 24-week regimen (29–48%).[55–57] Published data for HCV genotypes other than 1–4 are very limited. Fung *et al.* demonstrated that 48 weeks of peg-IFN plus standard dose of RBV could achieve the significantly higher SVR rate of 86% in HCV-6 patients than that of 52% in HCV-1 patients.[58] However, the optimal regimen for HCV-6 patients has not been well established. A retrospective study observed that 48 weeks of peg-IFN plus standard dose of RBV could achieve a significantly higher SVR rate (75%) than the 24-week regimen did (39%) in HCV-6 patients.[59] Nevertheless, a prospective randomized trial conducted by the same study group demonstrated that the SVR rates in HCV-6 patients were similar between 24 (70%) and 48 (79%) weeks of peg-IFN/RBV.[60] Both studies were small scaled trials. Further, larger prospective studies for HCV-6 patients are needed to confirm the optimal regimen. Dev *et al.* observed that 79% of HCV-7, -8, or -9 Southeast Asians could achieve an SVR with 48–52 weeks of conventional IFN plus ribavirin.[61] However, no data are available for the treatment response to peg-IFN/ribavirin for HCV-7, -8, and -9, because these strains are currently classified as HCV-6 variants. Taken together, the current available data indicate that the standard 48-week regimen is an appropriate approach for HCV-4 and -6 patients, whose response to treatment is at an intermediate level compared to those for HCV-1 and HCV-2/3 cases.

Response-guided Therapy

Abbreviated Treatment Duration for Super-Responders

Although HCV genotype has been considered to be the most important baseline predictor of SVR, a recent study with multivariate logistic regression analysis demonstrated that an RVR at week 4 is the single best predictor of SVR, regardless of viral genotype. Patients achieving a RVR have a high probability of SVR, from 86–100% for genotype 1 to 4 with current SOC.[62] Interestingly, an earlier study in Taiwan demonstrated that 78% of HCV-1 patients who had an RVR could attain an SVR with only 24 weeks of IFN/ribavirin.[34] In addition, Yu *et al.* followed 32 patients who required early termination from peg-IFN/ribavirin among 617 patients.[63] None of the 16 HCV-1 patients with treatment duration of <20 weeks achieved an SVR. By contrast, 38.5% (5/13) HCV-2 patients, including two of four with 8–15 weeks and all of the three patients with 16 weeks of treatment, attained an SVR. Notably, all the five sustained responders achieved an RVR. These observations raise the question of whether RVR might be a marker for tailoring treatment regimen to a shorter duration without compromising the efficacy of these super-responders. Several studies have tried to answer the question since 2004 and are discussed below.

HCV-1/4 Patients with an RVR

Several European and North American studies have provided data to support the concept that HCV-1/4 infected patients who have an RVR during peg-IFN/ribavirin have high rates of SVR when the treatment course is shortened to 24 weeks, in particular patients with lower baseline viral loads. First, Zeuzem *et al.* conducted a single-arm study for HCV-1 patients with low baseline viral load (≤600,000 IU/mL), treated with peg-IFN plus weight-based ribavirin at 800–1,400 mg/day for 24 weeks, and compared to a 48-week historical control.[36] The overall SVR rates were significantly lower in the 24-week regimen (50%) than in the 48-week historical controls (71%). However, a subset of patients who achieved an RVR attained similar SVR rates with 24-week regimen (89%) to those in the 48-week control group (85%).[36] Another retrospective analysis of data from the trial by Hadziyannis *et al.*[28] observed that among HCV-1 patients with an RVR, the SVR rates were comparable between 24-week (89–90%) and 48-week (73–91%) groups.[35] A recent Italian study demonstrated that for HCV-1 patients with an RVR, the SVR rate in patients with 24-week treatment was substantial (77.2%), which is lower than that in those with 48-week treatment (87.1%).[64] RVR patients with baseline viremia <400,000 IU/mL achieved an SVR rate of 84.4% with 24-week treatment, which was comparable to the 83.3% rate in the 48-week group; however, RVR patients with baseline viremia >400,000 IU/mL achieved lower SVR rates when treated for 24 weeks than for 48 weeks (73.1% vs. 86.8%, $P = 0.14$).[64] Yu *et al.* conducted the first randomized controlled study to evaluate the utility of an RVR in determining a shorter, 24-week versus standard 48-week treatment duration for HCV-1 infected Asian patients.[52] They observed that, although the results were better when looking at the group of patients achieving RVR, SVR appeared to remain lower in the 24-week arm than in the 48-week arm (89% vs. 100%, $P = 0.056$). Nevertheless, for patients with low viremia (<400,000 IU/mL) and

RVR, the 24-week group had a comparable SVR rate (96%) with the 48-week group (100%). By contrast, for patients with high viremia and RVR, the SVR rate was significantly lower in the 24-week group than in the 48-week group (77% vs. 100%, $P = 0.045$).[52] Liu *et al.* accumulated additional evidence to support the abbreviated 24-week regimen in HCV-1 Taiwanese patients with low viral load and RVR.[51] The issue was further reinforced in another four studies for HCV-1 and HCV-4 patients.[64–67] These findings confirmed the RVR and baseline viral load as important markers in individualized treatment for HCV-1 patients.

HCV-2/3 Patients with a RVR

Studies investigating shorter treatment duration for HCV-2 or -3 patients have been conducted since 2001. The hypothesis was first tested in three European studies.[29,30,37] Dalgard *et al.* conducted the first trial to determine the efficacy of a 14-week treatment duration for HCV-2/3 patients with an RVR.[37] In the non-randomized controlled study, SVR was obtained in 90% of RVR patients with a 14-week regimen of peg-IFNα-2b 1.5 μg/kg/w plus ribavirin at a dose of 800–1,400 mg/d, while only 56% of non-RVR patients achieve an SVR with a 24-week regimen. Mangia *et al.* reported the first randomized, controlled trial with peg-IFNα-2b 1.0 μg/kg/w plus ribavirin of 1,000–1,200 mg/d.[29] HCV-2/3 patients were randomly allocated to either the 24-week regimen (standard-duration group) or a variable regimen (variable-duration group) of 12 weeks for RVR patients and 24 weeks for non-RVR patients. The SVR rate in the standard-duration group (76%) was comparable with that in the variable-duration group (77%). Von Wagner *et al.* randomly assigned 153 patients who had HCV RNA below 600 IU/mL after four weeks to receive either 16 or 24 weeks of peg-IFN plus ribavirin of 800–1,200 mg/day. Patients with HCV RNA ≥ 600 IU/mL at week 4 were treated for 24 weeks.[30] The SVR was obtained in 82% of patients achieving an RVR with an abbreviated 16-week regimen, which was similar to 80% of those with the standard 24-week regimen. By contrast, only 36% of non-RVR patients could achieve an SVR after 24 weeks of treatment.

Yu *et al.* evaluated the utility of an RVR in determining an abbreviated, 16-week regimen for HCV-2 patients in Asia.[38] They observed that both 16 and 24 weeks of treatment could achieve a high SVR, 100% and 98%, respectively, among patients with RVR. These results observed that for HCV-2/3 patients with an RVR, shorter treatment duration of 12–16 weeks peg-IFN plus ribavirin could achieve comparable SVR rates (82–100%) when compared to the 24-week regimen (80–98%). One proviso of these studies is that doses of ribavirin were the standard weight-based ones.

Notably, another two studies investigating the efficacy of shorter treatment duration for HCV-2/3 patients with an RVR had inconsistent observation.[68,69] These two studies, using a fixed, low dose of ribavirin at 800 mg/d, demonstrated that the abbreviated 12-week[69] or 16-week[68] regimen was inferior to the 24-week regimen, even among patients with an RVR. The SVR rates were significantly lower in the shorter treatment arm (71%[69] and 82%[68]) than in the standard treatment arm (91%[69] and 90%[68]) for HCV-2/3 patients with an RVR. The results raised the question of whether a fixed, lower dose of ribavirin may be suboptimal for shorter treatment duration in treating HCV-2/3 patients with an RVR. At the same time, the inferiority of an abbreviated 14-week regimen to the 24-week regimen among HCV-2/3 patients with an RVR was observed with peg-IFN plus weight-based ribavirin at 800–1,400 mg/d in another study by Dalgard *et al.* (81.1% vs. 90.7%, difference 9.6%, 95% CI 1.7–17.7, intention-to-treat analysis).[70] Nevertheless, the difference became less evident when the analysis included only patients with an HCV RNA test at the end of follow-up (86.3% vs. 93.2%, difference 6.9%; 95% CI, −0.1 to + 13.9).

To clarify the controversial issues, Di Martino *et al.* included data from seven randomized clinical trials of HCV-2/3 studies for meta-analysis.[71] Shorter therapy with 12- or 14-week regimens had an 11% lower SVR rate relative to the 24-week standard regimen, but the difference was not observed between 16-week and 24-week regimens. Shorter therapy with fixed-dose 800-mg/d ribavirin also had a 13% lower SVR rate relative to the 24-week regimen, but the difference did not exist between the shorter treatment regimen with standard dose of ribavirin and standard 24-week regimen. When the shorter therapy was further divided into two groups — optimal short therapy, defined as 16-week therapy with weight-based, standard dose of ribavirin, and suboptimal short therapy, defined as 12- or 14-week regimens and/or fixed, low-dose ribavirin — they found that only optimal shorter therapy could achieve a comparable SVR rate when compared with a standard 24-week regimen for HCV-2/3 patients with an RVR. These data provide evidence to support an optimal shorter therapy, 16-week peg-IFN plus weight-based standard dose of ribavirin at 800–1,400 mg/day for HCV-2/3 patients in whom RVR is observed, without compromising treatment efficacy.

Optimal Treatment Duration for Slow-Responders

HCV-1/4 Patients without an RVR, but with an EVR

Several studies have investigated the benefits of extending the treatment duration to 72 weeks, especially for the subgroup of HCV-1/4 and patients without an RVR.[64,72–76] Studies by Berg *et al.*[72] and Sanchez-Tapias *et al.*[73] were conducted at a time when an optional dose of ribavirin for HCV-1/4 had not been established, so that they used a fixed, low dose of 800 mg/d for all patients. Data from the study by Sanchez-Tapias *et al.* demonstrated that a 72-week regimen could improve the SVR rate from 28% in the 48-week regimen to 44% ($P = 0.003$) for HCV-1 patients without an RVR.[73] Interestingly, the benefits of extended treatment duration were observed mainly in patients with lower baseline viral loads (< 800,000 IU/mL) (27% and 51%, respectively, for 48- and 72-week regimens, $P = 0.007$), but not among those with high viral loads. Berg *et al.* observed that a significantly higher SVR rate in the 72-week arm as compared to the 48-week arm was observed only among HCV-1 patients who did not achieve a cEVR (29% vs. 17%, $P = 0.04$), but not among those who achieved a cEVR (77% vs. 75%).[72] Furthermore, the SVR rate was substantially higher in the 72-week arm than in the 48-week arm among HCV-1 patients with a pEVR (48% vs. 33%), although the difference did not reach significance (P = 0.154). The subsequent study by Pearlman *et al.* observed consistent and significant findings in a more strictly defined population.[74] They included only HCV-1 patients who did not attain an RVR and a cEVR, but who achieved a pEVR and HCV RNA seronegative at week 24. Patients were randomized into either 48 or 72 weeks of peg-IFN plus standard dose of ribavirin. The SVR rate was significantly improved from 18% in the 48-week arm to 38% in the extended 72-week arm (P = 0.026). More recently, Ferenci *et al.*[76] randomized HCV-1/4 patients who did not achieve an RVR but achieved an EVR into a standard 48-week regimen or an extended 72-week regimen. A lower dose of peg-IFN-α 2a, 135 μg/w was used in the extended 49th to 72th weeks. They observed that extended 72-week therapy could decrease only the rate of relapse, but not that of SVR, in patients with an EVR. The SVR rate was 51.1% and 58.6%, respectively, in the 48- and 72-week arms (P = 0.1). The relapse rate was 33.6% and 18.5%, respectively, in the 48- and 72-week arms (P = 0.0115). The benefit of extending therapy in reducing the relapse rate was observed mainly in those with viral loads <400,000 IU/mL (29% and 8%, respectively, for 48- and 72-week regimens, $P < 0.05$) and those achieving a pEVR (57% and 31%, respectively, for 48- and 72-week regimens, $P < 0.05$), but not observed among those with viral load ≥400,000 IU/mL and those achieving a cEVR. These findings suggested favorable results in extending therapy to 72 weeks for HCV-1 patients with a pEVR, but not those with a cEVR.

By contrast, another two studies investigating individualized therapy for HCV-1 patients observed conflicting results. Mangia *et al.*[64] demonstrated that a substantially higher SVR rate was attained in the 72-week arm than in the 48-week arm (64% vs. 38%, P = 0.068) among HCV-1 patients whose serum HCV RNA became negative during week 8 and 12. Ide *et al.*[75] assigned patients into the standard 48-week regimen or an extended regimen according to the first-time HCV RNA negativity. They observed that the extended group (60–68 weeks of therapy) patients who became negative for HCV RNA between weeks 16 and 24 had a significantly higher SVR rate (78%) than those in the standard group (9%; P = 0.005).

Since the study design, patient selection criteria, ribavirin dose, and subgroups for outcome analysis have varied across these trials of published data, it is difficult to obtain conclusive information from outcomes across the published data. To clarify the controversial issues, Di Martino *et al.* included data from five randomized clinical trials of HCV-1/4 studies for meta-analysis.[71] They found that a 12% greater SVR rate in the 72-week regimen versus the 48-week regimen was observed only among HCV-1 patients with a pEVR and treated with a weight-based, standard dose of ribavirin. The results accumulated additional evidence to support the finding that the benefits of extending therapy to 72 weeks in increasing SVR rate was observed only among HCV-1/4 slow responders with weight-based, standard dose of ribavirin, but not for those with a cEVR or those treated with fixed, lower dose of ribavirin.

HCV-1/4 Patients without an EVR

HCV-1/4 patients without an EVR have little chance of achieving an SVR with a standard 48-week regimen (< 2%).[39,40,52,68,69] Extending therapy to 72 weeks also adds little chance to achieve an SVR for HCV-1/4 patients without an EVR.[76] Therefore, the current treatment guidelines recommend the 12-week stopping rule for HCV-1/4 patients without an EVR.[21,40]

HCV-2/3 Patients without a RVR

The SVR rate in HCV-2/3 patients without an RVR was consistently lower among patients treated with shorter treatment duration than in those with a standard 24-week regimen.[38,68] Nevertheless, the treatment efficacy of the standard 24-week regimen for this subgroup remains unsatisfactory. Many studies are attempting to improve the SVR rate for HCV-2/3 patients without an RVR by extending the therapy beyond 24 weeks. Therefore, the current recommendation for treatment of HCV-2/3, non-RVR patients is 24 weeks of peg-IFN plus ribavirin.

Measurement of HCV RNA at week 12 is not currently recommended for HCV-2/3 patients, because almost all HCV-2/3 patients achieved an EVR. Nevertheless, in the subgroup analysis in the HCV-2 study by Huang *et al.*[77] all of the HCV-2 patients who achieved a cEVR at week 12 attained an SVR; while among those without an EVR, none attained an SVR. The findings provide evidence suggesting the measurement of HCV RNA at week 12 for HCV-2/3 patients without an RVR.

Roadmap for Treating Chronic Hepatitis C

According to the published literature, a treatment roadmap of individualized therapy for chronic hepatitis C can be constructed (Fig. 1). The current standard of care in treating CHC remains 48 weeks of

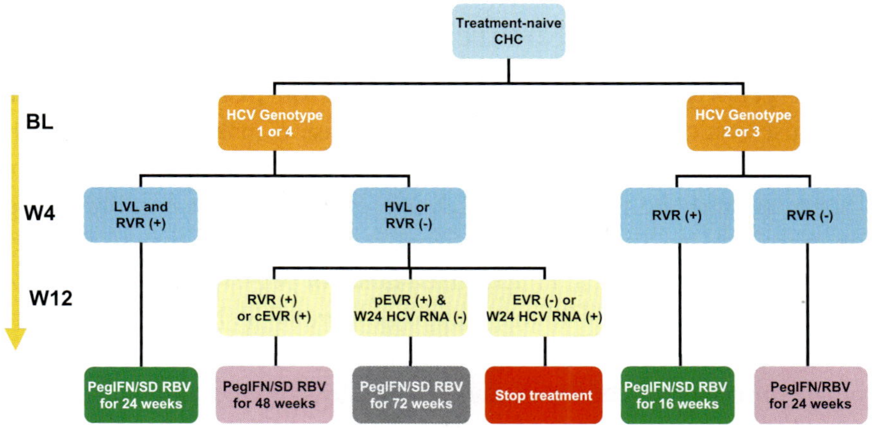

Fig. 1. Roadmap for individualized therapy for chronic hepatitis C: BL: baseline; W4: treatment week 4; W12: treatment week 12; W24: treatment week 24; LVL: low viral loads, HCV RNA <400,000 IU/mL at baseline; HVL: high viral loads, HCV RNA ≥ 400,000 IU/mL at baseline; RVR: rapid virological response, HCV RNA <50 IU/mL at week 4; EVR: early virological response, seronegative of HCV RNA or decrease of HCV RNA by 2 logs from baseline values after 12 weeks of therapy; pegIFN: pegylated interferon; SD RBV: standard dose of ribavirin, 1,000–1,200 mg/day; cEVR: complete EVR, non RVR, but seronegative of HCV RNA after 12 weeks of therapy; pEVR: partial EVR, no RVR, seropositive of HCV RNA at week 12 , but decrease of HCV RNA by 2 logs from baseline values after 12 weeks of therapy.

peg-IFN plus standard dose of ribavirin for HCV-1/4–6; and 24 weeks of combination therapy for HCV-2/3 patients. For HCV-1/4 patients without an EVR at week 12, treatment should be stopped.[26] Nevertheless, an abbreviated 24-week regimen of peg-IFN plus standard dose of ribavirin could be applied for HCV-1/4 patients with lower baseline viral loads and an RVR; whereas an abbreviated 16-week regimen of peg-IFN plus standard dose of ribavirin could be applied for HCV-2/3 patients with an RVR, without compromising the treatment efficacy. Using the treatment strategy, there is an estimated cost saving of 11.8% per SVR for HCV-1/4 patients and 29% per SVR for HCV-2/3 patients based on the response rates in Taiwan.[38,52] For HCV-1/4 slow responders, defined as patients achieving a pEVR and in whom HCV RNA test becomes negative at week 24, consideration should be given to extend therapy to 72 weeks.[21]

Conclusions and Future Directions

HCV genotype, baseline viral load, and on-treatment virological response could provide information for decision-making in individualized treatment. The tailored treatment regimen can reduce the cost of treatment and the incidence of adverse events without compromising on efficacy. However, a number of patients remain refractory to the current treatment regimens. Higher fixed doses of peg-IFN-α-2a (270 μg/week) and ribavirin (1,600 mg/day) may increase SVR rates in difficult-to-treat patients.[78] Recent progress in the development of novel recombinant interferon, albuferon, and specifically targeted antiviral therapy for HCV, such as protease, polymerase, and cyclophilin inhibitors, might further optimize the antiviral treatment for CHC.[79–85]

References

1. Lauer GM, Walker BD. (2001) Hepatitis C virus infection. *N Engl J Med* **345**: 41–52.
2. [No author listed] (2002) National Institutes of Health Consensus Development Conference Statement: management of hepatitis C: 10–12 June 2002. *Hepatology* **36**: S3–20.

3. Freeman AJ, Dore GJ, Law MG, *et al.* (2001) Estimating progression to cirrhosis in chronic hepatitis C virus infection. *Hepatology* **34**: 809–816.
4. Fattovich G, Stroffolini T, Zagni I, Donato F. (2004) Hepatocellular carcinoma in cirrhosis: Incidence and risk factors. *Gastroenterology* **127**: S35–50.
5. McCaughan GW, Omata M, Amarapurkar D, *et al.* (2007) Asian Pacific Association for the study of the liver consensus statements on the diagnosis, management and treatment of hepatitis C virus infection. *J Gastroenterol Hepatol* **22**: 615–633.
6. Maylin S, Martinot-Peignoux M, Moucari R, *et al.* (2008) Eradication of hepatitis C virus in patients successfully treated for chronic hepatitis C. *Gastroenterology* **135**: 821–829.
7. Huang JF, Yu ML, Lee CM, *et al.* (2007) Sustained virological response to interferon reduces cirrhosis in chronic hepatitis C: a 1,386-patient study from Taiwan. *Aliment Pharmacol Ther* **25**: 1029–1037.
8. Yu ML, Lin SM, Chuang WL, *et al.* (2006) A sustained virological response to interferon or interferon/ribavirin reduces hepatocellular carcinoma and improves survival in chronic hepatitis C: a nationwide, multicentre study in Taiwan. *Antivir Ther* **11**: 985–994.
9. Yu ML, Lin SM, Lee CM, *et al.* (2006) A simple noninvasive index for predicting long-term outcome of chronic hepatitis C after interferon-based therapy. *Hepatology* **44**: 1086–1097.
10. Shiratori Y, Ito Y, Yokosuka O, *et al.* (2005) Antiviral therapy for cirrhotic hepatitis C: association with reduced hepatocellular carcinoma development and improved survival. *Ann Intern Med* **142**: 105–114.
11. Yu ML, Huang CF, Dai CY, *et al.* (2007) Long-term effects of interferon-based therapy for chronic hepatitis C. *Oncology* **72**(Suppl 1): 16–23.
12. Hoofnagle JH, Mullen KD, Jones DB, *et al.* (1986) Treatment of chronic non-A, non-B hepatitis with recombinant human alpha interferon. A preliminary report. *N Engl J Med* **315**: 1575–1578.
13. Di Bisceglie AM, Martin P, Kassianides C, *et al.* (1989) Recombinant interferon alfa therapy for chronic hepatitis C. A randomized, double-blind, placebo-controlled trial. *N Engl J Med* **321**: 1506–1510.
14. McHutchison JG, Gordon SC, Schiff ER, *et al.* (1998) Interferon alfa-2b alone or in combination with ribavirin as initial treatment for chronic hepatitis C. Hepatitis Interventional Therapy Group. *N Engl J Med* **339**: 1485–1492.
15. Poynard T, Marcellin P, Lee SS, *et al.* (1998) Randomised trial of interferon alpha2b plus ribavirin for 48 weeks or for 24 weeks versus interferon alpha2b plus placebo for 48 weeks for treatment of chronic infection with hepatitis C virus. International Hepatitis Interventional Therapy Group (IHIT). *Lancet* **352**: 1426–1432.
16. Feld JJ, Hoofnagle JH. (2005) Mechanism of action of interferon and ribavirin in treatment of hepatitis C. *Nature* **436**: 967–972.

17. Bodenheimer HC Jr., Lindsay KL, Davis GL, *et al.* (1997) Tolerance and efficacy of oral ribavirin treatment of chronic hepatitis C: a multicenter trial. *Hepatology* **26**: 473–477.
18. Lai MY, Kao JH, Yang PM, *et al.* (1996) Long-term efficacy of ribavirin plus interferon alfa in the treatment of chronic hepatitis C. *Gastroenterology* **111**: 1307–1312.
19. Manns MP, McHutchison JG, Gordon SC, *et al.* (2001) Peginterferon alfa-2b plus ribavirin compared with interferon alfa-2b plus ribavirin for initial treatment of chronic hepatitis C: a randomised trial. *Lancet* **358**: 958–965.
20. Fried MW, Shiffman ML, Reddy KR, *et al.* (2002) Peginterferon alfa-2a plus ribavirin for chronic hepatitis C virus infection. *N Engl J Med* **347**: 975–982.
21. Ghany MG, Strader DB, Thomas DL, Seeff LB. (2009) Diagnosis, management, and treatment of hepatitis C: an update. *Hepatology* **49**: 1335–1374.
22. Farrell GC. (2007) New hepatitis C guidelines for the Asia-Pacific region: APASL consensus statements on the diagnosis, management and treatment of hepatitis C virus infection. *J Gastroenterol Hepatol* **22**: 607–610.
23. Ge D, Fellay J, Thompson AJ, *et al.* (2009) Genetic variation in IL28B predicts hepatitis C treatment-induced viral clearance. *Nature* **461**: 399–401.
24. Tanaka Y, Nishida N, Sugiyama M, *et al.* (2009) Genome-wide association of IL28B with response to pegylated interferon-alpha and ribavirin therapy for chronic hepatitis C. *Nat Genet* **41**: 1105–1109.
25. McHutchison JG, Poynard T. (1999) Combination therapy with interferon plus ribavirin for the initial treatment of chronic hepatitis C. *Semin Liver Dis* **19**(Suppl 1): 57–65.
26. Strader DB, Wright T, Thomas DL, Seeff LB. (2004) Diagnosis, management, and treatment of hepatitis C. *Hepatology* **39**: 1147–1171.
27. Zeuzem S, Diago M, Gane E, *et al.* (2004) Peginterferon alfa-2a (40 kilodaltons) and ribavirin in patients with chronic hepatitis C and normal aminotransferase levels. *Gastroenterology* **127**: 1724–1732.
28. Hadziyannis SJ, Sette H Jr., Morgan TR, *et al.* (2004) Peginterferon-alpha2a and ribavirin combination therapy in chronic hepatitis C: a randomized study of treatment duration and ribavirin dose. *Ann Intern Med* **140**: 346–355.
29. Mangia A, Santoro R, Minerva N, *et al.* (2005) Peginterferon alfa-2b and ribavirin for 12 vs. 24 weeks in HCV genotype 2 or 3. *N Engl J Med* **352**: 2609–2617.
30. Von Wagner M, Huber M, Berg T, *et al.* (2005) Peginterferon-alpha-2a (40KD) and ribavirin for 16 or 24 weeks in patients with genotype 2 or 3 chronic hepatitis C. *Gastroenterology* **129**: 522–527.
31. Neumann AU, Lam NP, Dahari H, *et al.* (1998) Hepatitis C viral dynamics *in vivo* and the antiviral efficacy of interferon-alpha therapy. *Science* **282**: 103–107.

32. Layden JE, Layden TJ. (2001) How can mathematics help us understand HCV? *Gastroenterology* **120**: 1546–1549.
33. Zeuzem S, Herrmann E, Lee JH, *et al.* (2001) Viral kinetics in patients with chronic hepatitis C treated with standard or peginterferon alpha2a. *Gastroenterology* **120**: 1438–1447.
34. Yu ML, Chuang WL, Dai CY, *et al.* (2006) Different viral kinetics between hepatitis C virus genotype 1 and 2 as on-treatment predictors of response to a 24-week course of high-dose interferon-alpha plus ribavirin combination therapy. *Transl Res* **148**: 120–127.
35. Jensen DM, Morgan TR, Marcellin P, *et al.* (2006) Early identification of HCV genotype 1 patients responding to 24 weeks peginterferon alpha-2a (40 kd)/ ribavirin therapy. *Hepatology* **43**: 954–960.
36. Zeuzem S, Buti M, Ferenci P, *et al.* (2006) Efficacy of 24 weeks treatment with peginterferon alfa-2b plus ribavirin in patients with chronic hepatitis C infected with genotype 1 and low pretreatment viremia. *J Hepatol* **44**: 97–103.
37. Dalgard O, Bjoro K, Hellum KB, *et al.* (2004) Treatment with pegylated interferon and ribavirin in HCV infection with genotype 2 or 3 for 14 weeks: a pilot study. *Hepatology* **40**: 1260–1265.
38. Yu ML, Dai CY, Huang JF, *et al.* (2007) A randomised study of peginterferon and ribavirin for 16 versus 24 weeks in patients with genotype 2 chronic hepatitis C. *Gut* **56**: 553–559.
39. Davis GL. (2002) Monitoring of viral levels during therapy of hepatitis C. *Hepatology* **36**: S145–151.
40. Davis GL, Wong JB, McHutchison JG, *et al.* (2003) Early virologic response to treatment with peginterferon alfa-2b plus ribavirin in patients with chronic hepatitis C. *Hepatology* **38**: 645–652.
41. Lee SS, Ferenci P. (2008) Optimizing outcomes in patients with hepatitis C virus genotype 1 or 4. *Antivir Ther* **13**(Suppl 1): 9–16.
42. Berg T, Carosi G. (2008) Optimizing outcomes in patients with hepatitis C virus genotype 2 or 3. *Antivir Ther* **13**(Suppl 1): 17–22.
43. Huang CF, Yang JF, Huang JF, *et al.* (2010) Early identification of achieving a sustained virological response in chronic hepatitis C patients without a rapid virological response. *J Gastroenterol Hepatol* **25**: 758–765.
44. Foster G, Mathurin P. (2008) Hepatitis C virus therapy to date. *Antivir Ther* **13**(Suppl 1): 3–8.
45. [No author listed] (1999) EASL International Consensus Conference on hepatitis C. Paris, 26–27 February 1999. Consensus statement. *J Hepatol* **31**(Suppl 1): 3–8.
46. Zeuzem S, Hultcrantz R, Bourliere M, *et al.* (2004) Peginterferon alfa-2b plus ribavirin for treatment of chronic hepatitis C in previously untreated patients infected with HCV genotypes 2 or 3. *J Hepatol* **40**: 993–999.

47. Kuboki M, Iino S, Okuno T, *et al.* (2007) Peginterferon alpha-2a (40 KD) plus ribavirin for the treatment of chronic hepatitis C in Japanese patients. *J Gastroenterol Hepatol* **22**: 645–652.
48. Lee HJ, Eun JR, Choi JW, *et al.* (2008) [Comparison of therapeutic results between combination therapy of peginterferon alpha-2a plus ribavirin and interferon alpha-2b plus ribavirin according to treatment duration in patients with chronic hepatitis C]. *Korean J Hepatol* **14**: 46–57.
49. Lee H, Choi MS, Paik SW, *et al.* (2006) [Peginterferon alfa-2a plus ribavirin for initial treatment of chronic hepatitis C in Korea]. *Korean J Hepatol* **12**: 31–40.
50. Liu CJ, Chuang WL, Lee CM, *et al.* (2009) Peginterferon alfa-2a plus ribavirin for the treatment of dual chronic infection with hepatitis B and C viruses. *Gastroenterology* **136**: 496–504.
51. Liu CH, Liu CJ, Lin CL, *et al.* (2008) Pegylated interferon-alpha-2a plus ribavirin for treatment-naive Asian patients with hepatitis C virus genotype 1 infection: a multicenter, randomized controlled trial. *Clin Infect Dis* **47**: 1260–1269.
52. Yu ML, Dai CY, Huang JF, *et al.* (2008) Rapid virological response and treatment duration for chronic hepatitis C genotype 1 patients: a randomized trial. *Hepatology* **47**: 1884–1893.
53. Yu ML, Dai CY, Lin ZY, *et al.* (2006) A randomized trial of 24- vs. 48-week courses of PEG interferon alpha-2b plus ribavirin for genotype-1b-infected chronic hepatitis C patients: a pilot study in Taiwan. *Liver Int* **26**: 73–81.
54. Kim KT, Han SY, Kim JH, *et al.* (2008) [Clinical outcome of pegylated interferon and ribavirin therapy for chronic hepatitis C]. *Korean J Hepatol* **14**: 36–45.
55. Kamal SM, El Tawil AA, Nakano T, *et al.* (2005) Peginterferon {alpha}-2b and ribavirin therapy in chronic hepatitis C genotype 4: impact of treatment duration and viral kinetics on sustained virological response. *Gut* **54**: 858–866.
56. El-Zayadi AR, Attia M, Barakat EM, *et al.* (2005) Response of hepatitis C genotype-4 naive patients to 24 weeks of Peg-interferon-alpha2b/ribavirin or induction-dose interferon-alpha2b/ribavirin/amantadine: a non-randomized controlled study. *Am J Gastroenterol* **100**: 2447–2452.
57. Hasan F, Asker H, Al-Khaldi J, *et al.* (2004) Peginterferon alfa-2b plus ribavirin for the treatment of chronic hepatitis C genotype 4. *Am J Gastroenterol* **99**: 1733–1737.
58. Fung J, Lai CL, Hung I, *et al.* (2008) Chronic hepatitis C virus genotype 6 infection: response to pegylated interferon and ribavirin. *J Infect Dis* **198**: 808–812.
59. Nguyen MH, Trinh HN, Garcia R, *et al.* (2008) Higher rate of sustained virologic response in chronic hepatitis C genotype 6 treated with 48 weeks versus 24 weeks of peginterferon plus ribavirin. *Am J Gastroenterol* **103**: 1131–1135.
60. Lam KD, Trinh HN, Do ST, *et al.* (2010) Randomized controlled trial of pegylated interferon-alfa 2a and ribavirin in treatment-naive chronic hepatitis C genotype 6. *Hepatology* **52**: 1573–1580.

61. Dev AT, McCaw R, Sundararajan V, *et al.* (2002) Southeast Asian patients with chronic hepatitis C: the impact of novel genotypes and race on treatment outcome. *Hepatology* **36**: 1259–1265.
62. Fried MW, Hadziyannis SJ, Shiffman M, *et al.* (2008) Rapid virological response is a more important predictor of sustained virological response (SVR) than genotype in patients with chronic hepatitis C virus infection. *Journal of Hepatology* **48**: 7.
63. Yu ML, Dai CY, Lee LP, *et al.* (2006) Outcome of chronic hepatitis C patients who required early termination of peginterferon-alfa plus ribavirin combination therapy. *Antivir Ther* **11**: 1015–1019.
64. Mangia A, Minerva N, Bacca D, *et al.* (2008) Individualized treatment duration for hepatitis C genotype 1 patients: a randomized controlled trial. *Hepatology* **47**: 43–50.
65. Kamal SM, El Kamary SS, Shardell MD, *et al.* (2007) Pegylated interferon alpha-2b plus ribavirin in patients with genotype 4 chronic hepatitis C: the role of rapid and early virologic response. *Hepatology* **46**: 1732–1740.
66. Ferenci P, Laferl H, Scherzer TM, *et al.* (2008) Peginterferon alfa-2a and ribavirin for 24 weeks in hepatitis C type 1 and 4 patients with rapid virological response. *Gastroenterology* **135**: 451–458.
67. Berg T, Weich V, Teuber G, *et al.* (2009) Individualized treatment strategy according to early viral kinetics in hepatitis C virus type 1-infected patients. *Hepatology* **50**: 369–377.
68. Shiffman ML, Suter F, Bacon BR, *et al.* (2007) Peginterferon alfa-2a and ribavirin for 16 or 24 weeks in HCV genotype 2 or 3. *N Engl J Med* **357**: 124–134.
69. Lagging M, Langeland N, Pedersen C, *et al.* (2008) Randomized comparison of 12 or 24 weeks of peginterferon alpha-2a and ribavirin in chronic hepatitis C virus genotype 2/3 infection. *Hepatology* **47**: 1837–1845.
70. Dalgard O, Bjoro K, Ring-Larsen H, *et al.* (2008) Pegylated interferon alfa and ribavirin for 14 versus 24 weeks in patients with hepatitis C virus genotype 2 or 3 and rapid virological response. *Hepatology* **47**: 35–42.
71. Di Martino V, Richou C, Thevenot T, *et al.* (2008) Modulations of peg-interferon plus ribavirin duration according to hcv-genotype and virologic response at w4 and w12: meta-analyses of rcts with individual data. *Hepatology* **48**: 213.
72. Berg T, von Wagner M, Nasser S, *et al.* (2006) Extended treatment duration for hepatitis C virus type 1: comparing 48 versus 72 weeks of peginterferon-alfa-2a plus ribavirin. *Gastroenterology* **130**: 1086–1097.
73. Sanchez-Tapias JM, Diago M, Escartin P, *et al.* (2006) Peginterferon-alfa2a plus ribavirin for 48 versus 72 weeks in patients with detectable hepatitis C virus RNA at week 4 of treatment. *Gastroenterology* **131**: 451–460.
74. Pearlman BL, Ehleben C, Saifee S. (2007) Treatment extension to 72 weeks of peginterferon and ribavirin in hepatitis c genotype 1-infected slow responders. *Hepatology* **46**: 1688–1694.

75. Ide T, Hino T, Ogata K, *et al.* (2009) A randomized study of extended treatment with peginterferon alpha-2b plus ribavirin based on time to HCV RNA negative-status in patients with genotype 1b chronic hepatitis C. *Am J Gastroenterol* **104**: 70–75.
76. Ferenci P, Laferl H, Scherzer TM, *et al.* (2010) Peginterferon alfa-2a/ribavirin for 48 or 72 weeks in hepatitis C genotypes 1 and 4 patients with slow virologic response. *Gastroenterology* **138**: 503–512, 512 e501.
77. Huang C-F, Yang J-F, Huang J-F, *et al.* (2010) Early identification of achieving a sustained virological response in chronic hepatitis C patients without a rapid virological response. *Journal of Gastroenterology and Hepatology* **25**: 758–765.
78. Fried MW, Jensen DM, Rodriguez-Torres M, *et al.* (2008) Improved outcomes in patients with hepatitis C with difficult-to-treat characteristics: randomized study of higher doses of peginterferon alpha-2a and ribavirin. *Hepatology* **48**: 1033–1043.
79. Kwo PY, Lawitz EJ, McCone J, *et al.* (2010) Efficacy of boceprevir, an NS3 protease inhibitor, in combination with peginterferon alfa-2b and ribavirin in treatment-naive patients with genotype 1 hepatitis C infection (SPRINT-1): an open-label, randomised, multicentre phase 2 trial. *Lancet* **376**: 705–716.
80. McHutchison JG, Manns MP, Muir AJ, *et al.* (2010) Telaprevir for previously treated chronic HCV infection. *N Engl J Med* **362**: 1292–1303.
81. Hezode C, Forestier N, Dusheiko G, *et al.* (2009) Telaprevir and peginterferon with or without ribavirin for chronic HCV infection. *N Engl J Med* **360**: 1839–1850.
82. Moucari R, Marcellin P. (2009) Hepatitis C: the role of new interferons in the era of STAT-C. *Nat Rev Gastroenterol Hepatol* **6**: 509–511.
83. McHutchison JG, Everson GT, Gordon SC, *et al.* (2009) Telaprevir with peginterferon and ribavirin for chronic HCV genotype 1 infection. *N Engl J Med* **360**: 1827–1838.
84. Gane EJ, Roberts SK, Stedman CA, *et al.* (2010) Oral combination therapy with a nucleoside polymerase inhibitor (RG7128) and danoprevir for chronic hepatitis C genotype 1 infection (INFORM-1): a randomised, double-blind, placebo-controlled, dose-escalation trial. *Lancet* **376**: 1467–1475.
85. Vermehren J, Sarrazin C. (2011) New HCV therapies on the horizon. *Clin Microbiol Infect* **17**: 122–134.

Index